# The Hubble Space Telescope
From Concept to Success

David J. Shayler with David M. Harland

# The Hubble Space Telescope

## From Concept to Success

Published in association with
Praxis Publishing
Chichester, UK

David J. Shayler, F.B.I.S.  
Astronautical Historian  
Astro Info Service Ltd.  
Halesowen  
West Midlands  
UK

David M. Harland  
Space Historian  
Kelvinbridge  
Glasgow  
UK

## SPRINGER-PRAXIS BOOKS IN SPACE EXPLORATION

Springer Praxis Books  
ISBN 978-1-4939-2826-2     ISBN 978-1-4939-2827-9   (eBook)  
DOI 10.1007/978-1-4939-2827-9

Library of Congress Control Number: 2015945584

Springer New York Heidelberg Dordrecht London  
© Springer Science+Business Media New York 2016  
This work is subject to copyright. All rights are reserved by the Publisher, whether the whole or part of the material is concerned, specifically the rights of translation, reprinting, reuse of illustrations, recitation, broadcasting, reproduction on microfilms or in any other physical way, and transmission or information storage and retrieval, electronic adaptation, computer software, or by similar or dissimilar methodology now known or hereafter developed.  
The use of general descriptive names, registered names, trademarks, service marks, etc. in this publication does not imply, even in the absence of a specific statement, that such names are exempt from the relevant protective laws and regulations and therefore free for general use.  
The publisher, the authors and the editors are safe to assume that the advice and information in this book are believed to be true and accurate at the date of publication. Neither the publisher nor the authors or the editors give a warranty, express or implied, with respect to the material contained herein or for any errors or omissions that may have been made.

Front cover: An IMAX film still from STS-31, showing the Hubble Space Telescope released from the Remote Manipulator System. A cloudy blue Earth is reflected in the closed Aperture Door.

Rear cover: Left: STS-31 deployment mission emblem; Center, STS-61 Service Mission 1 emblem; Right, the cover of the companion book *Enhancing Hubble's Vision: Service Missions That Expanded Our View Of The Universe*.

Cover design: Jim Wilkie  
Project Editor: David M. Harland

Printed on acid-free paper

Praxis is a brand of Springer  
Springer Science+Business Media LLC New York is part of Springer Science+Business Media (www.springer.com)

# Contents

| | |
|---|---|
| **Preface** | ix |
| **Acknowledgements** | xiii |
| **Foreword** | xv |
| **Dedication** | xx |
| **Prologue** | xxi |

**1 Deployment** ... 1
    Deploying a telescope ... 1
    An astronomer deploys Hubble ... 7
    Preparing to fly ... 12
    Building the 'stack' ... 23
    Letting Hubble go ... 31
    High, but not too high ... 39
    Trouble with Hubble ... 44

**2 The dream** ... 50
    Out where stars don't twinkle ... 50
    Stratoscope: Small steps to space ... 54
    Astronomy from space before Hubble ... 55
    Orbiting astronomical observatories ... 58
    Laying the groundwork ... 61
    A manned orbital telescope ... 64
    The large space telescope is born ... 72
    Nimbus: A candidate for orbital servicing? ... 73
    Astronomy mission board report, July 1969 ... 77

## vi Contents

| | | |
|---|---|---|
| **3** | **A dream becomes reality** | 83 |
| | After the Moon | 84 |
| | Pieces of a jigsaw | 85 |
| | Getting started | 87 |
| | Bringing the pieces together | 98 |
| | A space shuttle system | 99 |
| | The remote manipulator system | 103 |
| | A meeting in space | 107 |
| | Managing Hubble | 115 |
| | A European partner | 121 |
| | Protection against contamination | 124 |
| | A broad and involved development | 127 |
| **4** | **LST becomes ST, becomes HST** | 131 |
| | The 1972 Large Space Telescope Phase A design study | 131 |
| | LST Phase A design study update | 141 |
| | Maintaining maintainability | 146 |
| | EVA at LST | 152 |
| | The Hubble Space Telescope: a brief description | 153 |
| | The journey from LOT to HST | 181 |
| **5** | **Simulating servicing** | 190 |
| | Creating a concept | 191 |
| | Establishing the guidelines | 195 |
| | Astronaut Office involvement | 198 |
| | EVA support equipment for servicing | 200 |
| | Crew aids | 203 |
| | Baseline equipment | 204 |
| | EVA profile | 207 |
| | Ground support | 207 |
| | EVA training | 208 |
| | Taking Hubble underwater | 213 |
| | Progressive systems analysis | 216 |
| | The NBS test facility | 218 |
| | Full scale simulations | 225 |
| | NBS tests 1983–1990 | 235 |
| | Gaining EVA experience | 239 |
| | DTO-1210 | 240 |
| **6** | **Tools of the trade** | 246 |
| | The beginning | 247 |
| | The HST servicing tool kit | 251 |
| | Russell Werneth, managing the tools | 257 |
| | Mechanisms engineer Paul Richards | 263 |
| | No manned maneuvering unit | 268 |
| | Tools of the trade | 271 |

| | | |
|---|---|---|
| 7 | **Behind the scenes** | 275 |
| | Training a shuttle crew | 276 |
| | Ground teams | 282 |
| | Hubble program office | 283 |
| | Other service mission team members | 300 |
| | Astronaut office support | 312 |
| | Mission Control | 313 |
| | The role of a Hubble flight director/mission director | 322 |
| 8 | **Service Mission 1** | 330 |
| | The Hubble comeback | 330 |
| | Choosing a crew | 332 |
| | Planning the mission | 344 |
| | First house call at Hubble | 353 |
| | Passing the baton | 376 |
| | **Closing comments** | 382 |
| | **Afterword** | 386 |
| | **Abbreviations** | 389 |
| | **Bibliography** | 397 |
| | **About the author** | 401 |
| | **Other works by the author** | 403 |
| | **Index** | 405 |

# Preface

It always fascinates me how my various book projects evolve from ideas, sometime going back years, even decades. During the late 1960s, I, like many of my generation, became fascinated with space exploration and the race to the Moon. As a young teenager I soaked up all I could on "the space age," which was at the start of its second decade. As the early Apollo missions reached for the Moon, I became more interested in what the astronauts would do on the surface, using the equipment provided and the procedures developed, to enable them to venture across that alien world. I suppose it was the sense of wonder and magic at that time, the grainy images on a black and white television, and the artistic impressions of research bases on the Moon and projections towards a far distant future—the 1980s—and exciting expeditions to Mars.

This period is fondly remembered as the start of my interest in extravehicular activity, EVA, commonly known as spacewalking. In addition to following the progress towards the lunar landings, the early months of 1969 included my first introduction to Soviet space flight with the EVA transfer from Soyuz 5 to Soyuz 4, followed soon thereafter by the first Apollo EVA. There were media forecasts of extensive spacewalking activities in plans for the rest of the century, of a new spacecraft called the space shuttle, of large space stations, and the repair and servicing of satellites in space. Wow, what an adventure the future held.

Well, one thing I learnt quickly in following the space program was patience and optimism: patience in waiting for things to happen, and optimism that they would become a reality—eventually. While investigating EVA in more depth, I became aware of grand plans for Apollo, spacewalking techniques from space stations, and how the shuttle would support EVAs to service and maintain various payloads and satellites, including an astronomical observatory called the Large Space Telescope. Over the next two decades, I gained further insight into EVA involving the shuttle and what was expected to be achieved by astronauts visiting the telescope. The years between 1971 and 1981 saw many changes from the heady days of my youth to the reality of understanding the complexity of the space program. The shuttle suffered from many delays and setbacks, as did the payloads it was planned to carry, including the Large Space Telescope—which for a while was known simply as the Space Telescope prior to its being named in honor of acclaimed astronomer Edwin P. Hubble in 1983.

By the mid-1980s there was more detail available about the shuttle program and the Space Telescope, including reports on how astronauts would maintain and service the instruments, not on Earth as first reported, but in space. Now that caught my attention! Just how are they going to manage that, I asked myself. No one had yet performed an EVA from the shuttle—in fact no American had walked in space for a decade since Skylab, but now teams of astronauts were going to do "space age home improvements" a few hundred miles above the Earth flying at 5 miles per second, this clearly required more study. And so the research began on what became known as the Hubble Space Telescope Service Missions and the genesis of this book.

Across the next three decades my research followed many tracks connected with this project, and I soon became aware of the huge infrastructure that was required to support humans in space, the shuttle, and the telescope in particular. There was the challenge of how all three elements were put together. Thrown into this mix was the complication of carrying out useful work while wearing a bulky pressure suit and thick cumbersome gloves in order to improve or repair the delicate parts of the telescope, without disturbing its science work, or breaking it.

Of course the telescope had first to be launched, and that presented its own problems and setbacks. In researching the service missions that were to follow, it became very clear that an immense amount of work had to be conducted to put each spacewalk together. But it was not just the EVAs, there had to be the replacement items prepared and tested, the tools for the astronauts to complete the task, and the crews trained to achieve the objectives. Researching all of this also took me into other areas, such as creating the facility to allow for servicing the telescope, ensuring the safety of hardware, crews and equipment at all times, organizing the ground teams, and understanding the environment in which the telescope was operating and the astronauts would work. It also took me into the realms of materials science, human factors engineering, flight control dynamics, orbital operations, and systems engineering. I quickly decided that, not being an astronomer, I would leave the pure science of Hubble to others. Also, much had been written over the years on the politics and management of getting the idea of Hubble from the drawing board to orbit. Though I knew I would have to touch on this subject, I decided I would not delve too deeply into it. What was *not* really covered was the network of small things which together made up the service missions—the hundreds of hours spent on the ground preparing for each mission, and more than anything the devotion, dedication, belief, and tenacity of everyone involved from the worker who put together the smallest components, to the teams who prepared and tested the hardware, the launch team, the flight controllers, the managers, scientists, engineers, technicians, and, last but not least, the astronauts "at the sharp end." There were, over the period of Hubble operations literally thousands involved in keeping the telescope flying. Many of these people were not directly involved in Hubble's scientific activities, but every one of them nevertheless contributed to enabling the telescope to obtain the stunning results that the instruments have returned and rightly proud they should be.

This was the story I set out to tell. It is not simply a detailed account of six space shuttle flights, nor is it a historical narrative of the people involved, but a blend of both, together with background information of the hardware and preparations, a jigsaw puzzle of small items which when put together presents the finished result. And that result has been flying around our planet for over 25 years, altering the way that we look at the universe, our understanding of that infinite depth, and how we see ourselves as part of that infinity.

This story has been spread across two titles. Firstly, in *The Hubble Space Telescope: From Concept To Success*, I return to April 1990 to recall the deployment of the telescope in space by STS-31, and the challenges addressed to achieve that feat. Then the background to the Hubble story unfolds, from its origin and the birth of satellite servicing, to developing the techniques and tools to achieve that capability at the telescope, and of the huge infrastructure on Earth to support such mammoth undertakings. This work closes with the huge success of the first servicing mission and restoration of its vision.

The second title, *Enhancing Hubble's Vision: Service Missions That Expanded Our View Of The Universe*, takes up the story with the development of the series of servicing missions required to keep the telescope flying and at the forefront of science, despite inflight failures and a second tragic blow to the shuttle program. The story closes with the often overlooked work on post-flight analysis of returned items of Hubble hardware, and to the fate of the telescope as the 25th anniversary of its launch was celebrated in 2015.

I have enjoyed the complexity of putting this book together and continue to be fascinated in the deeper story of each mission. This work has generated follow-on projects and new ideas that will appear in other titles, so enjoy the journey as I continue to do so.

David J. Shayler, FBIS
Director, Astro Info Service Ltd
www.astroinfoservice.co.uk
Halesowen, West Midlands, UK
May 2015

# Acknowledgements

This was a far reaching project involving the support and cooperation of a number of individuals whose names are etched in the history books of the Hubble Space Telescope program. Firstly I must extend my personal thanks and appreciation to all who have offered their help and assistance in compiling both books, from those who supplied information or offered their recollections and experiences to those who pointed me in the right directions. There are also a number of people who worked tirelessly on the production side, which is never an easy task.

My thanks go to a number of former astronauts who went out of their way to provide at times some very personal recollections of their time working on the Hubble service missions, as well as their insights into the "real" workings of what it means to be an astronaut and all that this entails. Specific to the Hubble missions, my thanks go to Steve Hawley who, in addition to providing valuable explanations of what it was like to "be the arm man," also crafted the Foreword to the book. With Bruce McCandless, Steve also offered personal recollections of the mission to deploy the telescope. From the crew of STS-61 my thanks go to Dick Covey, Tom Akers, Jeff Hoffman and Story Musgrave; from STS-82, Steve Smith, Joe Tanner and once again Steve Hawley supplied useful information in response to my queries; Mike Foale and European astronaut Jean-François "Billy-Bob" Clervoy provided generous support; and from STS-109 Jim Newman and "Digger" Carey gave fascinating insights into their roles and experiences on the third and fourth service missions. Story Musgrave is to be thanked for providing the Afterword to the book.

Other astronauts who helped in my research included Bob "Crip" Crippen and George "Pinky" Nelson who explained both the early years of shuttle rendezvous and the servicing of the Solar Max satellite, a precursor to the Hubble missions. Thanks also to Paul Richards, who explained his role in developing tools for Hubble in the years *before* he became an astronaut and used those same tools on the ISS during 2001.

Significant and important support came from the Public Affairs Office at Goddard Space Flight Center in Maryland, in particular Susan Hendrix, Lynn Chandler and Adrienne Alessandro. Also from Goddard, my thanks go to several individuals who provided insights into the background world of Hubble servicing: Preston Burch, Joyce King, Ben Reed, Ed Rezac, Al Vernacchio and Russ Werneth.

At the Johnson Space Center, my thanks go to Robert Trevino for his explanations of how its EVA support team functioned. Former Flight Director and Hubble Mission Director Chuck Shaw explained in great detail the working of Mission Control in Houston, and his role in support of Hubble servicing. The Public Affairs staff at JSC, and former employees who worked at what used to be the History Office at JSC, now Clear Lake University, the Collections held at Rice University then at NARA in Fort Worth, together with those at the Still Photo Library and Audio Library at JSC and Media Services at KSC have, over a period of many years, supported my research, including the early days of this project. They include: Eileen Hawley, Barbara Schwartz, Dave Portree, Glenn Swanson, Jeff Carr, James Hartsfield, Janet Kovacevich, Joey Pellerin, Joan Ferry, Margaret Persinger, Lisa Vazquez, Diana Ormsbee, Jody Russell, Mike Gentry, and Kay Grinter. And Ed Hengeveld is to be thanked for supplying some of the illustrations. In addition, my thanks go to Lee Saegesser, Roger Launius and Bill Barry at the History Office in NASA Headquarters for years of support and interest in my work.

At Lockheed, my thanks go to Andrea Greenan, Buddy Nelson and Ron Sheffield. At the European Space Agency, I must thank Carl Walker and Lother Gerlach, with appreciation to Claude Nicollier for his offer of assistance. I am grateful to John Davis at Hawker Siddley Dynamics/BAe Systems for information on a proposed orbiting astronomical observatory. And I thank once again Suzann Parry and the staff of the British Interplanetary Society in London for access to their library archives. I must also express my appreciation to Joachim Becker of *Spacefacts.de* and Mark Wade of *Astronautix.com* for permission to use some of their images. All images are courtesy of NASA and from the AIS collection unless otherwise stated.

On the production side, I must thank Clive Horwood at Praxis in England, Nora Rawn and Maury Solomon at Springer in New York, and project editor David M. Harland for his expert guidance, additional efforts (and patience!). These projects are never easy. I must also thank Jim Wilkie for his skills in turning my cover ideas into the finished product.

Love and appreciation go to my wife Bel for all the effort spent transcribing the numerous audio taped interviews in the AIS collection, and scanning numerous images for the book, and to my mother Jean Shayler for the hours spent reading the whole document and for her helpful suggestions to improve the manuscript. Apologies must also go to both of them for the weeks spent away from all our home improvements, days out, and cooking nice meals. Finally, I express my apologies to our wonderful German Shepherd Jenna for having missed out on more than a few long walks!

To one and all, a huge thank-you.

# Foreword

The Hubble Space Telescope is arguably the single most important scientific instrument ever developed. Astronomers knew that a large telescope in orbit would fundamentally change our understanding of the universe. However, in conjunction with the advances over the last two decades in the technology of ground-based telescopes, HST has initiated a revolution in our understanding of the universe unprecedented since the time of Galileo. Astronomers now know the age of the universe to a few percent, have confirmed the existence of black holes, have imaged planets around other stars, and in the Hubble Deep Field and Ultra Deep Field have found a rich population of galaxies of different types and different ages, some of which apparently formed less than a billion years after the Big Bang. HST has helped confirm the existence of "dark matter" and "dark energy" which together make up roughly 96 percent of the universe and about which we know almost nothing. As a wise person once pointed out, we are still confused, just confused at a more sophisticated level.

Crucial to the success of HST was the ability, envisioned right from the beginning, for the telescope to be repaired and upgraded while on-orbit. That capability was provided by the space shuttle and astronauts using sophisticated techniques in robotics and EVA. The first service mission in 1993 was the most complex shuttle mission attempted up to that time, and probably the most important NASA mission since Apollo 11. At stake was not only the very future of HST, but NASA's reputation. Service Mission 1 was to replace the original solar arrays that were impairing HST's pointing stability and to install hardware that would compensate for the spherical aberration in the main mirror that was preventing the telescope from simultaneously attaining both its design sensitivity and spatial resolution. This success was followed by four more challenging but successful service missions that replaced failed components and upgraded the science instruments.

In addition to HST's impact on science, developing the methods necessary to successfully execute the deployment and service missions provided valuable experience and confidence in the numerous operational techniques needed to assemble the International Space Station—perhaps the single greatest engineering accomplishment in history. Lessons learned by my crew on the HST deployment mission in 1990 led to improvements

Dr. Steve Hawley, NASA astronaut 1978–2008. (Courtesy Steve Hawley)

in our ability to handle large, massive payloads using the robot arm of the shuttle, increased the EVA time available for maintenance or assembly tasks, motivated important improvements in crewmember situational awareness during rendezvous and robot arm operations, and instigated better training for integrated robotics and EVA activities.

Perhaps not well-known is that HST data are archived and ultimately made available for the general researcher community. This allows for even greater use of the data for projects other than those for which it was originally obtained. New computer processing has been used to reveal previously unknown information in existing observations. One example is the direct detection of a planet orbiting another star, made possible by enhanced data processing capabilities and the remarkable stability of HST imagery. More scientific publications are now being written using HST archived observations than are published using newly obtained data. The science legacy of HST will increase for decades after the telescope is no longer in orbit.

During my astronaut career, I met many people who thought that the shuttle launched from Houston and flew to the Moon. However, the vast majority had heard of the Hubble Space Telescope and were amazed at the images. HST imagery is commonplace in class rooms. It inspires new generations of scientists, engineers and explorers. I was privileged to be one of the few people to get to work on HST in space twice. I may also be the only person to have been blessed to have had the opportunity to work on HST in space and then the chance to use it for research. My career as an astronaut and as an astronomer has been closely tied to the Hubble Space Telescope. Now that the shuttle and I are both done flying in space, the most significant accomplishment for which we will ultimately be remembered may be our long and successful association with the Hubble Space Telescope.

Dr. Steve Hawley<br>
Professor, Physics and Astronomy<br>
Director, Engineering Physics<br>
Adjunct Professor Aerospace Engineering<br>
University of Kansas<br>
Former NASA Astronaut (Mission Specialist STS 41-D, 61-C, -31, -82 and -93)

The Hubble Space Telescope in orbit.

*This project is dedicated to Hubble Huggers everywhere. In particular recalling the work, skills, and dedication of all who worked, from the ground up, on the Hubble servicing program, and to their families for allowing them to devote time to work when they really should have been at home.*

*Also to the memory of
Andrew Salmon
(1961–2013)*

*Fellow author and amateur astronomer who would have loved this project, and who would have offered countless suggestions and guidance.*

# Prologue

The year 2009 marked the 400th anniversary of the first use of a telescope to look skywards. Italian physicist and astronomer Galileo, used a crude instrument to make ink renderings of the Moon, recording mountains and craters, as well as a ribbon of defused light stretching across the night sky, a region that we know as the Milky Way. To celebrate this event the International Astronomical Union (IAU) and United Nations Educational, Scientific and Cultural Organization (UNESCO) created, in 2003, the International Year of Astronomy as a global effort to engage "citizens of the world" and encourage them to rediscover their place in the universe by observing the day and night sky and communicate their finding across the world. The program was endorsed by the United Nations and the International Council for Science (ICSU).

When the final report on the project was released, it was revealed that at least 815 million people in 148 countries across the world participated in one of the world's largest scientific projects. Astronomy was at the forefront of news, and one of the major driving forces of this popularity was a telescope, not on Earth but orbiting it at an altitude of 380 miles (600 km). This telescope, called Hubble, has changed the face of astronomy and our understanding of the universe far beyond anything that could have been envisaged by Galileo. Over the last four centuries, improvements to telescopes and our knowledge of the sciences has enabled us to make observations that penetrated progressively deeper into space and time. However, it was soon recognized that the distorting effects of our own atmosphere impose a limit on the clarity of the image provided by a telescope on Earth. The next step was to place a telescope in orbit to observe from above the atmosphere. The full story of Hubble has yet to be written, detailing its successes and difficulties in delivering stunning images and fascinating insights into the nature of the universe. Some people have said that Hubble does not deliver ground-breaking science, but merely confirms what we already suspected. Others maintain that this is nonsense. Arguing the case for or against Hubble science is not the purpose of this current volume.

In fact, this is not a typical book on the Hubble Space Telescope (HST) because it is not devoted to the science and images generated by the facility. Instead, this book examines the vast infrastructure and effort that was created to keep the Hubble flying and how the various techniques were brought together to build, launch, repair, service, and maintain it, thereby learning how to fly such an intricate instrument in space, and keep it operating for years in order to obtain, at the very least, some of the most breathtaking images ever seen of the universe around us.

Originally intended as a one-book project, in-depth research and generous cooperation by many individuals personally involved in the project resulted in the inspired decision by Springer to divide the project across two titles. This first part of the story focuses upon the development of the Hubble Space Telescope concept and the ability to service it to facilitate prolonged research over many years. The story of how an idea for a large optical telescope in orbit evolved over four decades into what became the Hubble Space Telescope is matched by the long period of development which resulted in the ability to service the telescope on-orbit using specially designed tools and procedures.

The story opens with the deployment of Hubble by the STS-31 mission in April 1990. The milestone of placing the telescope into space drew to a conclusion decades of proposals that such a facility would be of immense benefit to astronomy and space sciences, followed by debates about how that feat should be achieved. Astronomers were jubilant. But just a few weeks later came the shocking discovery that the optical system of the telescope was not as precise as expected, and was unable to focus correctly.

With the telescope on-orbit, the challenge was to deliver not only what had be proposed for almost as long as the instrument had been suggested, that of maintenance and servicing, but also to repair what had gone wrong. Five additional shuttle missions over the next two decades met and kept that promise, allowing Hubble to deliver its science and imagery for a decade longer than its planned 15 year lifetime, and, with a little of the Hubble luck, it will surpass its 30th anniversary in space, still delivering first class science and great images. The rest of the opening chapter looks in more detail at STS-31, and the efforts involved in taking the telescope into orbit.

The second chapter travels back in time and describes how the original idea for placing an optical telescope in orbit evolved and was slotted into the wider astronomical program of the early American space program. This chapter recalls early plans for satellite servicing, reviews the proposal for man-tended astronomical platforms, and the role envisaged for astronauts in astronomical research from space.

The third chapter summarizes the turbulent years of the 1970s and early 1980s, and picks out several key developments in technology and procedures that would prove crucial to the later service missions. These include the decision to launch on the shuttle with its capability for on-orbit servicing, the sizing of the shuttle's payload bay to suit the requirements of the US Air Force, which was responsible for launching the nation's classified satellites, the need for and development of a remote manipulator system, and various rendezvous techniques. This chapter also summarizes the management of the project, which, along with the never-ending battle for funding, featured a number of key decisions that affected the manner in which the telescope would be maintained. It concludes with a review of the participation of a European partner in the program, and the environment in which the service missions would have to operate.

The next two chapters review the plans for the predecessor of Hubble, called the Large Space Telescope, and how this emerged as the Space Telescope prior to its being named the HST. These chapters also give a brief summary of the HST hardware at the time of launch in 1990. They are a guide to the approaches to servicing and a useful reference for later chapters on the individual service missions. While the science instruments are mentioned and briefly explained, the science program of the telescope is only mentioned in passing as the main focus remains the on-orbit servicing of such a spacecraft rather than its utilization between shuttle service missions. The development of underwater EVA simulations years before Hubble flew was the key to mastering the service missions. These water tank exercises, begun in the late 1970s, ended only a few weeks before the final service mission in 2009. The fifth chapter also describes how the EVA servicing of the telescope evolved and the techniques of maintaining and repairing the telescope were developed mainly underwater, although there were exercises involving models, mockups and other 1-g simulators.

The sixth chapter reflects on the equipment developed to support the servicing objectives on the missions. As would any professional craftsman here on Earth, the astronaut servicing crewmember on-orbit required certain "tools of the trade," and these are detailed along with first-hand accounts by people who were at the cutting edge of developing such tools and the procedures for their use.

The public face of the shuttle service missions were the astronauts plying their trade in space, but on the ground there were several important and vital teams of engineers, flight controllers, and scientists who were the often unseen backup team on every mission. The seventh chapter explains the support team infrastructure and the roles they fulfilled on each mission. The astronauts who would fly a mission, the Mission Control team in Houston, the Hubble team at Goddard, and the launch team at Kennedy, all worked together as one huge team.

With the HST safely in space with defective vision, and with an impressive operational infrastructure in place, the eighth chapter recalls the heady days of the first service mission. During December 1993, STS-61 saved Hubble and NASA from disaster. The series of five EVAs installed the corrective optics to restore Hubble's vision and completed other repairs and upgrades to the telescope in what was arguably the most important mission by the space agency since the historic lunar landing by Apollo 11, restoring its reputation to a level last seen on Skylab or perhaps Apollo 13 over two decades before.

This first part of the story spans the years from the birth of a concept through many years of uncertainty, delays, frustration and hope, to a final launch, then more disappointment with a flawed mirror, to the plans to overcome what appeared to be a major setback, and to the first servicing mission that restored the telescope to a fully operational observatory.

But this was not the end of the Hubble story. Now that the telescope was up and running, the challenge became to keep it so: to provide new instruments and hardware to improve and extend its capabilities and potential far beyond that initially envisaged, to enhance its vision. The planning, preparation and creation of an infrastructure to support a protracted scientific program was in place. It was now time to execute that plan. This part of the Hubble story is told in the companion volume *Enhancing Hubble's Vision: Service Missions That Expanded Our View Of The Universe*. It details the next four servicing missions. This period of almost 15 years also includes the recovery by NASA from the second tragedy to hit the shuttle program—the loss of Columbia. As the 25th anniversary of the Hubble Space Telescope is celebrated, the story is brought up to date with details of activities after the servicing missions were completed—a time where the Hubble Space Telescope enjoys the status of a national treasure.

# 1

# Deployment

> *I really did not want to mess this up.*
>
> Steve Hawley, RMS operator, STS-31

On April 25, 1990 professional astronomer, NASA astronaut and STS-31 Mission Specialist Steven A. Hawley, was looking intently out of the aft flight deck windows of the space shuttle Discovery orbiting 380 miles (600 km) above the Earth, at the large payload on the end of the robotic arm that he was controlling. To his left was Mission Commander Loren J. Shriver, flying the orbiter. With them on this the 35th mission of the shuttle series were Mission Specialists Bruce McCandless II and Kathryn D. Sullivan, both of whom were on the middeck, preparing for a possible spacewalk if things went wrong with the payload. Floating between decks was Pilot Charles F. Bolden Jr., who was helping both pairs of colleagues. It was a tense time.

For Hawley, it seemed all attention was on his actions over the next few minutes. He was also monitoring the small TV screens which displayed views of the payload that he was about to deploy: the Hubble Space Telescope. It seemed appropriate that the responsibility to place the long-awaited 'Great Observatory' into orbit should fall to a professional astronomer. For many who had worked on the project, Hubble had dominated their entire professional career, and it carried the hopes and expectations of the astronomical community, together with scores of designers, engineers, contractors, scientists, controllers, managers, politicians and even the general public at large.

## DEPLOYING A TELESCOPE

The responsibility of his next actions was not lost on Hawley. "I remember, throughout the mission, but in particular on deploy day, thinking about all of the people and all the years of effort that had led up to this moment. To some extent, what I was doing was just representing the community of scientists and engineers who had taken [Lyman] Spitzer's idea

2 **Deployment**

for an optical telescope on-orbit and made it a reality. So I really did not want to mess this up. I did feel, particularly as a professional astronomer, the weight of the astronomical community. They had done all their jobs, and it's up to me to finish it off. I remember thinking about that quite a lot."

Looking out the overhead aft flight deck windows [left to right] Bruce McCandless, Steve Hawley and Loren Shriver.

**Unwanted motions and collision avoidance**

Hawley, an astronaut for 12 years, was on his third flight into space, fully proficient and experienced in operating the Remote Manipulator System (RMS). Although he had operated the arm in simulators and had gained insights from colleagues who had already operated it in space, he had yet to 'fly' the arm in space. But the task was daunting. Hubble was by far the largest mass that the arm had been required to hoist out of the payload bay to date, and the tolerances were tight, so the trick was to recognize the momentum of movement and be able to react quickly enough to do something about it.

Hawley explained this as "the worse failure you can have, when you are trying to unberth Hubble and it is very near the orbiter structure [where] tolerances are very small. With a run-away motor you can have your first hypothetical failure. As there was no collision avoidance software available, the collision avoidance is the RMS operator and you can make [all kinds] of estimates as to how long it would take the operator to recognize a run-away and take [the appropriate corrective] action. Obviously, when the telescope is that low in the bay, and that happens, you have a concern that it is going to take longer to recognize and stop it than the [time and distance] you have. In order to protect for that failure, what they did was, through the software, to limit the amount of current that you can send to the motors and that way, if a motor failed 'on' for some reason it would drive slowly enough that the operator, in principle at least, would be able to recognize that and take action before there was an impact."[1]

That is what was planned for STS-31. Hawley could control the rate of motion, or drive, through the software. As he unberthed Hubble, the software load would be changed so that when he had lifted the payload high enough, greater current was sent to the motors slowing the rate down. But what happened on-orbit was that these loads became lost in the general 'noise' of the system, to the extent that the signal 'noise' that Hawley was sending from the controller to the arm was not that much bigger than the 'noise' in the system. This combined noise then acted like a command, as far as the motors were concerned, confusing the system. Normally the ground simulators did not model this effect, and the crew didn't notice it until they were actually deploying Hubble. As Hawley explained, "The consensus was, when I would command pure 'up' motion, it wouldn't move purely up, it would wallow around. I remember that it was really confusing when it was doing that, because it wasn't the way it behaved in the sims and made the deployment task [about] 50 percent longer, because we kept having it stop and take out the commands that we weren't requesting and the different axis [that the arm was heading in]."

"Once we came back after STS-31," Hawley continued, "we went over to the Shuttle Engineering Simulator and we modeled the 'noise', and my recollection is that we very accurately reproduced what we saw in deployment. Ultimately, what we did was develop a control mode for the arm called POHS [pronounced POSH] for Position Orientation Hold Submode [in which] you could select that mode, then command in the minus Z orbiter axis, purely straight up out of the bay, with the software only allowing motion in that one axis. It would cancel out the noise in the system that we had on STS-31 that was trying to rotate the telescope in pitch and yaw, and send opposite commands. That made it quite a bit easier."[2]

## Hubble on the arm

The training for the mission had rehearsed a number of contingency and backup modes, including two worse-case scenarios that involved deploying the telescope in the backup mode and the total loss of the RMS. A failure of the orbiter's Main Bus A or Manipulator Controller Interface Unit (MCIU) could disable all the modes reliant on software. It had also been found that in the event of a failure of the systems management general purpose computer this could be replaced by a guidance navigation and control computer in order to support the operation of the arm. Should it be required, the crew had trained to remove and replace a MCIU with a spare. The individual joint motors could be driven in the backup mode, but without computer support or information displays. Reflecting on these contingency procedures, Hawley wrote in 2014 that had such a failure occurred, it would have been preferable to have delayed the deployment of Hubble until the MCIU backup unit had been installed.[3]

Although the Earth return mode for servicing needs had been abandoned in 1985 owing to the cost and the risk to the integrity of the telescope, the option for a contingency return was discussed in the event that Discovery was unable to attain the minimum deployment altitude. The overriding concerns which fuelled the decision to pursue orbital servicing remained, and as a result the team developed a means of deploying the telescope even if the RMS was not functional. This concept, which became known as "backaway deploy", would have involved turning the orbiter to the tail-to-Sun attitude to allow the telescope's Sun sensors to lock onto the Sun after its release. The disconnection of the umbilical would then have been followed by the opening of the four payload retention latch assemblies. With Hubble essentially free of the orbiter, Discovery would then have flown out from under the telescope. During flight techniques meetings in 1989, the question of providing the procedures in the STS-31 flight plan were discussed several times. There were many issues to be resolved with this proposal, and of course if anything had gone wrong in the post-deployment period, such as the failure of solar array or high gain antenna deployment, there would have been very little the crew could have done without an operational RMS. In 2014 Hawley recalled that some standalone training was done in the shuttle mission simulator for this mode, but they did not progress to the integrated simulation level. The backaway deployment procedures were available on the checklists for the mission, but fortunately neither this nor any other contingency plans were needed in deploying the telescope.

As he maneuvered Hubble, Hawley realized that the window view was not helpful once he had the telescope very high above the bay, because the aperture door was blocking the rear windows and was highly reflective, forcing a reliance on the TV cameras. He suspected that he over-drove the cameras trying to get a better view, "I remember the end effector camera was useless and I thought it had failed actually, because once the telescope was gone, it was operating again, and so what I concluded was that it was just unable to handle the sunlight reflecting off the aluminum surface. One of the recommendations that I made once we got back was that when we [went] back to Hubble, to rendezvous with it, we ought to do that at 'night' because I found the reflecting sunlight made it really hard to capture it, [as] the TV cameras would not work very well."

Hawley's next task was to tip Hubble over into its solar array deploy attitude, tilting the telescope over the top of the crew cabin, with the aperture door pointing to the rear of the payload bay. Described as resembling a butterfly leaving its cocoon, the release of the aft mast latches was the next step in transforming the inert cargo to a very active, free flying satellite. During the deployment of the mast, Discovery's thrusters were inhibited to prevent firings that could damage the load bearing pivots on the masts. With the orbiter in free drift, the 'go' was given to release the masts. But first, in order to prevent Discovery drifting off the desired attitude, Shriver gave a little burst to the orbiter's rate of roll.

After 4 minutes of watching the sides of the telescope for any signs of movement of the masts, without result, the astronauts were becoming concerned. Even McCandless was now on the crowded flight deck. Knowing that each array would move slowly, but not this slowly, he queried Mission Control whether any movement had been recorded by telemetry. It was almost 10 minutes into the maneuver when the masts finally rotated into position, but there was no signal to confirm that they had locked firmly into place. This news further increased tension in the Space Telescope Operations Control Center (STOCC) as they worked to issue new commands to the telescope to overcome the problem. In the meantime, McCandless and Sullivan were directed to continue their preparations for a

Hubble is moved by the RMS over the crew module. Note folded solar arrays and stowed high gain antenna.

contingency EVA, to preserve an Orbit 20 backup deployment opportunity. The manual openings of the masts, unfurling of the solar arrays, and deployment of the high gain antennas were three of the primary tasks which they had trained to carry out. However, shortly after they had resumed their suiting up in the airlock, STOCC received confirmation that both masts had been locked into position.

## Solar arrays and high gain antennas

The next step was to unfurl the solar arrays. With events running 30 minutes behind schedule, and with Discovery moving into darkness, it was decided to postpone the deployment of the arrays and to advance the deployment of the high gain antennas to make up some of the lost time.

When news came in that the first array was ready for unfurling, the EVA crew decided to remain on the flight deck and help, since they had a greater feel for nominal deployment of the arrays than their colleagues. As the first array rolled out of its canister all looked fine, but when the signal was sent to halt its motion, this cut off the data which would have confirmed that the array was locked in its fully deployed position. More delays ensued as further checks were required prior to deploying the second array. McCandless and Sullivan were informed that they should resume their EVA preparations on the middeck as the second array began to be unfurled, then suddenly stopped. It appeared that the built-in safety measures designed to halt the deployment if the tension on the array exceeded 10 pounds (4.53 kg) had intervened, indicating they might require a little manual assistance to unfurl fully.

McCandless and Sullivan finished donning their pressure suits and closed the inner hatch of the airlock, ready to depressurize it. While waiting, Sullivan reviewed her cuff checklist to mentally rehearse the manual deployment of the array, a task that she was trained as prime for. She reasoned that the problem she was hearing about over the radio did not seem to be a mechanical issue—nothing was jammed or broken. The problem was likely to be a software issue, and they should be able to overcome the difficulty if they proceeded with the EVA and used a manual tool. Meanwhile, the crew on the flight deck carefully inspected the partially deployed array and reported that there were no visible problems. Some 30 minutes after the first attempt, the second array was commanded to continue unfurling. It did briefly, but then stopped as the tension warning system again intervened. The plan, by now, was to have had both arrays extend and locked in position, generating a "power positive" situation in which Hubble could generate more power than it actually needed, but the reality was one and a half arrays out and a long way from the desired power status.

The EVA preparations had reached the point where the pre-breathing had to be stopped in order to retain sufficient rapid response time to support an EVA. If the third attempt with the array was unsuccessful, the airlock would be depressed but McCandless and Sullivan would remain inside, ready, if needed, to open the outer hatch and venture into the payload bay to manually deploy the solar array. Orbital darkness was looming. Not only was there a power constraint, there was also a temperature time limit to ensure the telescope could survive on its own. It did not help to ease the tension for the crew that they had been told beforehand that it was important to have both solar arrays out in order to ensure the survival of Hubble. The fact that this was not the case notched up the concern inside Discovery, as well as on the ground.

Meanwhile the high gain antennas were also commanded to deploy, and were confirmed locked into position. In images taken of the solar arrays, it was noted that wires near the dish on the number 2 antenna were bowed out of their normal position and had become hung up. There is often criticism that the space program spends a lot of money upon the simplest of devices, but this is not always true. In this case, engineers reverted to using a children's toy model of the telescope to visualize the antenna motions and as a result, on April 30, the day after Discovery landed, they were able to free the dish. However, its future movement would remain restricted in order to ensure that it would not again become tangled in the noose-like wiring.

After the final simulation, completed some weeks previously, the crew could not believe what they were seeing. McCandless suggested it was the tension monitoring module that had halted the deployment in order to protect the array. If there was undue strain on the structure, then the crew should be allowed to go out and fix it by manually pulling it out. This untried exercise worried Bolden, who had the responsibility of ensuring the two EVA crewmembers were correctly kitted out prior to allowing them to venture out. He knew they were extremely well trained, their equipment was more than adequate, and their procedures were sound, but the proposal really bothered him.

On the ground, engineers at Goddard continued to ponder the problem but data revealed that there was not enough tension to have triggered the halt. The suspicion then fell on the tension check sensor; perhaps it was simply too sensitive. The decision was to try again, but this time override the tension check command. The decision was easy, but the paperwork to authorize this option took longer. Eventually, the command was passed up to the telescope. Not wishing to waste more time in waiting for full sunrise, the controllers commanded the array motors to start the motion in darkness, with several onboard cameras monitoring the event. This time the array reached its full length, and at 1 day, 6 hours, 30 minutes into the mission both arrays were confirmed locked. All considerations for an immediate EVA were put on hold, with McCandless and Sullivan fully suited floating in the airlock, which was almost depressurized. Although McCandless knew the attempt would work—that was why the procedures were devised and the tension monitoring module was installed on Hubble in the first place—he was disappointed that he and Sullivan wouldn't get to perform the EVA for which they had trained over so many years.

## AN ASTRONOMER DEPLOYS HUBBLE

The series of delays had cost the team the intended deployment on Orbit 19, but the backup deployment on Orbit 20 seemed a plausible opportunity. With the opening of the deployment window just 15 minutes away, and the release opportunity fast approaching, there was not much time to orientate Hubble in its release attitude. Hawley recalled the situation after the flight, "When the solar array that had been giving us trouble was finally unfurled properly, we actually did not have very much time until the release opportunity. There were a number of activities that Loren [Shriver] and I had to do, with Charlie [Bolden]'s help." In particular, Hawley had to position the arm in such a way that it could be promptly moved out of the way when Hubble was released.

8  **Deployment**

Discovery's Remote Manipulator System (RMS) is shown attached to Hubble high above the Earth. Yellow handrails, foot restraint attachment points and insulation covering are clearly visible.

With everyone tied up in their own duties, and after an intense year of choreographing their activities for exactly this point, none of the three astronauts on the flight deck actually had thought to grab a camera to record the event. It was at this point that the advantages of cross training crewmembers became apparent in an unusual way. Hawley casually pointed out to Bolden that both McCandless and Sullivan, currently in the airlock, were the prime crewmembers for taking still images and IMAX footage. In 2004 Bolden recalled this episode as being a "nightmare", but it also gave rise to his "one moment of fame" on the mission. All members of the crew had trained to use the IMAX equipment, both the bulky cabin unit and the one installed in the payload bay, together with the other photographic equipment, "so we were ambidextrous; we could all do what was needed to be done. As it turned out, this was fortuitous because the two primary camera operators were locked in

Solar panels are unfurled while the RMS remains attached.

the airlock." Hawley was operating the arm, and Shriver the vehicle. Bolden was backing them both up. They were "playing musical chairs, trying to get the cameras set up and document everything we did ... Between the three of us we managed to capture everything there was to capture on the deploy and [got] some absolutely spectacular footage of Hubble." Bolden did manage to get an interior shot with IMAX once his colleagues were out of the airlock. The unit flown had no automatic mode so all of its settings were manual, and it amazed Bolden that he captured a spectacular sequence coming up from the middeck featuring a view of Shriver at the aft controls flying the vehicle, and then a view outside the window at Hubble. It remained in focus for the whole sequence. As Bolden admitted 24 years later, "Everybody said that was absolutely phenomenal, but I didn't have a clue what I was doing, it was just luck."

There were only minutes left to the release point, but for the crew primed to let go of the telescope the wait seemed much longer. Finally, as the window opened, the 'go' was given for deployment, which would occur a minute later than planned. The release from the RMS was confirmed by an indication on the Payload Deployment and Retrieval System (PDRS) console at Mission Control. The telescope was finally flying solo in Earth orbit as the shuttle passed high over the Pacific approaching Ecuador in South America.

Hawley later recalled the surprisingly small amount of clearance between the arm and one of the solar arrays at separation; it seemed less than the simulations had led them to believe. "However, the control of the arm was very precise", he noted, "and the control of the orbiter at separation was very precise. We had no concerns about contacting the arm with the solar arrays, but it was something that Loren and I kept a very close eye on."

Two small separation burns were completed to move the shuttle away from the telescope. McCandless and Sullivan had remained in the airlock until Hawley had cradled the RMS, as another planned precaution in case they had been required to manually latch the arm down. Because this was the first opportunity since STS-61B in December 1985 that an EVA team had the opportunity to gain valuable inflight data from running tests on the pressure suits, by remaining in the airlock to finish these tests the two astronauts had passed up the experience of seeing the telescope deployed.

## A pretty impressive sight

Shriver thought the sight of the telescope out of the window with all its appendages deployed looked "awesome, a pretty impressive sight... we had spent [years] training for these events. Until you actually get to see it happen with the real piece of equipment, there is always some tendency to underestimate what it's going to look like, and I think that was probably the case here." Even seeing the same solar arrays deployed in the factory in England, was different to seeing them stretched out from the sides of the huge telescope at 330 miles (531 km) above the Earth. "It adds quite a bit to the picture," Shriver recalled.

For Sullivan there was a tinge of disappointment at not witnessing the actual deployment, stuck inside the airlock. She had resigned herself to the fact that this was one of the scenarios they had prepared for, but simulations are not space flights. On later seeing film coverage of the deployment, she mused wryly that although she had been just 8 feet (2.44 meters) from the action, she had missed it all. She had had a "really rotten view" in the airlock and couldn't hear (off radio) the excitement on the flight deck.[4] In his 2004 oral history, Bolden recalled, "We're just oohing and aahing, and Bruce and Kathy are going crazy, because now they have lost their EVA and they didn't get to see their telescope [deployed]."

In this IMAX still frame, Hubble flies free with its solar arrays outstretched but with its aperture door still closed.

For astronomer-astronaut Steve Hawley, the achievement of personally releasing the telescope was a big moment in his career, "although it was tempered a bit by the fact that we had not opened the aperture door and so we would remain on standby for the next 48 hours if the door did not open. So I don't remember feeling relaxed until we got notice two days later that the aperture door had worked, and that was when we put the sign on the middeck saying that Hubble is open for business."

12  Deployment

**Minimum deployment altitude**

One of the overriding issues in the planning for STS-31 was the deployment altitude of the telescope. As Hubble had no onboard propulsion system, it relied upon the shuttle to perform re-boost maneuvers to prevent premature orbital decay. The concerns were not for re-entry, as the shuttle was planned to revisit the telescope for servicing and possible re-boost every 3 to 3.5 years, but, as Steve Hawley wrote in 2014, "as the atmospheric density increased, the reaction wheel assemblies would not be able to overcome the increased drag". The end result would have been a loss of precision pointing that compromised the science. Had Hubble been placed in an orbit that required more frequent re-boosts, more servicing missions would have been necessary. By raising the overall cost of the program, this would have undermined the logic of planning for more missions.

By 1990 the shuttle had been flying for 9 years and normally its orbit was between 172 and 231 miles (276 to 371 km), so 380 miles (600 km) was a challenge to its capabilities. The altitude of deployment was dependent upon a number of factors including the initial altitude and an estimate of the timing and intensity of the solar cycle relative to the launch date, since the timing of the launch would affect the number of re-boosts required during the planned 15 year Hubble program.

In April 1985 the Marshall Space Flight Center conducted a pre-flight planning analysis for STS-61J which predicted a 1986 launch. The study suggested that if Hubble was deployed with the lowest point of its orbit in the range 264.67–307.25 miles (425.96–494.48 km) then it would require the first re-boost just 8 months after deployment, and six re-boosts by the middle of 1992. On the other hand, for 362.49–368.24 miles (583.38–592.64 km) a re-boost would not be needed for another 4.25 years, with only three required by mid-1992. The 1986 Challenger incident forced a delay to the launch of Hubble and changed the estimates of re-boosts as the solar cycle progressed. By 1987, it was estimated that 310.71 miles (500 km) would require two re-boosts, while 362.49 miles (583.38 km) would require only a single re-boost. When a Flight Operational Panel meeting on November 10, 1987, reviewed the launch of the Hubble deployment mission, now designated STS-31, the recommendation was for a June 1, 1989 launch and a deployment at 345.23 miles (555.6 km). Should the orbit be lower than that figure, a management decision would have been required either to release or return the telescope. As the months progressed, the data was revised but the minimum deployment altitude was not affected, even when the actual launch occurred 11 months later than planned at the meeting in 1987.

According to Steve Hawley, "At that time there was a discussion about accepting a lower than targeted deploy altitude, and of investing propellant to raise the orbit by giving up the contingency rendezvous option. The need to achieve the minimum deployment altitude was one factor in the selection of Discovery over the more massive Columbia for the HST deployment mission."[5]

**PREPARING TO FLY**

It had been a long and at times difficult path leading up to this point. From the first realistic suggestions during the 1940s for an optical telescope in space, through numerous proposals, studies, and budget wrangling, to authorization of the project in the late 1970s. Then there was a wait of 13 years before the telescope was launched; the result of several delays, most notably the loss of the shuttle Challenger and her crew of seven in January 1986, just 9 months before Hubble had been scheduled for launch. Following the shuttle's return to flight in 1988 there was optimism for the resumption of regular flights and catching up with the manifest and the backlog of payloads, in particular the Hubble telescope.

**Naming the crew**

During the 1970s and early 1980s several astronauts held technical assignments supporting the development of the Space Telescope, mostly related to developing EVA and servicing techniques. On April 5, 1985, Bruce McCandless, Steve Hawley and Kathryn Sullivan were the first astronauts named to the Hubble deployment mission, which was then manifested as STS-61J and scheduled for the third quarter of 1986.[6] They were the Mission Specialists on the crew. McCandless (MS1) and Sullivan (MS3) were also to train to make a contingency EVA to support the deployment or re-stowing of the telescope in the payload bay in the event of a problem. Steve Hawley (MS2) would serve as primary RMS operator and place Hubble on-orbit; he would also serve as launch and entry Flight Engineer on the flight deck, assisting the still to be named Commander and Pilot during ascent and landing.

Bruce McCandless, as part of his technical assignments in the Astronaut Office, had been working on Space Telescope EVA issues for over 6 years when he was named to the crew that would deploy it. In February 1983 he had been assigned to the crew of STS-41B, which flew in February 1984, and he also assisted in developing and selling the concept of the Solar Max retrieval and repair that was carried out by STS-41C in April 1984.

A short time before the public announcement, McCandless was called to the office of the Director of Flight Crew Operations at Johnson Space Center in Houston, Texas, who, at the time was George Abbey. McCandless was asked whether he was interested in flying on the telescope deployment mission which, Abbey explained, would have a contingency EVA requirement on Flight Day 2 to assist in the event that the telescope failed to deploy properly. This was an important consideration. Abbey noted that a crew was being put together where everybody had flown before, in order to take the maximum advantage of their experience in training and in flight. Another selection factor was that an astronaut must not have previously shown any susceptibility to Space Adaption Syndrome (SAS), a form of space sickness. This was important because most of the primary objectives of the mission would occur in the first 48 hours, and SAS could have disabling effects on an individual while the body was adapting to weightlessness during the first 2 or 3 days of a flight. Some astronauts suffered from this affliction, others did not. As there was no way to predict who would be susceptible, it had been decided to create a crew of veterans who had proven immune. McCandless told Abbey that he was eager to participate in the mission, and assured him he could handle the EVA if required. Prior to this discussion in Abbey's office, McCandless had no prior knowledge of his pending assignment to the deployment mission. "The whole flight crew assignment saga was [still] thrown up in the air darkly, with not much transparency," he recalled in 2006, but he had noticed that Abbey did seem to credit what previous assignments had been held, and then try to match these to logical flight crew assignments.[7]

At the time of the announcement, Hawley was in training to fly STS-61C. This was not unprecedented, but it was uncommon to be assigned to two missions at once. For his first mission on STS-41D the group of astronauts who were to be assigned to that flight were all called over to George Abbey's office and told of their assignment. But for Hubble, Hawley recalled, he was called to Abbey's office on his own. "He just called me in and said, 'We're going to assign you to the Hubble launch' and I think I said 'Great.' Then I said, 'Who else is on the crew?' I think George said, 'Does it matter? And I quickly said, 'No'".[8] Hawley was aware that if he were to say that he was not prepared to fly with someone, Abbey might have second thoughts on assigning him. But he was pleased to get another assignment, and time to train for it. "Frankly, [it's] not so much that you get trained," he explained, "but that there is an actual crew who can work the [various] mission issues. Hubble was obviously a big deal, and I think George wanted to have an assigned crew to be able to go to meetings and engage with Hubble science teams and things like that. I am looking at it now from the perspective [many years later] of having been a manager and that would be far more important—getting the flight crew involved in mission issues at the appropriate time."

**Change of command**

It was not until September 19 that NASA announced John Young as Commander and Charles Bolden as Pilot.[9] This completed a very experienced crew of five astronauts, all of whom had previously flown on the shuttle. Young had been an astronaut since 1962 (Group 2) and had served as the Chief of the Astronaut Office since 1974. He was by far the most experienced astronaut still in the office, having flown two Gemini and two Apollo missions including the fifth lunar landing. In addition to commanding the first shuttle mission in April 1981 he had commanded the first Spacelab scientific research mission in December 1983.

Bruce McCandless's career as an astronaut was almost as long as Young's, having been selected in 1966 (Group 5), but with far fewer missions. Having served as a member of the support crew for Apollo 14 and the backup crew for Skylab 2, he had worked for years to develop EVA techniques and equipment for the shuttle, notably the Manned Maneuvering Unit, and in support of the development of EVA methods for servicing the telescope. As a crewmember on STS-41B in 1984 he became the first person to make an untethered EVA, flying the MMU a distance of 328 feet (100 meters) from Challenger.

The other Mission Specialists were members of the first shuttle era astronaut selection in 1978 (Group 8). After Steve Hawley flew on STS-41D, the maiden mission of Discovery in 1984, he was scheduled to fly on STS-61C later that year (subsequently delayed to January 1986). Sullivan had previously flown STS-41G, also in 1984, becoming the first American female to conduct an EVA, and that experience probably led to her position on the telescope deployment flight.

Pilot Charles Bolden was selected as astronaut in 1980 (Group 9) and at the time of being selected for the Hubble deployment mission had not yet flown in space, although he was in training as Pilot, together with Hawley, for STS-61C. Bolden was elated at the assignments, not only because the mission was going to deploy Hubble but also because he would get the opportunity to fly as Pilot to the legendary John Young.

**Preparing to fly** 15

STS-61C landed on January 18, 1986, and for the next few days both Bolden and Hawley were busy readjusting to life on Earth and participating in the formal post-flight debriefing in advance of joining the STS-61J crew to prepare for launching aboard Atlantis on August 18, 1986. But Challenger and her crew of seven were lost on January 28 in the STS-51L launch accident, and the program ground to a halt. All assigned crews were stood down and placed into generic (basic proficiency) mission training, pending a new flight manifest. Astronauts also supported the Accident Investigation Team and Review Board in their analyses of what happened to Challenger, submitting recommendations to the Return to Flight program.

For the most part, the selected Hubble crew remained together and continued very basic refresher and familiarization work, but no specific mission training. A year after the loss of Challenger things began to change. On April 15, 1987 it was reported that Dan Brandenstein would take over as Chief of the Astronaut Office, with Young moving to a more managerial role.[10] Then on March 17, 1988, NASA revealed that the Hubble deployment mission, now STS-31, would be commanded by Loren Shriver instead of the grounded Young. The other members of the crew were the same: Pilot Charles Bolden, MS1 Bruce McCandless, MS2 Steve Hawley, and MS3 Kathryn Sullivan.[11] According to Hawley, formal crew training for STS-31 did not begin until July 1989.

A proud crew displays the mission emblem [*left* to *right*] Loren Shriver, Charles Bolden, Kathryn Sullivan, Bruce McCandless and Steve Hawley who is holding a model of the HST.

In 2002 Loren Shriver was interviewed for the JSC Oral History Project, and mentioned that he was asked by John Young whether he was interested in taking the command seat on STS-31. "I was up in Canada, participating in one of the IMAX opening events, John wanted to know if I was interested, and I said I most certainly was. I didn't get much information on the phone, other than the number STS-31 and that it was the Hubble mission."[12] The fact that the call came from Young, rather than Brandenstein, who had replaced him in the Astronaut Office, probably reflects the desire of Young to personally hand over his last command seat now that he was moving to more managerial and administrative roles. Although the veteran astronaut officially kept his 'active' status for several years, the prospect of a seventh mission was extremely slim. Unsurprisingly, the transfer of command from Young to Shriver wasn't discussed in *Forever Young*, John Young's 2012 biography, written together with James R. Hansen. Flying as commander of the Hubble deployment mission would have been a fitting end to an illustrious astronaut career, but it was not to be.

STS-31 was a contrast to Shriver's first mission in 1985. The first flight of an astronaut usually grabs the attention of the media, but for Shriver as Pilot-51C, and his other rookie colleagues, their first flight had been a Department of Defense mission and was cloaked in secrecy. Shriver could not even tell his family about the mission, so when he was assigned to the Hubble mission there was a lot of publicity anticipating the deployment of the telescope.

"If there were ever two missions that were completely opposite in terms of public attention given to them, it would be my first and second missions," Shriver said in 2002, reflecting on the huge interest generated in the Hubble missions, not only the pending deployment, but in the forthcoming service missions as well. This pushed the crew into the spotlight rather more than recent missions, and was a hint of how important public outreach was to become for the Hubble related missions.

It had been a challenge to keep the Hubble team intact for 3 years and maintain their proficiency with such a protracted training cycle. When STS-61J became STS-31 it had not been certain that they would in fact remain as a crew. Steve Hawley recalled a memo from late 1986 or early 1987 saying that all previously assigned crews were dissolved, and at that point he did not take for granted that he would be on the Hubble crew when it was reformed. Nevertheless, he did think it would make sense to keep the core crew together, and certainly McCandless and Sullivan, in order to track the status of Hubble and work on interfaces and EVA procedures. In 1987 Hawley became Dan Brandenstein's deputy, thereby establishing the tradition within the Astronaut Office during the shuttle era of the Deputy Chief being a Mission Specialist. Although his new job involved a lot of non-Hubble issues, Hawley kept up his skills on the RMS whenever he could find the time.

## Training

When the new crew announcement for the re-manifested deployment mission was made, a number of astronauts including Mark Lee, George 'Pinky' Nelson and Bruce McCandless had been working on various EVA issues for some time, mostly related to tool development interfaces and to assessing whether the item of apparatus supplied by the payload sponsor or the procedures devised by the flight operations team really worked. The liaison between the payload sponsor and the flight operations team could at times result in differences of opinion and procedure. Support astronauts assigned to track a task of developing a system for a given mission could often suggest the best way to do things, and this could be completely different from the way that a flight-assigned crewmember may wish the same task to be approached or worked on. It was natural, at least early in the program, that the flight-assigned members of the office should have greater influence because it was they who would have to carry out the task. As Steve Hawley noted, "I think that is what George [Abbey] had in mind, to have the guys who're going to fly identified so that they had some leverage in some of the meetings."

Following the loss of Challenger, in April 1987 Hawley was assigned as Deputy Chief of the Astronaut Office and therefore became deeply involved with administrative assignments. Like many of his colleagues, for most of 1986 and 1987 he was involved in the accident investigation. As a result, he was not able to devote much time to Hubble-related training or meetings, although as the robotics lead on STS-61J, he did manage to continue work on a number of issues which were related to the Hubble deployment mission, including using a laptop-based RMS simulator which had a model of Hubble on it and could be linked into two hand controllers. "When I had an hour free during the day, I would go downstairs and use that just to keep [my hand in]. It was a pretty good model and I could run through procedures and get familiar with the procedures, but it was not as good as some of the other model trainers."

McCandless did not keep a record of his EVA training for STS-31. In fact, at the time of his selection for STS-61J he didn't put in a lot of training time at all because it was 12 to 18 months away. With Hawley assigned to STS-61C, both McCandless and Sullivan conducted some basic simulations and refreshed themselves on a few things. As McCandless explained, "The consensus was that each of us had flown in the shuttle fairly recently, so we thought we were reasonably well trained. With respect to the Hubble contingency EVAs, both Kathy and I had done a lot of work up at the Huntsville [water] tank. When we got the Hubble mockup down in Houston we couldn't stand it up, despite what the supervisor of the water tank there had promised us. It had to lie on the bottom of the pool with part of the forward bulkhead of the shuttle in the water too. We didn't really have good spatial relationships, so it was just a matter of principally demonstrating that we could deploy a solar array if necessary, jettison one, the same with the high gain antennas, the electrical connectors, and the aperture door." Following the experiences of previous on-orbit servicing missions where unexpected satellite configurations were encountered, McCandless and Sullivan, using data obtained during the integration and testing of the telescope, embarked upon a comprehensive review of tool fit-checks and clearance levels. They also reconfirmed the required torques for each tool, just to ensure they were fully prepared for any eventuality. Most of the generic EVA training for the two astronauts focused on the latches that secured the payload and using the winch and ropes for closing the payload bay doors, but for issues related to the solar arrays more specialist training was required.

There were three trips to the manufacturer of the solar arrays. "One inordinate amount of effort was to get British Aerospace in Bristol, England, to agree to make the Marmon clamp that held each solar array, so that it was reversible—meaning that it was not just for release and jettison of the solar array but could be used for installation of our new arrays. The spec had said for release, jettison, or removal, but nothing about installation, and by golly they weren't about to back off from that. There were all sorts of obtuse rationales, but we finally demonstrated that it could be done out at the clean room at Lockheed, Sunnyvale, and so it got done. It was totally necessary, as we went on to replace the solar arrays twice and if we hadn't had that capability the telescope would be out of business," McCandless explained.

When Shriver was named to command the mission in 1988, the four other astronauts had been together in theory as a crew for 3 years. In fact, McCandless and Sullivan had been working on Hubble deployment issues for longer than that, although not in constant mission related training. Joining the crew after such a length of time could have been challenging, but Shriver didn't recall any real problems. "I think coming on into the crew went very well", he reflected, "I didn't notice even any ripples in my training or theirs. There were just five of us. It was a joy working with them, everybody felt they had a defined purpose and basically got right to the training… it all went very well and smooth."

In his 2004 NASA oral history Charlie Bolden reflected that training for the deployment mission benefited from its having three of the most Hubble-aware astronauts in the office—McCandless, Sullivan and Story Musgrave (as one of the Capcoms). With their extensive background on the telescope, its workings, EVA procedures, and past flight experiences, "they knew everything there was to know".[13] He said they were prepared to "expect the unexpected", and for McCandless and Sullivan to evaluate and overcome the problems.

For 6 months the crew proceeded through generic training to bring them up to speed on developments and upgrades since their last sessions over 2 years earlier and refresh their knowledge in some systems. It resumed the crew ethos of working as a team. In the final 6 months prior to launch they carried out integrated training with the assigned controller teams at Mission Control. The focus of this phase was on the link between the mission controllers, the flight crew, and the flight plan—both the nominal one for routine operations and various off-nominal plans rehearsing all manner of contingencies and incidents. As Bolden recalled, "I always tell people, its 'catastrophic training', getting ready to go fly in space. You'd love to get in the simulator and everything goes right. Never happens. The training team's life is designed to make you miserable all through the training, but prepare you for everything that can go wrong. And they generally do. They do a superb job of imagining every conceivable thing that can go wrong, and exposing it to you at least once."

For familiarization, the astronauts visited Lockheed Martin in California to view the final assembly processes and system checkout. They also made a trip over to British Aerospace to inspect the first set of solar arrays and the method of deployment, both automatically and, if required, manually during an EVA. As Shriver has observed, "It was deemed pretty crucial—and, indeed, it turned out to be pretty critical—to know what the solar array looked like and how the mechanism functioned as it was being unrolled simultaneously in both directions. I'm very glad, looking back on the deployment mission, that we had the opportunity to go do that."

Bolden also recalled that their final integrated simulation exercise included a solar panel failure mode which proved to be a critical factor in their preparations. The simulated failure would require sending McCandless and Sullivan out to manually wind out the array. Whilst they were confident of success if called upon to attempt this task, there was an element of doubt. During such an EVA the telescope would have to be taken out of its automatic mode and would no longer be capable of taking care of itself; if things went from bad to worse, a second mission might have had to be mounted early in order to restore the telescope to full operation capability. The worry was the time between the deployment mission and such a 'rescue' service mission, and whether the incapacitated telescope would still be in a good enough condition to be repaired. Trying to prepare for every eventuality once the telescope was in space was a nerve wracking experience for the entire team.

When the crew visited British Aerospace to inspect the flight solar arrays, the company's engineers had devised a training mockup of a long tank filled with water that could suspend the solar array when it rolled out. Each crewmember had the chance to roll the array out and back to observe what it would be like for real. Both McCandless and Sullivan became quite competent in this task.

Because the mission focused upon the deployment of the telescope, most of the training for McCandless and Sullivan centered on EVA options in support of that process. They kept up to speed with the latest developments on preparing the telescope for launch, and with the nominal and contingency sequences for deploying the telescope out of the payload bay. The main function for Hawley as the primary RMS operator, was the physical placement of the telescope on-orbit. In addition to their orbiter duties, Shriver and Bolden also played a vital role in commanding Hubble. As Shriver recalled in 2002, "We didn't have much insight into the actual systems, the internal working of the [telescope]. We were keypunch operators and command relayers and things like that. But being able to get into the software that deployed the telescope or the solar arrays and actually bypass that module, that had to be done on the ground."

Kathryn Sullivan and Bruce McCandless train underwater for HST contingency EVAs in the Neutral Buoyancy Simulator (NBS) at Marshall Space Flight Center, Huntsville, Alabama.

20 **Deployment**

Emergency egress training at the 195 feet level of LC 39B, KSC, during the Terminal Countdown Demonstration Test on March 20, 1990.

## Developing an EVA plan

Even though this was only the deployment mission, the delay resulting from the Challenger accident allowed additional time to re-evaluate the list of on-orbit servicing tasks and refine planning for contingencies. Some of the inputs from the crew during this period included a suggestion by Kathy Sullivan that the bracket which secured the low gain antenna should be modified to allow replacement in space. It was also the suggestion of the crew that clusters of red and yellow reflectors of the type used on bicycles should be installed on the aft bulkhead of the telescope to provide orientation cues during the rendezvous.[14]

Prior to resuming shuttle flights in 1988, changes were made to the way that the astronauts were to conduct EVAs based upon work done by Bruce McCandless and Kathy

Sullivan. Up to 1986, flight rules allowed for just two scheduled and one contingency EVA per mission, with each lasting no more than 6 hours. Detailed evaluations of the ability to complete the proposed servicing tasks within these rules revealed that there would simply not be enough time available to complete the work. Even though McCandless and Sullivan were only tasked with potential contingency tasks related to the deployment of the telescope on STS-31, they were involved in the initial planning for the first servicing mission. As a direct result of their work the procedures, timelines and consumables on the flight had been revised by the time the first servicing mission flew to permit up to four, 7 hour EVAs.

Another EVA subject reviewed was when to schedule the contingency EVAs in the Crew Activity Plan (CAP). With Hubble intended to be deployed on FD 2 there was a need for a quick-response EVA capability in the event of anything being amiss. The major consideration centered upon the power supply to Hubble. Secure in the payload bay the telescope received shuttle power via an umbilical, but during the deployment sequence this umbilical had to be disconnected before the solar arrays were unfurled, leaving the telescope reliant on internal batteries until the arrays could lock onto the Sun. If the arrays were unable to be deployed automatically, then with an estimated supply of only 6 hours there would be insufficient time to make the EVA preparations, exit the airlock, and manually deploy the arrays. Therefore, it was decided to begin EVA preparations soon after entering orbit so that the astronauts would be ready to conduct an EVA as the telescope was being deployed.

Under the new flight rules introduced prior to the resumption of shuttle flights in 1988, it was decided that an EVA would not be scheduled before FD 4 and that any contingency EVA related to a payload could not occur before FD 3. These changes were to allow crewmembers to fully adapt to space flight conditions, including any effects of Space Adaptation Syndrome. As Steve Hawley wrote, "There was substantial discussion of whether to impose that new constraint on the HST deployment mission."

During an HST Payload Operations Working Group meeting on July 18, 1989, discussions were held on whether, under normal timeline rules, the EVA crew would have completed the pre-breathing protocol by the time they entered the airlock. In the end, it was reasoned that planning for the deployment mission as STS-61J had been prior to the new rules and included a deployment on FD 2 with quick-response EVA capability, as indicated when George Abbey assigned McCandless to the mission. What helped the decision to allow a FD 2 EVA option was the fact that both McCandless and Sullivan had flown previously, had known adaptation patterns, and had experience of EVA. "Consequently," Hawley wrote, "STS-31 was the only time in the shuttle program when a flight day 2 EVA was an approved option."*

Unlike the subsequent servicing missions, where Hubble was held in the Flight Support System and the spacewalking astronauts were able to mount a platform on the RMS, all of the EVA tasks assigned to STS-31 were designed "non-RMS" because the arm would be

---

*On STS-37 in April 1991 a FD 3 contingency EVA was completed to deploy the high gain antenna of the Compton Gamma Ray Observatory, but this was consistent with the flight rules for supporting shuttle EVAs.

holding the telescope. This resulted in useful planning for potential RMS failures during the servicing missions.

**'Little black books'**

During the post-flight debrief of the STS-31 crew, it was said that "having their little black notebooks prior to flight" was advantageous. Although adding new procedures was never a concern, the fact they carried the books with them into orbit for reference had proven very useful.[15] These were personal handwritten notebooks containing information and reminders which accumulated during training. As Steve Hawley confirmed, "I think the original intent was so that you could take notes during the flight and in particular for debrief purposes; then someone had the idea that if you had the notebooks with you ahead of time, you could make notes during training that would then be useful to you during flight."[16] But it was essential that the data was kept up to date. The Flight Operations Team's concern was that something might have been missed or misinterpreted, which could cause a problem on-orbit.

"It would be helpful if a note reminded you of a certain fact or figure that may be useful," Hawley pointed out. "But for a procedure reminder, for example to throw a certain switch or a sequence of events that had to be followed in a particular way, the operations team would want to ensure that what was written down was exactly the procedure that was on the formal checklist, that things may be done out of sync with formal procedures or planning." In order to ensure that the notebooks, in particular procedures or sequences, were accurately recorded, Hawley remembered that "the ops team, before the books were stowed, made copies of what had been written, so that they could review [them] to make sure there were no discrepancies between what you had written down and the procedure that you were supposed to execute." On later missions these notebooks were superseded by personal computers, then laptops and more recently tablets.

**The Name Of the Game**

The original launch of the telescope was scheduled for August 1986, but was delayed several times over the next 4 years, most notably owing to the loss of Challenger in January 1986. Reworking the program to take account of the recommendations of the accident investigation added to the delays. Then when the shuttle returned to flight in September 1988, the backlog of payloads were re-prioritized and the Hubble deployment mission slipped to late 1989 and then into 1990.

For many of the astronauts who had been around in the program for several years, a fluid schedule and changing manifest became part of the game. In the aftermath of Challenger, the intense focus on flight safety, for both the hardware and the crew, along with technical issues or anomalies from other missions, slips due to processing of upcoming missions, and weather issues at the Cape all added to the mix for astronauts waiting to fly.

Reflecting in his oral history, Loren Shriver tells those who ask what it was like to be an astronaut, that the attributes of "patience and perseverance were the two things that helped me along, and it just doesn't pay to get too excited about changes like that."

## BUILDING THE 'STACK'

By 1989, the elements to launch Hubble were finally coming together at the Kennedy Space Center in Florida. After so many years of waiting for the telescope to be authorized and built there was a sense of relief that, finally, it would soon enter service. But it was not simply a case of placing Hubble on the shuttle and 'lighting the blue touch paper'! The sequence of events involved in readying a shuttle mission was lengthy and, as the preparations for this mission experienced, prone to delays and setbacks.

### Hubble slips into 1990

On October 22, 1987, NASA announced its intention to launch Hubble during 1989 as part of the schedule to catch up with delayed payloads stuck on the ground by the loss of Challenger. On 15 March 1988 the target date was clarified as June of 1989, but in early October, shortly after the somewhat delayed but ultimately successful Return to Flight mission of STS-26, this was postponed to December 11, 1989 in order to enable three planetary missions to achieve their launch windows. Then in April 1989, delays in making the modifications to Columbia meant that NASA had to bump one of its missions into 1990, and the most likely candidate was the Hubble deployment at a cost of about $7 million per month. Further delays resulted in a new updated manifest, issued on May 12, 1989, that listed Hubble's launch in February 1990. In yet other modification to the manifest on June 13, it was slipped to March 26, 1990, this time to protect the launch of the Galileo probe to Jupiter and outstanding DOD payloads.

### Steps towards launch

On October 6, 1990, the Hubble telescope arrived at KSC from Lockheed's manufacturing facility in Sunnyvale, California. It was then moved to the Vertical Processing Facility prior to insertion into the payload bay of the orbiter Discovery.

In November 1989, the launch processing for the STS-31 deployment mission saw the buildup of the Solid Rocket Booster segments in the Rotation Processing and Surge Facility. On November 30, as work continued to prepare the hardware that would launch Hubble, the Solar Max satellite which had been used to pioneer in-space servicing, re-entered the atmosphere, ending its mission of almost a decade. "It is conceivable that chunks of [Solar Max] debris as large as 400 pounds [181.4 kg] could survive [re-entry]," reported the Goddard Space Flight Center. This served as a warning to the Hubble program, because unless steps were taken to prevent it, the telescope too would rain down debris at the end of its mission.

Following the STS-33 mission, on December 4 Discovery, the orbiter assigned to carry Hubble into orbit, returned to KSC on the back of the Shuttle Carrier Aircraft (SCA). The next day, it was towed to Bay 2 in the Orbiter Processing Facility (OPF) for STS-33 post-flight operations and preparations for STS-31.[17] Most of this work centered upon removing the classified elements of the STS-33 payload and de-servicing the orbiter's systems and consumables, checking out the vehicle and removing its three main engines for inspection. Finally, on December 18, the work to create the STS-31 launch configuration

## 24 Deployment

Hubble undergoing pre-launch testing at the Vertical Processing Facility at KSC in February 1990.

began with stacking the left-hand aft booster segment on Mobile Launcher Platform 2 in High Bay 1 of the Vehicle Assembly Building (VAB), a process which was completed 2 days later. On December 21, with the seasonal holidays approaching, Discovery's payload bay doors were closed and the vehicle powered down until the New Year.

Discovery is towed into the Bay 2 of the Orbiter Processing Facility at KSC.

On January 2, 1990, power was restored to Discovery in the OPF and work resumed in preparing the orbiter for its new mission. During the third week of the month, this saw the three main engines (#1 2011, #2 2031 and #3 2107) fitted and a remote manipulator (serial number 301) installed on the port longeron (looking forward) of the payload bay.

Meanwhile, over in the VAB, the stacking of the right-hand SRB began on January 13, but was disrupted a few days later. Results from a 1989 standard leak test of the internal nozzle of the right-hand aft booster performed at Morton-Thiokol, the contractor, established that the internal joint could not be certified for flight and would require to be replaced. So the booster was de-stacked and replaced by one intended for STS-35.

## Deployment

In February a series of tests verified the orbiter to payload connections, the flight readiness of the three main engines, and the operational readiness of the RMS. Though the main focus was on the STS-31 launch, advanced planning was also underway for the *next* mission of this orbiter as STS-41 later in the year. Modifications to the orbiter included adding cooling lines along the fuselage which would assist in controlling the thermal environment from the extra heat produced by the radioisotope thermoelectric generator onboard the Ulysses solar probe that was to be carried by that mission. This type of work, often overlooked in media reports, was essential to keeping ahead in the overall processing flow for the coming year. However, it could also impact upon *current* planning. For instance, the hardware added to Discovery to support STS-41 also had to be taken into account during the weight and balance tests for the mass determination of the orbiter flying as STS-31. Though not directly associated with the Hubble deployment mission, such work illustrated the long lead time needed in preparing a shuttle for launch.

Stacking of the left-hand SRB booster was finished on February 7, followed 12 days later by its partner (resulting in SRB Set BI-037; RSRM #10). During the night of February 9/10, and prior to the transfer of Discovery from the OPF, a communication test via satellite was made between the orbiter and its payload interfaces at the Johnson Space Center and the Goddard Space Flight Center. By the end of the month, the External Tank (ET-34) had been transferred to the VAB ready for mating with the twin SRBs, and final tests were conducted prior accepting the orbiter in the VAB.

Discovery was towed from the OPF to the VAB on March 5, and the following day was hoisted vertically to be bolted to the SRB/ET stack, a task which was completed by March 7. Originally the plan was to roll out to the launch pad on March 15/16, then transfer Hubble to the pad on March 26 and install it into the payload bay 2 days later. However, in order to enable pad workers to get ahead in their tasks, and to obtain several contingency days in the countdown process, it was decided to advance the launch date from April 18 to April 12.

### Nuts!

Often it is the small things that hold up the process for launch. This was the case for STS-31, with the rollout to Pad 39B being delayed by 12 hours due to concerns about a small bearing nut on the nose wheel axle of the orbiter. During processing of both Columbia and Atlantis it was found that the bearing nut on each of these vehicles appeared cross threaded. Analysis of this problem and its possible impact on Discovery was conducted on the removed axle from Columbia, in preparation for STS-35. The axle was returned to the supplier in California for detailed examination. Fortunately, it was given a clean bill of health. If a serious problem had been found, this would have required Discovery to be demated from the stack and returned to the OPF in order to access the nose wheel, thereby significantly delaying the launch. But luck was on Hubble's side and processing resumed.

The long journey to space for Hubble continued with the rollout of the STS-31 stack in the evening of March 15. After a journey of 3 miles (5 km) that lasted eight and a half hours, the MLP bearing the shuttle was in place at Launch Complex 39B. Five hours later, the Rotating Servicing Structure (RSS) was wrapped around the combination.

The flight crew arrived at the Cape in their T-38 s on March 18 for the 2-day Terminal Countdown Demonstration Test, set to begin the next day. During their time at the Cape the crew received briefings on the condition of the payload and vehicle and on the overall status of launch preparations, and carried out an emergency egress training at the Pad 39B. The test itself commenced at the T-24 hour point at 8 am on the 19th and was wrapped up with a simulated Main Engine Cut Off (MECO) shortly after 11 am on the following day; after which the astronauts flew back to Houston.

**Bugs in the system**

The processing for launch resumed after the TCDT, with the March 21 loading of hypergolic propellants into storage tanks in Discovery. Then on March 25 Hubble was transferred in the Payload Transfer Canister to the Pad 39B Payload Changeout Room (PCR). When it came to placing the telescope in the payload bay of Discovery, the gremlins struck once again. As the RSS was retracted, three midges were found on the payload bay doors. For the next few days the number of midges found in traps prevented the transfer because they posed a threat to the extremely sensitive optics and electronics of the telescope, but fortunately Hubble was still in its protective covering inside the PCR. The traps consisted of lighted enclosures with a small vacuum device and dry ice, placed at several strategic positions in the clean room. The PCR lights were turned off to ensure the trap lights would attract the midges, which were drawn into the trap by the vacuum and killed by the dry ice. On March 27 the transfer was finally given the go ahead.

While Hubble was being held in the changeout room, other work continued on the orbiter, including resolving small issues involving Teflon coated seals on one of the main propulsion system turbo pumps, and internal circuitry problems with engine number 3 (2107).

On March 28 the management team held the Launch Readiness Review to determine the readiness of KSC to support the launch. This required a detailed review on the current status of Discovery, the SRBs and ET, payload operations, the readiness of the ground systems to support the launch, safety reliability, quality assurance and range support, including the status of the various abort sites across the globe. In summarizing the state of launch processing Jay Honeycutt, Director STS Management and Operations and Chair of the Management Team, reported there had been a "real fine team effort in getting ready for launch". He pointed out that although there had been some problems, and there remained a lot of work to do before they would be able to launch the vehicle, there were no significant issues posing a threat to the countdown.

**Hubble loaded for flight**

Later that day, the process to install Hubble into the payload bay resumed, starting with the removal of the protective cocoon around the telescope. At 10:40 am EDT on March 29, the process to physically install Hubble into the payload bay began, something that even under ideal conditions would take almost 6 hours. Once Hubble was safely in the payload bay, a series of tests were carried out to verify all connections between the payload and the orbiter. At the same time, a 2 day STS Flight Readiness Review (FRR) was conducted by mission management to again review the status of all components and the planned launch date. As a result of there being unused contingency time in the countdown, it was decided to move the launch forward 2 days to April 10. This was the first time in shuttle history that a launch had been advanced following the FRR.

All functional tests on the stack were completed 2 days before the planned launch, and once the pad had been cleared of non-essential personnel the final ordnance operations were conducted, then pressurization of vehicle propellant tanks and loading of cryogenics into the pad storage tanks. That day also saw the start of charging the telescope's batteries. This was completed 48 hours later, when further protective coverings were also removed from Hubble in preparation for closing the payload bay doors.

At 3 pm EDT April 7, the countdown commenced at the T-43 hour mark. The flight crew arrived from Houston in their T-38 jets later that day. After receiving an update on the status of the shuttle stack and its payload, and the weather forecast, which was 90 percent in favor of launch, they took a final look at Hubble. Later in the day the protective 'shower-cap' over the aperture door of the telescope and the covers of its low gain antennas were removed. The next day the payload bay doors were closed for flight.

**April 10, the first launch attempt**

The final phase of the April 10 countdown progressed well until just after T-5 minutes, when Pilot Charlie Bolden initiated the start-up process for the three APUs which were to provide hydraulic power during the launch and landing phases of the flight. Barely 1 minute after start-up, a problem was detected on the number 1 unit, necessitating a hold in the countdown at T-4 minutes. Early data indicated it was a problem that would require further analysis, so the launch attempt for the day was scrubbed because mission rules call for all three units to be functioning prior to launch. Post-scrub analysis revealed a hydrazine fuel valve inside the APU had failed, allowing fuel to enter the unit at a higher rate. What couldn't be determined was whether this was a faulty controller or the APU itself, so in order to isolate the problem the controller would be tested prior to removing the entire unit.

**Keeping the batteries charged**

The day after the abort, the controller of APU 1 was removed for shipping to the vendor and hypergolic propellants were drained from the vehicle. The flight crew returned to Houston to await the setting of a new launch date. After it was determined that the APU controller was not at fault, it became necessary to replace the APU before Discovery could be cleared for flight again. A new APU unit was fitted on April 12, and a new launch date of April 24 was set which offered a window of two and a half hours to get the vehicle off the ground.

Although setting a new launch date was not in itself a major concern, keeping Hubble's six nickel-hydrogen batteries at their peak of performance during the delay was. This could be done either by hooking the batteries up to a ground-based charging system or removing them from the telescope and taking them to another facility. The management team and the contractors decided to minimize the risk of contamination by removing all six batteries and relocating them to the battery facility in the VAB. Because this was not a normal situation, payload technicians spent April 13 practicing the removal and relocation using mockups of the real batteries and their stowage location in the telescope in order to identify any issues. The batteries were removed from the telescope the next day and taken to the VAB, where they were to be charged for 130 hours at 30 degrees Fahrenheit (minus 1.1 Centigrade). In the meantime, the payload bay doors of the orbiter were kept shut to protect the telescope, and devices were installed inside the vacant battery bay in order to monitor contamination levels. Over the weekend of April 14/15 the APU 1 unit was replaced, and was cleared for flight 3 days later. The charging of Hubble's batteries had been completed by April 19, and they were returned to the telescope the next day.

The countdown resumed at the T-43 hour mark on April 21 with the flight crew returning to the Cape on April 23. The 3 hour ET propellant loading operation began shortly after midnight on the 24th and later that morning the pad was re-opened to enable the ice team to inspect the vehicle and launch structure and the closeout crew to prepare to receive the flight crew. With the weather offering an 80 percent chance of launch, the prospects were looking good for the second attempt to place Hubble into orbit.

## April 24, the final countdown

The crew awoke around 3:35 am EDT on April 24 and, after breakfast, were briefed on the status of the shuttle. With a 70 to 80 percent prediction of favorable weather (later revised to 90 percent), the only concern was at Edwards Air Force Base in California, where recorded wind speeds were too high for an Abort Once Around (AOA) landing, so the AOA site was switched to White Sands, New Mexico, which had last been used 8 years earlier for the STS-3 landing.

Less than 2 hours after awakening, the five astronauts, dressed in their orange launch and entry 'pumpkin suits', departed the crew quarters at the Operations & Checkout Building to ride the 'Astrovan' out to the launch pad. Onboard Discovery, Mission Commander Loren Shriver occupied the flight deck front left-hand seat (#1) with Pilot Charles Bolden taking the front right-hand seat (#2). Between them in the aft center seat (#4) was MS2 Flight Engineer Steve Hawley. For launch, MS1 Bruce McCandless had the seat (#3) directly behind Bolden. MS3 Kathy Sullivan took the single mid-deck seat (#5), next to the side hatch. At the end of the mission, for entry and landing she was to swap positions with McCandless.

The countdown proceeded smoothly until T-31 seconds, when a fuel valve failed to shut properly. The problem was rapidly traced to faulty software and engineers overrode the error to allow the countdown to proceed. It was discovered that earlier in the month a burst water line had shorted electrical equipment that forced a temporary closure of a Launch Processing System control room. This had prompted a review of such incidents and the implications for the final stages of a countdown. The remedy was to change the computer software to protect the vehicle from damage, but the software was not tested completely and this resulted in the failure on April 24.[18]

The final seconds passed smoothly, and at T-6.6 seconds Discovery's three main engines ignited and reached 100 percent thrust in less than 4 seconds. With the ignition of the twin SRBs at 08:33:51 EDT, the launch hold down bolts were severed and Discovery, with Hubble nestled safely in its payload bay, lifted off the pad. The Space Telescope was airborne at last. The three main engines were throttle up to 104 percent at T+4 seconds, then at T+7 seconds the vehicle cleared the launch tower, the milestone that marked the transfer of flight control from the Kennedy Space Center to Mission Control at the Johnson Space

30  **Deployment**

Center in Houston for the remainder of the mission. One of those at the Cape to witness the historic launch was Professor Lyman Spitzer, who had the satisfaction of seeing his 40 year vision for sending a large telescope into space become a reality.

The Firing Room at KSC during the launch of STS-31. (which can be seen at top left through the windows).

## Hubble reaches space

The ascent profile was a Direct Insertion (DI) that dispensed with the nominal first firing of the Orbital Maneuvering System (OMS) known as OMS-1, owing to increased performance of the Main Propulsion System (MPS). This profile was designed to put Discovery into an initial orbit with an apogee of 325 nautical miles and a notional perigee of only 25 nautical miles (374×28.77 statute miles, 607.4×46.3 km) inclined at 28.45 degrees to the equator. At T+2 minutes 6 seconds into the flight, the twin SRBs were successfully jettisoned. They parachuted into the Atlantic Ocean for recovery, refurbishment and reuse. At 8 minutes 32 seconds, the three main engines were shut down. Then the spent ET was jettisoned and the orbiter fired its Reaction Control System (RCS) engines in order to separate from the tank, leaving that to pursue a destructive re-entry.

Hubble was in space, but not yet in a stable orbit. To achieve this, at T+48 minutes, near apogee, the twin OMS engines made a 4 minute 30 second burn that circularized the orbit at 311×331 nautical miles (357.89×389.9 miles, 575.97×613.0 km). Lasting 5 minutes 4.8 seconds, this OMS-2 was the longest of the shuttle program to date. It produced a change in velocity of 496.7 feet (151.39 meters) per second. Bolden later noted that the OMS engines gave a "smooth acceleration" and the RCS firings sounded like "cannons!"[19]

Less than 90 minutes later, as the crew began their second circuit of the Earth, Bolden commanded the payload bay doors to open to expose the radiators on their inside faces, at which point Mission Control gave the crew a 'go' for orbital operations.

From his vantage point at the front of the flight deck, Shriver had noticed at MECO that even the coast prior to the OMS maneuver was different to his first shuttle ascent 5 years previously. He knew they were going to end up much higher on this flight, which officially set a record for the shuttle; Hubble was already flying high.

## LETTING HUBBLE GO

It had taken decades from the earliest suggestion to put a large optical telescope into orbit to achieve the feat. The focus of the crew of Discovery and the controllers on the ground was now to deploy the telescope so that, after a series of tests, it would be able to begin its much anticipated science program.

## April 24, preparation day

Given the 'go' for orbital operations, Mission Specialist Steve Hawley on the aft flight deck powered up the RMS just 2 hours 54 minutes after leaving the launch pad. The first task was to ensure that the arm functioned as designed, by checking each of the various systems and modes of operation. With this done, Hawley activated the black and white camera at the end of the arm and conducted a visual survey of the telescope for any damage which might have been caused by the stresses and strains of the 8 minute ride to orbit. Fortunately, it appeared to be in good shape.

In the STS-31 Post-Flight Report, it was noted that additional time should be allocated to RMS checkout on Flight Day 1. After retrieving the LDEF satellite 3 months previously the crew of STS-32 had made a similar recommendation. Hawley reported that re-scheduling his activities in real time was not conducive to getting the RMS checkout completed. It was also reminiscent of the final Skylab mission, 16 years earlier, when mission planners added items to the flight plan on a daily basis for which the flight crew had little or no training; trying to get everything done within a defined time frame had imposed such stress that the astronauts had eventually rebelled. As Hawley explained in 2012, "In terms of the RMS checkout, I do remember the way it was. A timeline allows you to do all of the specific engineering things that you need to accomplish in order to certify that the arm was ready to do the task." This included checking the drive in the motors and in certain directions, checking out the snares and the software, but what it did not include was enabling the operator to 'fly' the arm. "So what I wanted was an opportunity to fly it [unloaded], just to get some experience before I actually released Hubble. I had an agreement with the Flight Director that part of the RMS checkout for me was that I was just going to fly the arm around the bay [and] maneuver it forward to the grapple fixture, like I was going to do the next day for the real deployment, just to get a sense of how it flies."

Although this was his third space mission, Hawley had not previously operated an arm in space. On his first flight, STS-41D in 1984, MS Judy Resnik had been prime on the arm and Commander Henry 'Hank' Hartsfield had backed her up owing to his experience on STS-4. And STS-61C, Hawley's second mission, had not carried an RMS. "In fact, in those days," Hawley explained, "the ground was really strict about that kind of stuff; if you hadn't been trained, then you weren't allowed to do it. Charlie [Bolden] was officially my backup arm operator on STS-31 and I had to spend some energy to get an agreement that after Hubble was released, Charlie could fly the arm around a little bit, so he could get some experience. The concern on the ground was that if something went wrong when the prime crewmember for the task was not in control, how is the justification made for allowing it to happen? The crew's argument against that, was that it is how a flight crewmember gains real experience. You can train in the simulator, and they were very good, but there is nothing like doing it for real, on-orbit. It was logical to allow a person to gain a few minutes of hands on experience, making it easier when that person is assigned a task for real. There were similar discussions in allowing a pilot some 'stick time' before he is tasked to help land the shuttle. It became a hard sell, convincing the higher echelons that it made sense to allow 'rookies' to experience real flight operations that they could compare to their time in the simulator, and raise their confidence on a real mission. As it was not an absolute requirement for the mission it was a difficult 'sell', but the argument was eventually won and real flight experience on various systems became a regular part of later missions, such as the shuttle pilots on ISS missions performing the undocking and fly around sequence to give them hands on experience of orbiter handling during proximity operations."

Back on Discovery, Hawley thought his RMS practice was an important element of their first full day in space. He couldn't remember specific events, but recalled being frustrated in being diverted from this primary task of trying to check out the arm and gain his confidence in actually using it before grasping the telescope, "which was a large, expensive, and highly visible payload to handle at your first attempt in RMS operations".

While the RMS was being checked out, the EVA crew of Bruce McCandless and Kathy Sullivan were on the middeck preparing their equipment and themselves for a possible EVA during deployment. Both astronauts had begun breathing pure oxygen through masks prior to lowering the pressure in the cabin from 14.7 to 10.2 psi some 4 hours into the mission, at which point they could come off the pure oxygen supply. This protocol had been developed over several shuttle missions and was intended to reduce the time that the EVA crew had to breathe pure oxygen prior to any spacewalk and gradually purge their system of nitrogen to alleviate the risk of contracting the bends.

During the rest of the first day in space, McCandless and Sullivan conducted a detailed checkout of the two full pressure garments and the spare upper torso unit that was available, and in completing the rigging of the tool caddy that had been designed specifically to stow the power tools, batteries, tethers, wrenches, and other apparatus that they might need for a variety of tasks to assist in the deployment of Hubble.

For the first time since the start of the countdown process, at 4 hours 30 minutes into the mission the crew sent power to the telescope through the connecting umbilical. As the first signals were received at the Space Telescope Operations Control Center at Goddard Space Flight Center, this first 'conversation' with Hubble was welcomed by a huge cheer and loud applause. Jean Olivier, Deputy HST Project Manager at the Marshall Space Flight Center, recalled the experience as "…fantastic; it gives you goose bumps when you send those first commands in the blind and the telescope talks back to you". It was the first of an enormous number of two way communications that would occur between the ground and the telescope over the coming years, some of which would not always deliver the best news, but on this day, early telemetry provided evidence that the telescope had survived the ride into orbit in good shape.

At 11 hours into the mission, the crew finally settled down for their first sleep period, but the controllers at Goddard were fully occupied over the next several hours in bringing Hubble back to life. A series of commands designed to activate the power circuits linking the support system module and the optical telescope assembly cleared the way for activating the onboard heaters that would maintain an even temperature across the sensitive optics.

**April 25, deployment day**

The crew awoke to their second day in orbit and first full day in space, eager to press on with their busy schedule. The morning teleprinter messages included a note of congratulations for Steve Hawley on the improvement of his launch average to 0.231 with 3 successful launches in a record 13 countdowns. After the routine post-sleep housekeeping chores, each member of the crew floated off to their individual work stations. Hawley took the right-hand position (when looking towards the rear of the shuttle) at the aft flight deck to control the RMS, with Loren Shriver at the aft flight controls to fly Discovery.

## 34 Deployment

The morning activities included a private medical conference between the Flight Surgeon and Bruce McCandless and Kathy Sullivan, following which both were certified fit for EVA. With the clearance to perform a contingency spacewalk, their main focus on the middeck of Discovery was to complete as much preparation work as possible to enable them to exit the spacecraft promptly should the need arise. This meant getting partly dressed in their coolant garment, bio-medical instrumentation harness, and the cotton clothing that would protect the water tubing and coolant garment from snagging and being damaged as they moved around the cabin. They also prepared the airlock to the point that they could quickly don the outer layers of their EVA suits, utilizing the hours of pre-breathing pure oxygen and lowering the cabin atmosphere to reduce the time spent in the airlock prior to opening the hatch.

By far the busiest crewmember during the deployment process was Charles Bolden, who not only aided McCandless and Sullivan in their EVA preparations on the middeck but also Hawley and Shriver on the flight deck during the deployment operation.

One of the challenges of the deployment day was the power management strategy of the telescope. During launch, and whilst in the payload bay, the telescope was powered via the orbiter. To deploy it, this power coupling had to be disconnected. However, until the solar arrays were deployed, the telescope would depend upon its internal batteries, which had a defined operating lifetime before requiring recharging via the solar arrays. Hawley called this the 'what if' phase of the deployment. What if the solar arrays do not open? What if an antenna does not deploy? What if the aperture door does not open? Fortunately, there were contingency EVA actions to overcome all of these potential problems.

Just over 2 hours into the second day Story Musgrave, who had a long involvement with Hubble EVA development and was duty Capcom in Mission Control, radioed the news that Discovery's crew were awaiting. "Good morning from the Orbit One team… you've got a 'go' for HST deploy operations." To which the reply was, "That's outstanding."

The timeline moved rapidly from this point, although the crew still had the opportunity to add their impressions of flying at the highest apogee yet in the shuttle program, with Shriver reporting that "the onboard astronomer [Hawley] said something to the effect that the 'blue marble' sure looks far away today".

### Hubble at 'high hover'

When the time for deployment arrived, Hawley commanded the arm to grasp the starboard (right-hand looking aft) grapple fixture of the telescope. When he had verified a strong grip on the instrument, the five latches that held Hubble in the payload bay were released. Once the umbilical connection to Discovery was disconnected, the internal power system on the payload began running its systems. It was now a race against time to raise the telescope to a point where its solar arrays could be unfurled and face the Sun before its onboard batteries ran down in about 8 hours.

Steve Hawley later gave a detailed explanation of his role as the prime RMS operator on STS-31 to *Spaceflight* magazine.[20] The astronaut explained that for mass balancing/center of gravity reasons, as well as raising the telescope for deployment, Hubble was launched with the mirror and service module at the aft end of the payload bay and its aperture door facing the crew module. But for release, the telescope had to be positioned so

Hubble on the end of the RMS seen through the overhead aft flight deck windows.

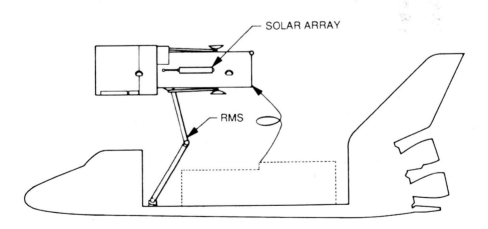

An artist's impression of the "High Hover" position.

that the aft bulkhead of the telescope was facing forward. "So what that means for the arm operator is that he has to lift the telescope out of the payload bay and get it high enough—about 15 feet [4.58 meters]—above the bay so you can rotate it end over end, then put it in a position that is a little bit forward and over the crew module so that you can look out of the big overhead windows and monitor the deployment of the appendages." The disadvantage of this evolution was that as the telescope was hoisted out of the payload bay the large aperture door blocked Hawley's view out of the aft flight deck windows. "That was probably the most difficult part of the operation. As soon as you start to lift the telescope up, all you can see out the aft window is basically your own reflection in the aperture door." The glare from the aperture door dazzled the aft flight deck as they concentrated on their actions. Conversation was very limited because the crew were told there was no need to acknowledge calls from the ground, to allow them to focus on the job in hand.

For better visual references, Hawley had to rely on camera and digital data relayed from the arm, and on the system of four cameras located at each corner of the payload bay which offered a view down each side of the payload as the telescope was raised, plus another view from a fifth camera located on the floor of the payload bay that viewed the underside of the telescope. The RMS coordinate system featured a digital display of the arm's position which provided further clues to what was happening out of direct line of sight for the operator. "All of that information together enables you to maintain the telescope in a position away from the orbiter structure. As soon as you get it 5 or 6 feet [1.5 or 1.8 meters] out of the bay, then it becomes a lot easier. You can see better and you have a lot more room to maneuver."

The first lift to the 'low hover' point some 15 feet (4.57 meters) above the payload bay progressed much slower than ground simulations had suggested. As Hawley recalled, "The view out of the window certainly degrades rapidly as the telescope… comes out of the bay." Bolden, who was backing up Hawley on the RMS, assisted by talking Hawley through the deployment operations, especially while the telescope was down low in the bay where the clearance was very restricted. As Hawley said of this phase, "Slow is good when you have two very large vehicles very close together." As Bolden recalled in his 2004 oral history, "There were characteristics of the arm that we didn't know at the time, and so we were making it up as we went along… What was supposed to take a few minutes took several hours."

As Hawley slowly raised Hubble to the highest point, called 'high hover', Goddard also commenced the step by step process that would culminated in the deployment of the twin solar arrays by activating the Deployment Control Electronics (DCE). At the 'high hover' position, Hawley pitched the telescope through 90 degrees, pointing its nose down into the payload bay, and Goddard issued a command to release the forward set of latches that had held the solar array masts against the sides of the telescope for launch.

In terms of orbiter coordinates, the 'Z axis' position of the point of resolution which was loaded into the RMS software varied between –416 when berthed in the bay to –750 at the 'high hover' point, which was a difference of 27.8 feet (8.48 meters) with the approximate center of mass located at the intersection of the aft and keel trunnions. The problem was to determine how 'high' was 'high hover'. As Steve Hawley explained, "The exact answer to that question depends on what part of the HST you want to refer to. The point of resolution for the maneuver, as I executed it [and explained above] was approximately the center of mass." Once lifted out of the payload bay the HST was rotated relative to the orientation it was in whilst berthed in the bay and, as Hawley pointed out, "not all parts of the telescope traveled the same Z distance".

Although the work on the middeck had become very intense, Sullivan, partly dressed in her EVA gear, came up to the flight deck to lend a hand, leaving McCandless to continue EVA preparations.

**Airlock cover up**

When problems arose in deploying the solar array on Hubble, both McCandless and Sullivan entered the airlock because it was thought it would be better to be ready to conduct an EVA at short notice, should the situation worsen. "This is where we wound up with a little bit of a disconnect," recalled McCandless in 2006, "in that in our trips to Bristol we were supposed to be the ones who knew what to look for to spot troubles, not only in the solar arrays but in the other systems as well. So the first array went out quite satisfactorily, and the second got part way out and at this point we thought we'd help and solve the problem by getting in the suits, climb into the airlock and start de-nitrogenation pre-breathing. So we were effectively taken out of the troubleshooting loop by the need to prepare for EVA.

"We had an agreement pre-launch with Bill Reeves, the Flight Director, that if he told us to depressurize the airlock we would go all the way to vacuum and open the door at least and qualify as quote 'EVA' unquote. But apparently after he gave us the 'go' to depressurization for EVA, the back row in the control center, the management row, descended on Reeves and asked him if that was really the thing to do, so we stopped at 5 psi and we didn't get to fully depressurize." Presumably the decision to proceed to full EVA would have been made only if the workarounds being devised at Goddard had not worked. So the pair waited in the airlock for almost a full 90 minute orbit until Hubble had been released, unable to contribute.

Apparently a piece of fabric on the outside that was attached by Velcro covered the small airlock window. Presumably it had been installed for the previous mission of Discovery, the military STS-33 mission, to prevent ground workers from viewing the classified cargo in the bay. As nothing is overlooked during ground processing, there must have been a decision to leave it in place for STS-31. Although a fortune was very likely saved by not generating the paperwork to have the item removed, the result was that McCandless and Sullivan, who had worked so hard to ensure that Hubble was able to be deployed, never actually got to see this accomplished.

Story Musgrave, the lead Capcom for the day's activities, recalled these challenges as "a very rough day, so very rough. I handed over to Goddard to manage the machine, so in terms of [crew] operations we were not so sophisticated at the time, but incredibly 'good to go', we just needed to warm up and get some experience. Operations on that day were really rough, but if you look at where we ended up, like on STS-61 servicing the telescope, the set up was perfect."[21] STS-31 marked the end of the preparation phase and the start of the operational phase, with the crew setting the standards for service missions to follow. "We didn't even need to tell them, they followed the procedures, they got their job done seamlessly without even talking about it."

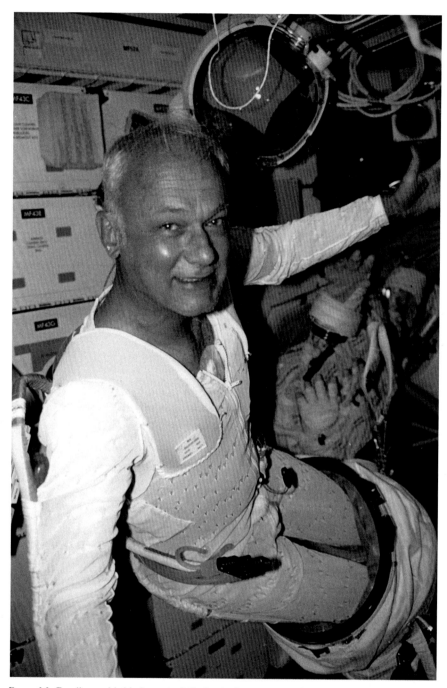

Bruce McCandless with his legs partially in the lower torso of an EMU, wearing the liquid coolant undergarment. His helmet and gloves float behind him.

# High, but not too high 39

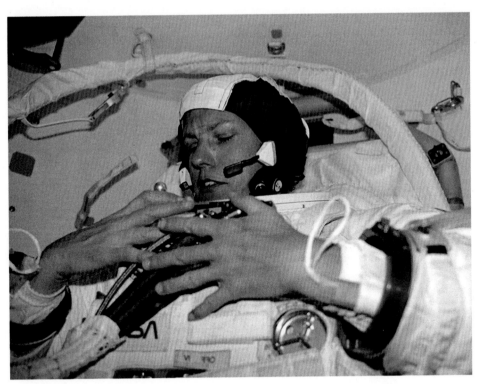

Kathy Sullivan in the airlock of Discovery, preparing for a spacewalk that neither she or Bruce McCandless would complete.

## HIGH, BUT NOT TOO HIGH

Hubble had to be placed in a high orbit above the distorting layers of the atmosphere to be able to view the farthest reaches of the universe clearly. But because the telescope was no longer to be returned to Earth periodically for maintenance and re-launched, instead being serviced in space, its altitude had to be within the operating range of the shuttle. The OMS maneuvering propellant of the orbiter was limited, so there were restrictions. In the case of STS-31, half this capacity was burned in climbing to the altitude at which Hubble was to be deployed, and most of the remainder would be consumed in getting the shuttle down again, so there was little scope for any additional maneuvers, especially if the telescope had to be revisited or recaptured. There was therefore a fine line between success and failure.

The operating orbit of the telescope was also influenced by atmospheric drag. The plan was to mount a service mission every 3 or 4 years, in between which the science program would take priority. The lower the operating altitude, the greater would be the atmospheric drag. Because Hubble did not have a system of attitude control jet, its gyros and associated systems would be influenced by the external forces acting upon them, complicating the task of holding the telescope stable to make long observations. The operating orbit had therefore to be as high as the shuttle could possibly attain in order to minimize atmospheric drag until the next shuttle arrived to boost the telescope back into a higher orbit.

**Flying a high orbit mission, the risk factor**

This requirement for the shuttle to place Hubble as high as possible answered operational requirements and attracted media headlines. It also placed additional risks on the crew and vehicle in terms of safety, reliability, and possible emergency contingency scenarios.

At 380 miles (600 km) STS-31 attained the highest apogee of any shuttle mission to date, or so the official records indicate. In her 2009 oral history Kathy Sullivan suggested that this may not have been the case. "The deployment altitude for Hubble was quite high for a space shuttle. I'm pretty sure it's the highest altitude *civilian* [author's italics] flight. Every time I said that, KT [Kathryn Thornton] squints at me as if she went higher than that at some point. Her first mission was a Department of Defense flight, so if she did, she can't say anything." That was STS-33 in November 1989, the last mission assigned to Discovery prior to flying the Hubble deployment mission.

In planning STS-31, consideration was given to the increased risks of flying the shuttle near to its design limits, particularly the possibility of an OMS propellant tank failure.

Formal Mission Safety Evaluation (MSE) reports formed part of the pre-flight and post-flight review process for every mission.[22] These reports were used by the NASA Associate Administrator, Office of Safety and Mission Quality (OSMQ) and by the Shuttle Program Director prior to each shuttle flight to document the changes, or potential changes, to the safety risk factors that were baselined in the Program Requirements Control Board (PRCB) element of the Space Shuttle Hazard Reports (SSHR). Any unresolved issues were included as part of the Flight Readiness Review (FRR Edition), the Launch Minus Two Days Review (L-2 Edition), and prior to the Launch Minus One Day Revision (L-1 Update). The final Post-Flight Edition was used to evaluate actual performance against safety risk factors previously identified in MSE editions for a given mission. Published on a mission-by-mission basis for use in each FRR, these were updated prior to the L-1 Review. For historical reference and archival purposes each MSE was issued in final report format following each shuttle flight.

Specifically for STS-31, the first shuttle mission (openly) planned to achieve such a high altitude, extracts from the post-flight mission safety evaluation report are reproduced here:

> Because of the extended OMS-2 burn required to reach a 330 nautical mile circular orbit, the OMS engine propellant reserves were depleted to a point that subsequent loss of access to OMS propellants (fuel or oxidizer in either OMS pod) would leave insufficient propellant to accomplish deorbit. *This is a known condition and is an acceptable risk of flying high orbit missions. The issue of sufficient OMS propellant can become a concern at any altitude.* [Authors italics] Pre-launch quantity, fuel usage, mission profile (including payload deployment maneuvers and rendezvous), payload mass, and center of gravity maintenance can vary and influence propellant reserves. OMS propellant tanks are made of titanium and operate at a

The view from orbit. Discovery attains the highest altitude for a shuttle to date.

maximum pressure of 313 psi. All tanks are proof tested prior to insulation and leak checked prior to each flight. Tank fracture control requirements require tank proof-pressure testing to be accomplished after five flights to protect against possible crack initiation, growth and tank rupture. There have been no propellant tank failures in flight. The STS-31 OMS propellant tanks successfully passed all acceptance and proof tests required by the tank fracture control plan.

In the 2009 documentary *When We Left the Earth*, the STS-31 mission was covered in some detail and there was a suggestion that it may have consumed too much propellant in reaching the high altitude. In her 2009 oral history Kathy Sullivan dismissed this, saying, "The artistic license the film guys took was that… we unexpectedly found ourselves with such low fuel remaining. That's invention, it just juiced up the story." She then noted that because 1990 was close to the maximum of the solar cycle, the envelope of the atmosphere would be much larger. The target altitude was 340 nautical miles (391 miles, 630 km). A lot of observations were made to calculate where in this cycle the telescope would be deployed. "When you put all those numbers together and run it against the orbiter's performance and consumables, it turns out that you arrive on-orbit with about 50 percent of your propellant already consumed. So you are less than 1 hour into a five-plus-day flight, you have got to release the telescope and back away, and you've got to station-keep nearby in case there are any failures. And if a failure happens you're going to have to rendezvous and capture again, service it, release, back away again, then deorbit. You need a margin for all of that, and half your propellant is already gone. That's a lot lower initial level of propellant [on] Day 1 than you typically see on a shuttle flight. That had everybody's attention."

## April 26–28, the standby days

For the remainder of the flight, at each day and night terminator the crew could see Hubble reflecting sunlight even at distances greater than 40 miles (64.36 km). It appeared star-like, strangely resembling one of its many targets in coming years.

The day after deployment, Hubble and Discovery were 54 nautical miles (62.14 miles, 100.0 km) apart. The main event of FD 3 was to open the aperture door to its full 105 degree position by controllers at STOCC, with McCandless and Sullivan again standing by in case an EVA were needed to manually open the door. In fact the first attempt was not successful. The onboard computer detected problems with the high gain antenna and placed Hubble into safe mode. The second attempt was more successful and Story Musgrave informed the crew, "Hubble is open for business." Hawley replied that the news was "outstanding" and the crew was proud to have been part of the team that put the observatory in business. "I don't suppose they'd want to give me any [observation time] on it, would they?" he joked.

During the 48 hours between deploying Hubble and opening its aperture door, Hawley had been on edge, mentally reviewing what he would need to do if the door failed to open. "We'd trained for a re-rendezvous, and I thought through my procedures for this and re-grappling of the telescope." The news that the door had opened was a great relief. As Musgrave informed the crew, "You've been released from Hubble support. It's on its own. Thanks a lot." At this point the prospect of McCandless and Sullivan making a contingency EVA in support of the telescope ended, and the cabin was re-pressurized to a normal 14.7 psi.

Engine burns placed Discovery into an orbit of 332 × 328 nautical miles (382.06 × 377.46 miles, 614 × 607.45 km). For the next several days the crew followed their flight plan, which included a series of small secondary experiments on the middeck, making Earth observations, recording more IMAX imagery, and general housekeeping duties in preparation for entry and landing. On April 27 Hubble experienced another problem arising from opening the aperture door. The movement had caused two of the four rate gyros to go out of limits. This triggered a 'software safe point' in which the onboard computer took one of the gyros offline and put the telescope in a second safing mode. This resulted in pointing the top of the telescope and its solar arrays towards the Sun to keep gathering power while the problem was investigated on the ground. It took the about 14 hours to return the telescope to its full four gyro control.

Sullivan noted in her oral history that the rationale to move the orbiter away from Hubble was to enable it to deploy its systems in a clean environment. "The shuttle is a comparatively dirty vehicle, plus you want to be far enough away that there is zero likelihood of collision. There would have to be an intention to go back to the telescope." It was also important to let the telescope finish outgassing before the aperture was opened, to prevent contaminating the mirror. For the shuttle to have returned to Earth shortly after deployment would have been a mistake, Sullivan explained. "The shuttle comes home and then you discover that the latch on the aperture door won't release; it's latched shut. Or the hinge motor won't drive. Those were the two final critical functions. If it won't unlatch or it won't hinge up, then no light gets into the telescope and you may as well not have done this. So the door would have been the main thing that could have brought us back."[23]

Loren Shriver, in his 2002 oral history, made an interesting comment on the prospect of returning to the telescope had something gone wrong in the hours after deployment. "I think originally there may have been some idea that if things hadn't been working out, [we] could have gone back and got it, but they gave up on that concept fairly early on. So I'm not exactly sure why we stuck around for two days, because we weren't going to go back and get it if it wasn't working correctly. They may have had us come back in and take a look at something just visually and take pictures, but that would have been about all we could do."

**April 29, landing day**

Following a descent that lasted about 15 minutes longer than normal as a result of its higher altitude, Discovery made a perfect landing on Runway 22 at Edwards AFB, in California, to finish a flight of 5 days, 1 hour, 16 minutes by testing a new design of carbon brakes during the rollout.

**Discovery returns to Florida**

After the crew disembarked Discovery, ground technicians made the vehicle safe for towing to the Mate/Demate Device (MDD), some 6 hours after wheel stop. The following day it was secured to the converted Boeing 747 Shuttle Carrier Aircraft for its flight back to KSC. The return journey, over 3 days, included a refueling stop at Kelly AFB in Texas and, due to marginal weather conditions in northern Florida, an overnight stop at Warner Robbins AFB, Georgia. The combination arrived at the Shuttle Landing Facility in the morning of May 7. That afternoon Discovery was offloaded and the next day was towed to Bay 1 of the OPF for post-flight processing, tests, and inspections. The payload bay doors were opened on May 14 for de-configuration from the Hubble deployment mission. At the end of the month, work began to prepare the orbiter to deploy the Ulysses solar probe during the STS-41 mission.

Meanwhile things were not going so well for the Hubble Space Telescope.

**TROUBLE WITH HUBBLE**

During the period of orbital verification that followed the deployment of Hubble, controllers working at STOCC successfully overcame a series of irritating anomalies which included a malfunctioning high gain antenna, aperture door closures, and several minor pointing issues. One complication was that data could not be sent to the ground immediately because one of the antennas was entangled with a power cable which prevented it from rotating to lock onto the geostationary Tracking and Data Relay Satellites (TDRS) used to relay information to the ground and send commands to the telescope. This slowed the calibration of instruments until the situation was resolved.

On May 2, a tiny vibration was discovered while sensors locked onto one of the 'guide stars' used to orientate the telescope. Every minute, a slight wobble of six hundredths of a degree up or down interfered with both the fine pointing mechanism and the clarity of star observations. Tests conducted two days later indicated that the vibration originate from the solar arrays as Hubble emerged from the cold of orbital night to the warmth of orbital day, and then as it slipped back into orbital darkness. The software of the guidance and pointing system was revised to compensate for this. On May 8, the telescope failed a test in finding stars generating a common brightness. This was resolved by sending up a program with an updated star chart for the computer guidance system.

The STS-31 Mission Report of May 20 stated, "The Wide Field/Planetary Camera shutter was opened and the HST experienced 'first light' when a photograph was taken of the open star cluster IC2602 in the constellation Carina. A preliminary evaluation indicated that, even though the telescope is not precisely focused, the quality of the images is far superior to that produced by the best ground-based telescope. Once the HST instruments have cooled to the design temperatures (in 3 to 6 months), the telescope should begin returning images that are orders of magnitude better than can be obtained using ground-based instruments."[24]

Then NASA revealed on June 14 that the problems with Hubble were more serious than first thought. The software update designed to cancel the solar array vibrations was actually making the situation worse, the telescope could still not lock on to stars, and it was proving unexpectedly sensitive to the South Atlantic Anomaly where the Van Allen radiation belts dip closer to the Earth. It was hoped that all three problems would be able to be solved by further modifying the onboard computer.

This optimism dramatically changed less than a week later, when NASA announced on June 27 that Hubble's main mirror had a flaw which made the telescope "near sighted". The instrument most affected was the Wide Field/Planetary Camera, which was considered to be the most important one. In an effort to allay fears of a crippled instrument, it was announced that the service missions scheduled for 1993 and 1996 would include new instruments that would not only upgrade those currently onboard but also compensate for the flawed mirror. There was talk of advancing at least one of these missions in order to speed up restoring the telescope to its full working capacity. It was also noted that checks to verify the accuracy of mirror focusing hadn't been included in the manufacturing process because that would have been far too expensive.

In July NASA set up the six-member Hubble Space Telescope Optical Systems Board of Investigation, chaired by Dr. Lew Allen, who was Director of the Jet Propulsion Laboratory, to investigate the "spherical aberration" of the mirror. As part of the investigation Leonard A. Fisk, NASA Associate Administrator for Space Science and Applications, headed a series of testimonies on the problem, and the reasons why full testing had not been conducted while the telescope was on the ground.

**Spherical aberration**

In depth investigation of the difficulty in focusing the telescope had nothing to do with the deployment mission, or the servicing capabilities, because replacing the mirrors had never been considered to be feasible tasks on-orbit. The saga of the faulty mirror falls beyond the scope of this book, but extensive details into the nature of the problem and its background can be found in the titles listed in the Bibliography of this book. Here is a summary from a pamphlet issued by NASA:[25]

> Controllers began moving the telescope's mirrors to better focus [the] images. Although the focus sharpened slightly during [those] six times the mirrors were moved, the best image achieved was a pin point of light encircled by a hazy ring or "halo". Controllers concluded that the telescope had a "spherical aberration", a mirror defect, only 1/25th the width of a human hair that prevented Hubble from focusing all light to a single point.
>
> At first some scientists believed the spherical aberration would cripple the telescope, but they were proved wrong. Engineers began running a battery of tests to determine which mirror—primary or secondary—was causing the spherical aberration. Pictures taken with the Faint Object Camera suggested the problem rested with the primary mirror. By late August, an investigation into the cause of the problem determined that a "null corrector", an optical device that was used as a guide in grinding and polishing the mirror contained a spacing discrepancy that caused the mirror to be ground too flat by two microns, an extremely small error in a mirror so large, but an error which resulted in the wrong prescription for the optics, preventing the Hubble from achieving the expected focus. The good news is that the error caused a "pure" spherical aberration, a problem relatively easy to correct much like the way an eye doctor corrects poorer vision with spectacles.

The effects of the spherical aberration on an early image from Hubble.

The pamphlet went on to explain that the spherical aberration would most impair the two cameras that used visible light to conduct their science, namely the Faint Object Camera and the Wide Field/Planetary Camera, but it also pointed out that "a great deal of science can be accomplished even before the spherical aberration is corrected". Computer "restoration" had been able to remove some of the blurring from around the center of the star images. Even in this flawed state, they were "10 times better than the best images produced through ground-based telescopes on a clear night". The other instruments, which were not so badly affected by the blurring, would continue to perform important studies, although some of the planned observing schedules would have to be revised.

Despite the upbeat reports, it was clear that NASA was considerably embarrassed by the error. This was not helped by fuel leaks on the shuttle posing serious delays to the manifest, and the escalating budget for the space station which was coming under increasing political pressure to be slashed. As a result of these difficulties, an advisory committee was set up by Vice President Dan Quale and the National Space Council to manage the nation's efforts in space and look into the grounding of the shuttle fleet by the ongoing fuelling problems, the serious cost overruns, design flaws on the space station, and Hubble's mirror flaw.

In an effort to limit excess spending in other areas, NASA was forced on June 7 to cancel a contract with TRW Corporation for the proposed and much delayed Orbiting

Maneuvering Vehicle (OMV) otherwise known as the "Space Tug", which was to have transferred various payloads between orbits, including Hubble. Then to add insult to injury, the following day a Titan 4 launch vehicle successfully placed a classified payload into orbit. In the wake of the Challenger accident, the Titan 4 had been chosen by the Air Force to supersede the shuttle as its primary launcher for military payloads.

What was clear was that the crew of STS-31 would not have been able to address any of the problems affecting Hubble, even if these had become evident while the shuttle was in space. There was certainly a lot to do, and it would take considerable time to develop the necessary fixes and incorporate them into the first service mission scheduled for 1993.

On learning of the flaw in the telescope mirror, Loren Shriver said, "You always come back from a mission on a real high, and especially a Space Telescope mission." All the good press the mission had received, and the high hopes of the astronomy community and Space Telescope Science Institute were now at risk. "The spherical aberration jumped right in the way of all that [and] the naysayers don't hesitate to come out of the woodwork and criticize anything and everything. As a crew, it was disappointing to have that happen. But of course [we] didn't have anything to do with that. The stuff that we did, we did properly."[26]

### New roles for the crew

With the STS-31 mission accomplished, it was time for the astronauts to move on to work on new goals in their individual careers.

Before the end of the year, Loren Shriver was assigned to a new flight as Commander of STS-46, the tethered satellite mission involving Italy. Charles Bolden undertook a series of administrative assignments, serving as technical assistant first to George Abbey, then to the outgoing JSC Director Jesse Moore and to Moore's successor, Gerald Griffin. After that, he received his first commander's seat on STS-60. Sullivan went back into the training cycle for her third crew assignment, the STS-45 Atmospheric Laboratory for Applications and Science (ATLAS-1) flight—the renamed Earth Observation Mission.

On June 7, Steve Hawley had been named as Associate Director Ames Research Center in California in order to widen his managerial experience in NASA. Effective August 31, 1990, Bruce McCandless retired from both NASA and the Navy in order to enter private industry. However, by joining Martin Marietta Astronautics Group in Denver, Colorado, he became a consultant for the first Hubble service mission.

### The long and difficult path to orbit

Though afflicted with difficulties, Hubble was at least in space, and a service mission was to attempt to restore it to full working order in 1993. Fortunately, it was the fact that Hubble had been designed to be serviced by shuttle crews which became its saving grace, a capability that was built into its design almost from the start of the program over 20 years earlier.

The concept for the telescope arose in the 1920s, and gained the enthusiastic support of a number of key individuals in the 1940s. Nevertheless, it would be another 20 years before a formal program began to take shape that suggested the telescope would remain in space for 15 years and be serviced by astronauts. To sustain this venture, a huge infrastructure on the ground would also be required to undertake a series of shuttle missions that would service, maintain, and update the telescope without the need to return it to Earth.

**REFERENCES**

1. AIS interview with Steve Hawley, March 1, 2012
2. Hawley, AIS interview 2012
3. How Launching Hubble Space Telescope Influenced Space Shuttle Mission Operations, Steven A. Hawley, *Journal of Spacecraft and Rockets,* March, 2014 Vol. 51, No. 2 : pp385–396, American Institute of Aeronautics and Astronautics, Inc.
4. Kathryn D. Sullivan Oral History, May 2009
5. Hawley 2014
6. NASA News: JSC 85-50
7. AIS interview with Bruce McCandless, August 17, 2006
8. AIS interview with Steve Hawley March 1, 2012
9. NASA News: JSC 85-131
10. NASA News JSC 87-018
11. NASA News JSC-88-008
12. Loren J. JSC Oral History Project, 18 December 2002, pp1–15
13. Charles F. Bolden JSC Oral History Project, 6 January 2004, p37
14. How Launching Hubble Space Telescope Influenced Space Shuttle Mission Operations, Steven A. Hawley, *Journal of Spacecraft and Rockets,* March, 2014 Vol. 51, No. 2 : pp385–396, American Institute of Aeronautics and Astronautics, Inc.
15. STS-31 Flight Data File Crew Debrief, memo from Richard D. Snyder, Chairman, Crew Procedures Control Board, May 21, 1990, reference DH4-90-139, in AIS archive
16. AIS interview Steve Hawley March 1, 2012
17. Launch processing references from Shuttle Status Reports various dates 1989–1990, KSC; Chronology of KSC and KSC Related Events for 1988, by Ken Nail, NASA KSC, KHR-13, March 1989; Chronology of KSC and KSC Related Events for 1989, by Ken Nail, NASA KSC, KHR-14, March 1990; and Chronology of KSC and KSC Related Events for 1990, by Ken Nail, NASA KSC, KHR-15, March 1991
18. STS-31 Mission Report, A New Window on the Universe, Roelof Schuiling and Steven Young, *Spaceflight* Volume 32, Number 6, June 1990 pp196–206, British Interplanetary Society; STS-31 Mission Status Reports, NASA April 24–29, 1990
19. STS-31 Post-flight Debriefing Guide, prepared by W.B.D. Jones, with annotations by Bolden, Charles F. Bolden Collection, NASA JSC History Archive, Clear Lake University, copy on file AIS Archive
20. A New Window on the Universe, Discovery crew deploys Hubble Space Telescope, *Spaceflight*, Vol. 32, No 6, June 1990 p202
21. AIS interview with Story Musgrave August 22, 2013

22. Mission Safety Evaluation Report for STS-31, Post-Flight Edition: October 15, 1990, Safety Division, Code QS, Office of Safety and Mission Quality, NASA HQ, Washington DC
23. Sullivan, Oral History 2009
24. STS-31 Space Shuttle Mission Report, May 1990, NSTS-08207, NASA
25. Hubble Space Telescope's First 125 days, NASA Fact Sheet, Marshall Space Flight Center, September 1990
26. Shriver, 2002 Oral History

# 2

# The dream

> ...a space telescope of very large diameter... and requiring the capability of man in space is becoming technically feasible, and will be uniquely important to the solution of central astronomical problems of our era.
>
> From the recommendations of the 1965 Woods Hole study group

For well over two decades the Hubble Space Telescope has provided stunning images and ground-breaking scientific data that has dramatically revised both our view of the universe and of public interest in space exploration. As with any of the world's largest space projects such as Apollo, the space shuttle, Mir, and the International Space Station, the real story of Hubble did not begin with the launch of the hardware into orbit, but many years—indeed decades—earlier in the minds of theorists, dreamers, and planners.

## OUT WHERE STARS DON'T TWINKLE

Humans have always been fascinated with the skies above them, both during the daylight hours and during the veil of darkness called night. Early observations using the naked eye gave birth to the science of visible astronomy. The wonder and mystery of the pin pricks of light set against a black background; the changing face of the Moon; and strange lights that streaked across the sky both fascinated and frightened early mankind, so these phenomena were attributed to Gods and demons.

In the early 17th century, Galileo Galilei turned the recently invented telescope to the sky and made remarkable discoveries, in particular revealing the Moon to be a world in its own right, that Jupiter was accompanied by a system of satellites and that Venus displayed a full range of illumination phases, proving the hypothesis of Nicolaus Copernicus that the planets travel around the Sun. Later, astronomers with better telescopes mapped the Moon in detail and discovered new planets. Also, atmospheric phenomena were monitored and measured, and in the late 18th century lighter-than-air balloons were invented. These were then used to carry scientific instruments into the upper reaches of the atmosphere to investigate not only the composition of the sky but also the celestial realm beyond. By the 20th century, the idea that a rocket could break free of most of the Earth's atmosphere and conduct science before falling back to the ground was perfected and utilized. This work supplemented high flying aircraft and ongoing ground observations throughout the early decades of the 20th century.

In 1957 the first payload to circle the globe above the atmosphere was launched, and a new means of exploration and discovery dawned—the Space Age. With the availability of more powerful rockets, and the knowledge and experience gained from the first flights into space, it became possible to conceive of developing sophisticated instruments which would operate in space to reveal what was really 'out there'.

Over the next three decades, space astronomy made ever greater steps towards placing a large telescope into space. In 1990 that dream finally became reality with the deployment on-orbit of the Hubble Space Telescope (HST). After a somewhat shaky start Hubble began to change our view of the universe around us, but it was not all about science. To ensure that the telescope continued to operate at its full potential and reveal the stunning images which it has now produced for over 25 years, it had to be regularly serviced in space and its capabilities improved.

**Origins of a telescope in space**

The origins of the Hubble Space Telescope can be traced back to the 1923 book *Die Rakete zu den Planetenräumen* (*By Rocket into Planetary Space*) by Hermann Oberth. In this work, Oberth suggested that an observatory in space would offer significant scientific advantages over those located on the ground.

This supported the well-known theory that a telescope in the vacuum of space would no longer be hindered by the Earth's atmosphere, and would obtain sharper views of the stars. Furthermore, a telescope would be able to sense a wider range of optical wavelengths than was possible from the ground. However, in Oberth's time the ability to place *anything* into space, let alone such a complicated instrument as a large telescope, was still many years in the future.

It was also evident that a telescope in space would require a significant infrastructure not only to manage the research but also to keep the telescope operational. As a result, few took the concept of astronomy from space seriously. Nevertheless, in February 1940 a proposal was published in the magazine *Astounding Science Fiction*. It was written by astronomer R. S. Richardson, who speculated on the future possibility of placing a 300 inch (7.62 meter) diameter telescope on the Moon. Published 30 years before mankind actually landed on the Moon, it was recognized then that the development of rocket technology would be the best way to support the exploration of space. However, with the Second World War raging, the idea of an expensive and difficult program of rocket development for anything other than a military role was far from the minds of the leaders of the few nations that were likely to be capable of developing such technology.

## Lyman Spitzer and an extra-terrestrial observatory

In 1946 astronomer Lyman Spitzer, in a report for the RAND Corporation that described the advantages of operating a 16.5 to 50 foot (5 to 15 meter) diameter telescope in space, said it would "uncover new phenomena not yet imagined, and perhaps modify profoundly our basic concepts of space and time".[1]

Spitzer recognized that such an observatory would be above the absorbing layers of our atmosphere, which blocked most of the ultraviolet and infrared light known to exist in the universe, but is difficult to study by ground-based observatories. He also explained that the ever-present turbulence of the atmosphere blurs our view. By placing a telescope above the atmosphere, the stars would not 'twinkle' and their appearance in images would be sharper. When this report was published, the 16.5 foot (5 meter) telescope on Mt. Palomar was still under construction. Although still firmly on Earth, this telescope would be the largest in the world for many years. In lay terms, its resolution of 0.02 seconds of arc meant the observer could, in theory, see a penny coin at a distance of 120 miles (193 km), but the atmosphere distorted the sharpness of the image to no better than 1 second of arc (which was 60 times better than the human eye nevertheless). However, if it were possible to place the Palomar telescope above most of the atmosphere, then its performance would be 3000 times better than on Earth! Spitzer calculated what could be attained by telescopes up to three times the size of the Palomar instrument. For the next 50 years Spitzer worked to see his vision turn into a reality. This was the genesis of what evolved into the Large Space Telescope (LST), then simply the Space Telescope (ST) before being named the Hubble Space Telescope.

Lyman Spitzer Jr. was born in 1914 in Toledo, Ohio. After graduating from Yale with a bachelor's in physics in 1935, he attended Princeton and earned his master's in 1937 and his doctorate in astrophysics in 1938. Then he accepted a teaching appointment at Yale. During the Second World War he conducted underwater sound research, and in 1947 was appointed chairman of Princeton's astrophysical science department. He made significant contributions to astrophysics, held a variety of positions at Princeton, and developed the case for placing a large telescope in space.

The RAND Project was an Air Force 'think tank' evaluating a 'secret study' of placing a large artificial satellite in orbit around Earth at an altitude of several hundred miles. Spitzer was asked by a friend, who was also a member of the RAND staff, to write a chapter in the report on the usefulness of astronomy from a satellite. As Spitzer later explained, "With my long and ardent background in science fiction, I found this invitation an exciting one and accepted with great enthusiasm. I spent some time analyzing a number of possible research programs for telescopes of different sizes above the atmosphere and wrote a brief description of these."[2] Although no orbital astronomical telescope resulted from this study, the exercise helped to convince Spitzer that such a large general-purpose optical telescope could offer the scientific community far greater discoveries than ground-based observatories.

Spitzer's paper indicated interesting targets for a telescope in space as:

- To push back the frontiers of the universe and determine the distances to galaxies by measuring very faint stars
- Analyze the structure of galaxies

Lyman Spitzer (1914–1997), pioneer of the concept of a large telescope in orbit.

- Explore systematically the structure of globular clusters
- Study the nature of planets, especially their surfaces and atmospheres.[3]

The continuing interest in astronomy from space in popular talks, science fiction journals, and formal project reports was encouraging, but it would be a further decade before the first satellite was launched into space, and that was far from a large astronomical payload. There was still clearly a long way to go.

## STRATOSCOPE: SMALL STEPS TO SPACE

After the war, America began launching captured German V-2 rockets carrying scientific payloads to study the upper atmosphere. By 1947 the results had convinced Spitzer that the orbiting of large instruments would soon be feasible. "In my thinking," he wrote, "personal association with the development and operation of such a large orbital telescope gradually became a major professional goal."

While pursuing his dream of placing a large telescope in space, Spitzer also collaborated on Project Stratoscope. This featured multiple flights high into the atmosphere of a 12 inch (30.48 cm) solar telescope slung beneath a balloon between 1957 and 1959. The mirror was made by the Perkin Elmer Corporation, which later won the contract to furnish the primary mirror for the Hubble Space Telescope. The project went on to obtain imagery from 80,000 feet (24,384 meters), which lies above most of the Earth's atmosphere. Stratoscope II carried a 36 inch (91.4 cm) reflecting telescope on a gondola that weighed 3.5 tons. It was a remote-controlled observatory, flown at night, and there were eight flights between 1963 and 1971. Instead of film cameras, the first two tests made infrared observations to prove the system worked. On two of the six imaging flights the telescope failed to unlatch from its vertical launch position and on a third the primary mirror door didn't open, in all cases preventing observations. The three successful flights observed planetary atmospheres, red giant stars, and external galaxies. Although the telescope never left the atmosphere, there was so little turbulence at its operating altitude that the high-resolution images were clear and sharp.

Notwithstanding the short duration and limited capability for long-term research of these balloon flights, Spitzer believed they were an important step towards his dream of placing a large telescope into orbit.

### Dawn of the Space Age

The Second World War led straight into a new and more sinister Cold War which became a race to create the most destructive weapons possible in order to demonstrate a technological superiority and psychological terror over the enemy and civilian populations. This arms race was waged primarily by the two superpowers of the United States and the Soviet Union. The ultimate 'high ground' in this contest was outer space, and one remarkable spin-off from all this flag-waving was a race to the Moon.

## Sputnik, Gagarin and the Moon

Under the guise of the International Geophysical Year, which actually ran for 18 months in 1957 and 1958, both the USA and the USSR strove to place a small 'scientific' satellite into orbit around the Earth.

On October 4, 1957 the Soviet Union became the first to achieve that goal, as the Space Age dawned with the "beep, beep, beep" sound of Sputnik. America was quick to catch up with a satellite in early 1958. Then it announced a plan to place a man into orbit, only to be beaten once again by the Soviets on April 12, 1961 with the single orbit of Vostok carrying cosmonaut Yuri Gagarin. Not to be out done, on May 25, 1961, President John F. Kennedy challenged his nation "to achieve the goal, before this decade is out, of landing a man on the Moon and returning him safely to Earth". He also pointed out that such a program would be both expensive and very difficult to accomplish, requiring the nation's technological might.

History records that this commitment was fulfilled on July 20, 1969 by the landing on the Moon of the Apollo 11 lunar module *Eagle*, carrying astronauts Neil Armstrong and Edwin 'Buzz' Aldrin. After completing a two hour excursion on the surface, they, along with their lunar orbiting colleague, Michael Collins aboard *Columbia*, safely returned to Earth on July 24. Just 4 months later, the lunar landing feat was repeated by the Apollo 12 crew. For a short time thereafter, the American space program shone in the glory of its achievements.

## ASTRONOMY FROM SPACE BEFORE HUBBLE

Whilst the effort to place men into space and on the Moon certainly attracted the headlines and a large portion of the funding within NASA, there was also a commitment to develop space science and technology, including satellites, launch vehicles, probes for deep space, space research, and, of course, placing astronomical instruments into space.

### The fields of astronomy

Quite naturally, the focus of space astronomy was on regions of the spectrum that cannot be seen by telescopes on the ground. Observations had been made by instruments lifted to high altitude by balloons, most notably Project Stratoscope, and some early ballistic missiles had been converted to carry scientific payloads as sounding rockets. But such instruments were limited, and offered only brief opportunities to observe. There was a consensus to fly more advanced instruments in space to make prolonged observations. But there were many fields within the realm of astronomy, characterized by wavelengths spanning the visible spectrum, radio, microwave, infrared, ultraviolet, X-rays and the high energy gamma rays. The debate was whether the effort should be directed at just one, or several parts of the spectrum.

## Creating an act and an agency for space exploration

Barely a month after launching Sputnik, the Soviet Union sent up a second satellite carrying a dog named Laika. The apparent superiority of their competitor's technology came as a shock to many Americans. In response, President Dwight D. Eisenhower created the Presidential Science Advisory Committee to determine a suitable approach and pace for a national space program which would place a strong focus upon scientific research. On March 26, 1958, the Committee recommended the establishment of a domestic science-based space program that would proceed at "a cautiously measured pace".[4]

A very simple timetable was listed under goals classified as 'Early', 'Later', 'Still Later', and 'Much Later Still'. Astronomy was the first priority in the 'Later' category and Human Flight in Orbit was rated sixth. Interestingly, Human Lunar Exploration was the third item in the 'Still Later' category, and Human Planetary Exploration was the single entry for 'Much Later Still'.

Following the successes of the early Sputniks, and in view of growing public concern in the United States, the White House moved quickly to establish an American civilian space agency. This would absorb the National Advisory Committee for Aeronautics (NACA) that had been established in 1915 but was far too small to cope with such an enormous task. The new agency would include rocket and space engineers then involved in a variety of defense programs. The National Aeronautics and Space Act of 1958 was signed by the President on July 29, and the National Aeronautics and Space Administration (NASA) became active on October 1.

The same Act also established the National Aeronautics and Space Council as a policy coordinating board at presidential level. Subsequently NASA, to improve its engagement with the scientific community, established a series of advisory committees, including the Astronomy Mission Board which offered commentary on the astronomical programs and plans. Things did not always go smoothly as this infrastructure developed, but generally it was recognized that, in appearance at least, most of the science plans of the agency were a close match to the objectives desired by the majority of American space scientists.

The objectives of the Space Act included (in part) the "expansion of human knowledge of phenomena... in space" and "the development and operations of vehicles capable of carrying instruments, equipment, supplies and living organisms through space".

Until then, astronomical research had focused mainly on using ground-based telescopes for visible-light studies, and using sounding rockets, balloons and aircraft for ultraviolet and X-ray studies. Identifying opportunities for space-based research beyond those resources, as well as trying to understand what leading astronomers were interested in learning, and then balancing such aspirations against the technical capability of available hardware became the main focus of the early years of astronomical research at NASA as it endeavored to create a sustainable and effective program.

By the end of the decade, studies were underway to identify the potential for conducting astronomy off the planet. In 1959 the American Astronomical Society (AAS) supported by the National Science Foundation (NSF) sponsored a symposium on space telescopes and the potential results they could contribute. It was realized that space telescopes would be vital to the future of astronomy over the next decade. Of course, this was also the decade in which the US gave reaching the Moon a higher political and national priority than any other space project. Nevertheless, a number of dedicated individuals believed in the promise that a large space observatory could offer the wider astronomical community and they quietly continued their studies, gradually gaining support and interest in their ideas.

**Creating a program for astronomy**

The civilian space agency's first astronomical program was conducted by sounding rockets under the auspices of the National Research Laboratory (NRL), and under the leadership of James E. Kupperian, formerly of NRL, who headed a team at the Beltsville Space Center in Maryland, renamed the Robert H. Goddard Space Flight Center. It is important to highlight this pioneering work at Goddard as part of NASA's early astronomy program because there was a direct link between that center and the subsequent development of the facilities which would operate the Hubble Space Telescope.

To determine the requirements for astronomical payloads on-orbit, Gerhardt Schilling, a former assistant to astronomer Fred Whipple at the Smithsonian Astrophysical Observatory (SAO), was appointed to lead the agency's astronomy program. He was assisted on a part-time basis by John O'Keefe of Goddard's Theoretical Division. The team set out to devise instruments and a spacecraft that would evolve into the Orbiting Astronomical Observatory (OAO), of which several were launched. In February 1959, Nancy G. Roman, formerly of NRL Radio Astronomy Branch joined the team to lead the Optical Astronomy Program, a position that also encompassed ultraviolet research. Within a year, Schilling left NASA and Roman took over the entire NASA astronomy program, which at that time included geodesy studies against the strong opposition of the Air Force, which believed this topic to be more their bailiwick than NASA's.

**A platform to build upon**

In addition to establishing the fields of research and experiments which could obtain data, significant work was required to develop a suitable spacecraft to carry the instrumentation and support hardware, and prove that a satellite could support a far longer period of study than was otherwise possible. The initial expectation was to obtain data for several hours to several days or even weeks, but as experience was gained it was expected that successful missions would last several years.

Taking into account the limited payload mass capability of launch vehicles at that time was only one of many hurdles to overcome. Ensuring that a satellite fitted within the tight confines of its aerodynamic shroud was another. And of course, for any given mass it was essential that the experiments and power supply were sophisticated enough to warrant the expense of launching the payload in the first place. The operating environment in space had to be considered, including the temperature variation between being in full sunlight and the Earth's shadow, and the effects of radiation on the materials of the satellite and how these might impair the scientific instruments. Other challenges that would have to be addressed included the method of keeping the satellite under control and in communication, how to keep the instruments pointing at the intended target, and how the data should be collected, stored onboard, and transmitted to Earth.

## ORBITING ASTRONOMICAL OBSERVATORIES

One of the earliest long range projects established by NASA's Office of Space Science was the Orbiting Astronomical Observatory (OAO) series. This began on May 15, 1958, with a primary study made by staff at what was then the NACA Langley Memorial Aeronautical Laboratory. In October 1958 Nancy Roman headed up a working group that, over the next several months, evaluated the feasibility of launching large astronomical observatories. The result was a proposal to use stable orbiting platforms as part of the long range space science program. A number of satellites would carry telescopes to make observations in the optical, infrared, ultraviolet, and X-ray regions of the spectrum.

The Orbiting Astronomical Observatory (OAO) was the forerunner of the Hubble Space Telescope (HST).

The OAO project joined the Orbiting Solar Observatory (OSO) and Orbiting Geophysical Observatory (OGO) projects in developing satellites to gather data on cosmic phenomena and physical objects from space. Following a briefing to potential industry partners, NASA issued Request for Proposal (RFP) documents on December 1, 1959, for both the spacecraft and the instruments to be flown on them. In February 1960, Goddard was appointed as the technical management field center for the project, then in October of that year Grumman was selected as prime contractor for the OAO series.[5]

**Follow-on to OAO?**

With the OAO project underway, it was time to look ahead and start planning the even more advanced spacecraft that would continue the development of astronomy from space.

In 1962, NASA asked the Space Science Board of the National Academy of Sciences to create two study groups to recommend possible follow-on programs to the OAO satellites. One of these study groups was led by Lyman Spitzer, still advocating the large observatory above the atmosphere.

The first study was conducted at the State University of Iowa, and the second at Woods Hole, Massachusetts. These two study sessions were the first to give serious consideration to the concept that would go on to become the Space Telescope. When the initial meetings were held in 1962, the first OAO had yet to be launched but already the focus was on what should follow that series. It was cautiously recognized that committing to a larger instrument with a diameter of 100 inches (2.5 meters) or more would result in "a truly enormous investment for astronomy. For this reason, it is vital that its scientific justification receive the most careful and comprehensive consideration by the astronomical and related scientific communities."[6]

There were frustrations within the astronomical community that a disproportionate amount of NASA's funding was being ploughed into the Apollo lunar program, undermining the case for developing ambitious astronomical programs.

One of many technical challenges was that the proposed orbital telescope would require to accurately track and capture precise images of stellar objects, and no one really knew whether it would be possible to develop a precise pointing system that would operate reliably. At that time, the first images from weather satellites were being received but the space telescope was something different, as it would have to travel at over 17,500 miles per hour (28,164 km per hour) whilst accurately pointing at a distant, faint pinprick of light. However, the engineers were encouraged by the emergence of classified military satellites and a growing number of successful scientific satellites that provided stable platforms for optical instruments.

Then there was the question of how to get data back from the telescope. Military satellites were demonstrating the option of a data return capsule, but this skill was still classified, very expensive, and limited by the number of such modules that could be carried and the amount of data that they could return. For a given launch mass, the heavier the instruments aboard a space telescope the smaller the mass available for capsules. Then the operational life of the mission was limited by the number of capsules. Weather satellites transmitted their images of Earth to ground stations, but their low-resolution television-scan views were not suitable for an astronomical observatory.

Another potential problem discussed in these sessions, centered on the possibility of such a large telescope having what was essentially a conventional optical mirror that was able to sustain high-resolution research for several decades, while new techniques and technologies might offer a more economical viewing option from the ground. It was recognized that the problem of atmospheric blurring might, in principle, be overcome in the future and improve ground-based observations of bright objects such as stars and planets, but for very faint stars and distant galaxies a telescope in space offered a better option. If the large telescope could be a general-purpose design with a variety of instruments onboard, then this would be even more preferable than operating several smaller telescopes. As a result of these discussions and additional independent studies, over the next three years the astronomical community became increasingly confident that a large space telescope was the logical successor to the OAO series.

The final 1962 report combined optimism at the results that could be obtained by a large telescope in space with due caution that the technology was still in its infancy. As yet there had not been a single image of any celestial body returned to Earth by an instrument on any satellite. It therefore seemed logical to await the outcome of the first OAO mission, before making a commitment to develop a large space observatory.

Over the next few years, significant advances were made in flying hardware in space on a regular and reliable basis. This was most notable in the manned Gemini program, created in 1961 and flown between 1964 and 1966 to provide experience prior to attempting the more ambitious Apollo lunar program. Ironically, the success of Gemini, in addition to the recent successes in the Ranger lunar-impact probes and the Mariner flyby missions to Venus and Mars, heightened confidence that the technology for advanced space programs, including a larger space telescope, was on the right track.

By 1965 the engineering concepts resulting from the 1962 studies led the Space Science Board to strongly recommend to NASA the development of a large space telescope, and to set up its own committee to define the scientific objectives that such an observatory might address. Lyman Spitzer was chosen to lead this committee, but despite his firm belief in the proposal and tireless drive, this would not be an easy road for the astronomer, or indeed for the idea of creating a space telescope. Many of his fellow astronomers were unsupportive, fearing that the cost of such a project would seriously hamper, if not threaten, their ground-based and small satellite astronomy programs. And of course scientists in other fields were alarmed at so much money being devoted to astronomy. Spitzer responded with a vigorous personal effort to convince not only the scientific community but also the politicians of the value of such a project. It would be well over a decade before he would see final fruition of his efforts. Meanwhile, work continued with the series of smaller satellites in investigating astronomy from space, and these results contributed to the argument for launching a larger instrument that would operate for a longer period of time.

## LAYING THE GROUNDWORK

Related to the development of the Space Telescope were the series of astronomical satellites launched between 1966 and 1978, each of which increased our understanding not only of the universe but also of how orbital platforms could supplement conventional research on Earth. These flights also increased confidence that operating a large observatory on-orbit for many years and with a variety of instruments would generate a vast scientific return that could not be achieved by any other means.[7]

### Orbiting Astronomical Observatories

There were four launches in the OAO series between 1966 and 1972, with a success rate of 50 percent.

Table 1  Orbiting astronomical observatories.

| Designation | Launched | Vehicle | Mass | Results |
|---|---|---|---|---|
| OAO-1 (A1) | April 8, 1966 | Atlas-Agena D | 3960 lbs (1769 kg) | Failure of spacecraft due to battery malfunction (overheating) after 1.5 days (22 orbits) |
| OAO-2 (A2) | December 7, 1968 | Atlas-Centaur | 4491 lbs (1996 kg) | Carried 11 UV experiments; operated successfully for over 4 years. Following the failure of the experiments power system it was turned off on February 13, 1973. The spacecraft far exceeded its expected operational lifetime |
| OAO-B (planned as OAO-3) | November 30, 1970 | Atlas-Centaur | 4644 lbs (2106 kg) | Failure of nose cone ejection sequence during launch, satellite lost before reaching orbit. |
| Copernicus | August 21, 1972 | Atlas-Centaur | 4851 lbs (2200 kg) | Highly successful science program returning data until 1980 (8 years) |

After all the hard work that went into its preparation, the loss of OAO 1 was immensely frustrating to all concerned. Its power supply blew a fuse during the checkout procedure. A review team examined the manner in which observatory-class spacecraft were powered, and recommended a redesign. The second loss occurred when the launcher failed to release the satellite. However, the other two were highly successful and helped to set the requirements for even larger telescopes.

### A desire for a standard 'streetcar' design

One of the priorities of the observatory-class program was to standardize the way spacecraft were built. Previously, scientific instruments and spacecraft were designed for the research that was to be undertaken and then adapted to fit onto the selected launch vehicle.

## 62 The dream

As many were one-flight designs, this was both expensive and time consuming. The development of a standard format made economic sense and would save considerable time in preparing a new spacecraft. The so-called 'streetcar' design approach was adopted for the OSOs and OGOs, where interchangeable scientific instruments could be carried on a standard framework that was known as a 'bus'. This concept was developed at Goddard Space Flight Center during 1959–1960. It would enable the shell of the satellite to be produced on almost a production line basis, independently of the specific scientific instruments to be carried. These satellites demonstrated that this type of design was well suited to the standardization of a framework that could support many different missions in a wide variety of fields. The next step was to make the 'bus' capable of on-orbit servicing.

### The Explorers

The work of the observatory-class missions supplemented that from the long-running series of small scientific satellites called Explorer. The Explorer program had already begun when NASA was created, because Explorer 1, launched on January 31, 1958, was America's first successful satellite. The wide range of studies conducted by the Explorer series included the Earth's environment, terrestrial-solar-interplanetary relationships, and various astronomical observations.

Table 2  Explorer astronomy satellites.

| Designation | Launched | Vehicle | Mass | Results |
|---|---|---|---|---|
| Explorer 38 | July 4, 1968 | Delta | 426 lbs (193 kg) | Radio Astronomy Explorer (RAE)-A, placed in high Earth orbit and found that our planet, like Jupiter emits radio waves; data transmission deteriorated after two months in orbit |
| Explorer 42 | December 12, 1970 | Scout | 313 lbs (142 kg) | Small Astronomy Satellite (SAS)-A '*Uhuru*', the first X-ray satellite; continued to survey until March 1973; decayed April 5, 1979 |
| Explorer 48 | November 15, 1972 | Scout | 366 lbs (166 kg) | SAS-B studied gamma rays until June 1973; Decayed August 19, 1980 |
| Explorer 49 | June 10, 1973 | Delta | 723 lbs (328 kg) | RAE-B placed in lunar orbit June 15, used the Moon as a shield again Earth's radio noise to study solar and galactic radio radiation |
| Explorer 53 | May 7, 1975 | Scout | 434 lbs (197 kg) | SAS-C X-ray telescope; decayed April 9, 1979 |

Developed prior to the observatory-class spacecraft, these missions were not of a standard design, with the experimentation and instrumentation being defined by the mission. Although they were inexpensive in comparison to the larger astronomical observatories, they provided valuable early data on the space environment as well as many new discoveries in astronomy, pioneering the work of the much larger observatories.

## A higher energy astronomical observatory

Intended as a follow-on to the work of several Explorer-class spacecraft, in the late 1960s studies began for a 'super explorer' which would obtain high quality data on high-energy celestial X-ray, gamma-ray and cosmic ray sources. Much heavier than the Explorer series, the heaviest of which was only 723 pounds (328 kg) and indeed the OAOs, the heaviest of which was 4800 pounds (2200 kg), the new payload would weigh in at about 21,388 pounds (9700 kg) and would consist of a pair of satellites, each of which would have an additional 28,665 pounds (13,000 kg) of experiments onboard. They would be launched together on one of the Titan III-class launch vehicles. There were also plans in development for two follow-on missions.

During the spring of 1969, the management of the High Energy Astronomy Observatory project was assigned to Marshall Space Flight Center and by September of that year NASA recommended to the Space Task Group (STG) that a high-energy astronomy capability be a high priority scientific goal for the new decade—a recommendation which the STG strongly endorsed in its report to President Richard M. Nixon. Although initially planned for launch in 1975, Congressional budget cuts imposed in January 1973 promoted a rethink of the program and instead of two large satellites that would be launched together it was decided to use three smaller satellites, each of about 6175 pounds (2800 kg) with the capacity to carry a further 2866 pounds (1300 kg) of experiments. These would be launched by Atlas-Centaur, on an annual basis starting in 1977.

Table 3  High energy astronomical observatories.

| Designation | Launched | Vehicle | Mass | Results |
|---|---|---|---|---|
| HEAO-1 | August 12, 1977 | Atlas-Centaur | 6002 lbs (2722 kg) | Carried four scanning X-ray experiments; exhausted control gas air supplies in January 1979; re-entered March 15, 1979 |
| HEAO-2 | November 13, 1978 | Atlas-Centaur | 6500 lbs (2948 kg) | Pointing X-ray telescope; achieved a highly successful 30 months of data return; re-entered March 25, 1982 |
| HEAO-3 | September 20, 1979 | Atlas-Centaur | 6002 lbs (2722 kg) | Gamma and cosmic ray investigations; completed 20 months of data return; re-entered December 7, 1981 |

## Recommendations from Woods Hole

Against the development and operation of these scientific missions, work continued to define the larger orbital telescope. Various technical problems were discussed by the Woods Hole Study Group in 1965 and clear parameters emerged for what was generally referred to as the Large Orbital Telescope (LOT). These included:

- The largest feasible aperture
- An operating location either orbiting Earth with an apogee of several hundred kilometers, a 24 hour synchronous orbit, or a site on the lunar surface
- The role of man for telescope maintenance and updating.

The inclusion of man for servicing the telescope was an important point in the future of the project. The suggestion to put the telescope on the Moon is interesting, because several years later the Apollo landings revealed that the ubiquitous lunar dust would be a hindrance to extensive surface operations for prolonged periods, especially for an optical telescope. In hindsight it is clear that the ultimate decision to place the telescope in low Earth orbit was a wise one.

The Woods Hole study group recommended "a space telescope of very large diameter, with resolution corresponding to an aperture of at least 120 inches (3.04 meters), detecting radiation between 800 Å and 1 mm, *and requiring the capability of man in space* [author's emphasis] is becoming technically feasible, and will be uniquely important to the solution of central astronomical problems of our era".[8]

The schedule was outlined as an 11 year development program between commencing the design in 1968 and launching the telescope in 1979. Compare this to what became the Large Space Telescope, and its development from about 1969 through to the initial expectation of launch by 1979. As events transpired, the renamed Space Telescope would not gain funding until 1977 and, for various reasons, would not be launched as the Hubble Space Telescope until 1990, which was some 13 years after authorization and 25 years after the study which recommended it.

## A MANNED ORBITAL TELESCOPE

In the early 1960s several companies undertook studies into the possibility of placing an observatory into orbit, leading to the proposal for a Large Orbital Telescope.

### A telescope for MORL?

In parallel with these studies, the debate about whether the telescope ought to be unmanned, man-tended, or fully occupied continued. Several groups were investigating the capabilities that would be enabled by utilizing Apollo hardware in missions other than President John F. Kennedy's goal of "landing a man on the Moon… and returning him safely to Earth". It was evident from studying the equipment being developed for Apollo that there was considerable scope for expanding the lunar exploration program beyond a few brief landings, and that by using surplus and adapted hardware and resources it would be possible not only to achieve a great deal in low Earth orbit but possibly also far beyond the Earth-Moon system.

This train of thought was certainly not new, and was foremost in long range studies prior to the national commitment to reach the Moon by 1970, particularly the desire to construct a manned space station. Adapting existing hardware to new concepts was attractive because it would likely save time and minimize costs.

One early study was the Manned Orbital Research Laboratory (MORL) proposed in 1963 by Douglas Aircraft engineers Carl M. Houson and Allen C. Gilbert. In its original form this was to be a 'wet' workshop in which the spent final stage of a Titan II or Atlas launch vehicle would be made safe and fitted out by astronauts ferried up by Gemini or Apollo spacecraft.[9] The following year this was revised to a 'dry' workshop configuration,

## A manned orbital telescope

An artist's impression of a Manned Orbiting Research Laboratory (MORL) with a large telescope facility. (An enlargement of a low resolution scan courtesy Mark Wade/Astronautix.com)

where the fully fitted laboratory would be prepared on the ground and then launched by a Saturn IB. Projected for operational use in the early 1970s, it was envisaged that successive crews would be launched in Apollo spacecraft to conduct a range of scientific observations and experiments, including astronomical research.

One of the many concepts developed for MORL included launching a telescope that was 13 feet (4 meters) in diameter and 15 feet (4.57 meters) long atop a separate Saturn IB. This would be docked to MORL and be operated by the resident crew, who would also carry out spacewalks to replenish film cassettes.

Further studies were initiated, but MORL remained only a study concept and was finally abandoned in the late 1960s in favor of the Apollo Applications Program (AAP) which then evolved into Skylab.

### A telescope for Apollo

Concurrent with the studies of a large man-tended telescope, there were studies envisaging the enhancement of existing Apollo hardware to fulfill a wide range of scientific missions, including astronomical telescopes.

On February 18, 1965, in a presentation to the House Committee on Science and Astronautics, Dr. George Mueller, NASA Associate Administrator for Manned Space Flight, outlined the Apollo Extension System (AES). This exploitation of Apollo hardware would be able to be realized subsequent to the initial lunar landings. Because a brand new program would need approval and specific funding, Mueller chose a name that implied a logical development of the systems and resources which were already in progress.[10] AES was envisaged as a broad program of at least 15 missions, and it would involve over 80 investigations with dedicated apparatus, including astronomy experiments. However, over the next year and a half, AES was merged into the grander AAP that was to carry out a variety of missions using Apollo hardware and expand NASA's future capabilities in space.

## 66 The dream

As the studies developed, so did plans for a third variant of the basic Apollo Command and Service Module to support these missions. The so-called Block III CSM came after the early Block I, which had no docking system and would be used only in test flights in Earth orbit, and the Block II that had a docking system and was intended for lunar missions. The Block III would be capable of Earth orbital missions lasting up to 45 days and lunar orbital flights of up to 35 days with resources to support a wide range of payloads. These payloads included a telescope called the Apollo Telescope Mount (ATM) that would be carried in a vacant bay of the Service Module.[11] This telescope hardware would evolve through many guises and amendments, being reassigned to a modified descent stage of the Lunar Module that would dock with the workshop and finally becoming an integral part of the unmanned Skylab space station which was inhabited by three crews of astronauts between May 1973 and February 1974. One aspect of the MORL plan which was carried over to AAP/Skylab was that spacewalking astronauts would exchange film cassettes on the ATM.

An interest in creating a telescope for a 'manned' spacecraft whereby astronauts, ideally *astronomer-astronauts*, could conduct astronomical observations was gaining support. From the mid-1960s there were literally dozens of plans, proposals, ideas and suggestions for what might possibly follow Apollo after it had reached the Moon. Many of these ideas remained firmly on the drawing board or in the pages of trade magazines, but others were developed in abbreviated form. One study from this period developed the idea from MORL of combining the concept of a large optical telescope in space and the services that could be provided by a space station and its crew.

**Costing a Manned Orbital Telescope**

Early in 1966 the Space Division of the Boeing Company of Seattle, Washington, produced for the NASA Langley Research Center a preliminary plan and costing report regarding the creation of a Manned Orbital Telescope (MOT).[12]

This report expanded upon concept studies, and provided a development and operation program plan for further study. It included contributions from a survey of related industries, including space divisions of General Electric and the American Optical Company, the Itek Corporation and the Kitt Peak National Laboratory. The plan addressed three major phases: the concept development phase, the project development phase, and the project operation phase. An on-orbit mission of 5 years was assumed in estimating costs. However, logistics and space station systems were not included, since these were expected to be developed "in other major programs".

The study envisaged launch by a Saturn V and docking to a space station. The hardware would also include MOT-specific apparatus stowed on the station, a universal aerodynamic shroud to protect the telescope during the launch phase, logistics to the station for periodic supply of spares, modification kits and supplies, astronaut training apparatus, and a control facility at the Manned Spacecraft Center (MSC) for the experiments. The projected cost of developing the flight model and transporting it to the Cape for processing and launch were also included, as were the costs for the envisaged operational life. The study estimated the cost of the MOT program over a period of 16 years, culminating with 5 years of service in space, as $1.3 billion (1966 dollars).

The 120 inch Manned Orbiting Telescope (MOT) concept. (Courtesy The Boeing Company, Aerospace Division, AIS collection)

68  The dream

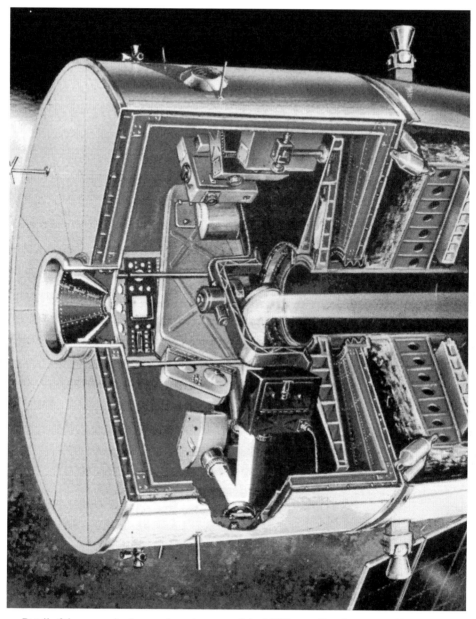

Detail of the pressurized crew observing cage of the MOT, revealing the proposed instrument and control layout. (Courtesy The Boeing Company, Aerospace Division, AIS collection)

The study also identified five new significant ground facilities that would be needed to support the program:

- Scientific Instrument Development Integration Complex
- Attitude Control and Rendezvous Development Complex
- Optical Development Complex
- MOT Training, Experiment Control and Data Center (at MSC)
- System Development Complex (at the Michoud Area Facility, New Orleans).

Existing facilities at MSC would require structural and subsystem development. At KSC both the barge transport system developed for Apollo hardware and LC-39 facilities would need to be adapted. Even incorporating as much current and proven technology and hardware as possible in order to minimize cost and delays, this would cost approximately $113 million (1966 dollars).

What could not be calculated at the time was the cost of developing and launching a space station and the means to support its logistics requirements, with or without a space telescope of the size envisaged in the study. The size of the MOT was interesting in comparison to the Hubble Space Telescope. The MOT design was 60 feet (18.28 meters) long and 15 feet (4.57 meters) in diameter with a mass of 28,000 pounds (12,712 kg). To fit into the shuttle, Hubble was 43.5 feet (13.3 meters) long and 14 feet (4.2 meters) in diameter, with a mass of 24,500 pounds (11,110 kg). The primary mirror of the MOT was 120 inches (3.04 meters), whereas Hubble's was 94.2 inches (2.4 meters). Their subsystems were similar, including the optics; science instruments; structures and mechanisms and miscellaneous flight support items. But unlike Hubble, the MOT included an attitude control system and the ability to rendezvous and dock with a space station.

**Forecasting the operational use of MOT**

Launched by Saturn V, the MOT would be remotely steered into a Hohmann transfer orbit trajectory to the synchronous orbit space station using two 1000 pound (453.60 kg) thrust multiple-start engines. Onboard batteries would supply electrical power for this flight, but after docking the MOT would draw power from the station.

The study envisaged a diffraction-limited Cassegrain telescope that combined a 120 inch (3.04 meter) primary concave mirror and a secondary convex mirror, equipped with a suite of suitable "astronomical instruments". In this design, the tube which contained the primary optical components was attached to a pressurized "cabin" that included instrumentation and operating equipment for the telescope, as well as offering a protective environment for the astronaut crew. Cylindrical in shape, it had flat pressure-type bulkheads at both ends. In the center of one of the bulkheads were an Apollo-type probe and drogue docking system and a tunnel for access from the space station. The other bulkhead provided an attachment for the telescope tube. A retractable pressure door would permit unobscured observations while the instruments were in use.

A brief summary of the hardware, instruments, and facilities of the MOT are reproduced below, followed by further details of the servicing and maintenance concept. As reported in the study, eight science instruments in two effective focal lengths were identified:

- Focal length ef/15
  - Wide field camera (0.5 degree) including an astigmatism corrector lens for the telescope optics
  - High dispersion ultraviolet spectrometer
  - Low dispersion ultraviolet spectrometer
- Low dispersion spectrograph focal length ef/30
  - Narrow field camera (<10 minutes of arc)
  - High dispersion spectrograph
  - High dispersion infrared spectrometer
  - Photoelectric photometer
  - Thermoelectric photometer.

The instruments were to be arranged inside the telescope cabin in two circular groupings, one for each focal length, and a TV camera was included in each group for real-time remote viewing. To support these operations the crew would work in a shirt-sleeve environment in a pressurized cabin during set up, servicing, and maintenance. However, because all apparatus would be accessible inside the pressurized compartment the MOT would not require an EVA to service or replace the instruments. Prior to initiating observations, the crew would exit the cabin, close the hatches, and depressurize it by storing the air in the environmental system of the space station.

The study noted that to fully support MOT operations, the creation of a space station still had to be formally authorized, and of course constructed. What was recognized was the need for a reliable logistics system to support not only the space station but also MOT operations, something which the space shuttle finally achieved over three decades later.

**Addressing the logistics problem**

The MOT study assumed that any logistics to support the program would already be in-situ at the space station. The logistics were identified as expendable supplies, spares, and the supply of equipment and materials to update the telescope and its instruments. It was also stated that any scientific data from the telescope would "be accomplished in conjunction with rotation of the space station and MOT crewmembers to the Earth".[13] Despite being planned as a *manned* optical telescope, with direct involvement by station crewmembers, the report noted that the development of "direct transmission of data from scientific instruments to the ground might reduce the need for ground-to-space logistics and man's space operational role".[14]

Any updates or modifications to the apparatus on the MOT once on-orbit would be either to improve the telescope and MOT-specific space station equipment operation, maintenance and performance, or to improve the flexibility of operations by modifying or exchanging the scientific instruments. In these cases, all prototype modifications would first be tested in the MOT qualification model. After qualification, flight kits would be developed to be installed on-orbit. As a precaution against a launch failure, the report's budget addressed the costs of constructing a backup MOT.

## The role of man on the manned orbital telescope

The report also recognized that additional studies would be needed in order to minimize the repetitive (and costly) logistics missions during the operational lifetime of 5 years, which was only one third of the duration that Hubble was expected to achieve and one sixth of what it will likely end up attaining. For the MOT, it was suggested that initially a ground to space logistics system be flown every 90 days, with physical data being returned at these times. It was thought that station crews would be involved mostly in activities designed to check out equipment and confirm nominal system operations. The system would then be placed into an automatic mode "with operations and control accomplished on the ground and scientific data transmitted to the ground", as was the case for Hubble a quarter of a century later. But crew would return to the MOT periodically in order "to perform maintenance and directly operate the systems for critical experiments requiring high resolution and physical return of data to the ground."[15]

The studies also recommended that consideration should be given to launching the MOT as an integral element of a dedicated telescope-space station concept, resembling one of the MORL configurations that was being evaluated at that time. This was a logical suggestion, since a self-contained station and telescope would reduce both the number of launches and conflicts with other disciplines for limited research time. Another recommendation was that thought should be given to installing the MOT's electrical and electronic equipment inside either the habitable part of that facility or in the space station itself. In this configuration, a shirt-sleeve environment would be continuously available for equipment maintenance and troubleshooting activities that would improve telescope operations.

During the development of the studies, it was envisaged that an ideal mission profile for the MOT would start with a three month period of manned operations that would check out and initially exercise the facility, prior to running the telescope in an automated free-flying mode for up to 9 months per year. Operations away from a space station were preferable due to the conflict of objectives, risk of contamination, independent stabilization and pointing requirements. It was determined that a man-tended operation for 3 months in 12 would provide both uninterrupted observations and a defined down-time for maintenance and modifications to equipment, whilst minimizing conflict with the complicated range of studies and operations aboard the space station. During the docked phase, the crew could also take the opportunity to conduct specific experiments and observations which required accuracies and resolution that couldn't be automated. This mode of operation would eliminate up to three of the four planned logistics flights per year, resulting in a considerable cost saving.

## Difficulties in defining a program

In addition to defining the basic requirements and program parameters, these studies were valuable for highlighting the serious hurdles that would have to be overcome before such a telescope could reach the launch pad, let alone become operational on-orbit. Foremost of all the challenges was funding. At that time, NASA was focused on the push to send Apollo to the Moon before the end of the decade and ahead of the Soviets. Though a space station was once seen as the logical next step, and in some circles it had a higher priority

72  **The dream**

than reaching the Moon just to beat a deadline and demonstrate technological and military might, NASA was worried that at such a crucial point in securing funds for the Moon, to request more to develop a space station, let alone a manned telescope, would not be welcomed by Congress or the general public. It was one thing to carry out related studies, but to start cutting metal was something else, particularly with the developing situation in south-east Asia consuming national treasure. For the foreseeable future NASA's focus was developing Apollo for the initial manned landing and up to nine further expeditionary missions, with the option of later adapting Apollo hardware to build a rudimentary space station. A manned telescope, a large space station and a logistics resupply system were far down the political and finance roads, and when they finally emerged they were very different from the original concepts.

## THE LARGE SPACE TELESCOPE IS BORN

To follow up the 1965 Woods Hole meeting the National Academy of Sciences, ever mindful of the difficulties of securing the funding for major projects, established an ad-hoc committee to study the concept of an unmanned telescope for a general purpose role. After assessing the engineering hurdles, the committee said there were no insurmountable difficulties preventing such a development. It was proposed that this Large Space Telescope (LST) should include:

- Occasional visits from astronauts for maintenance to sustain such a major and costly facility for at least a decade of operational use
- Capability to update the science package with improved instruments for changes to the scientific program.

It was from these basic points that the concept of telescope service missions evolved. In 1969 a Space Science Board report on the 'Scientific Uses of a Large Space Telescope' also recommended specific research programs. This helped to focus the discussions with scientists on the committee who would become either members or consultants, and publicized the great potential of research with the LST and potential close association with both conventional and other space-based astronomical research.

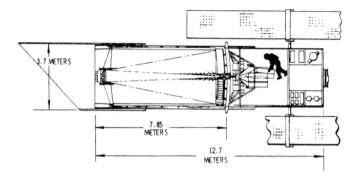

Cutaway of an early design for the Large Space Telescope (LST).

As a result of these efforts to define a balanced program, by the end of the decade support for the LST was growing as just one possible future direction of American space exploration after Apollo. However, there remained the matter of convincing Congress to fund such a long term project at a time when budgets were being slashed in order to help to fund the conflict in south-east Asia. Nevertheless, supporters continued to advance their ideas, and it was not just the LST that emphasized the potential of astronauts servicing and maintaining satellites.

## NIMBUS: A CANDIDATE FOR ORBITAL SERVICING?

With studies into the possibility of creating a space station and a reusable logistics system for "shuttling" cargo to and from Earth orbit, attention turned to developing the skills to maintain and service large space structures in space. Interestingly one early candidate for such a study was the Nimbus-class of metrological satellite.

A Nimbus satellite undergoing vibration tests in 1967.

## 74 The dream

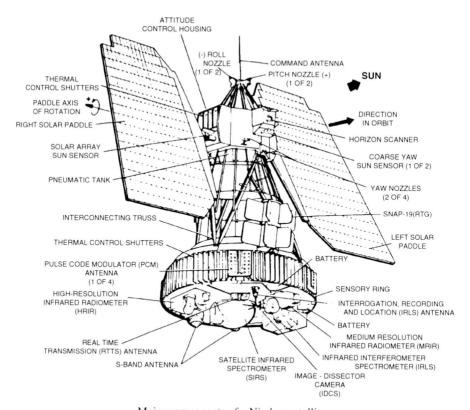

Major components of a Nimbus satellite.

On July 18, 1969, just 2 days before Apollo 11 landed on the Moon, a memo from the engineering consultancy that advised NASA, Bellcomm Inc., summarized this study into the effects of in-orbit servicing.[16]

Nimbus was representative of complex satellites suitable for servicing consideration. The study was fairly general and not that detailed, but it provided a 'straw man' case for further thought and discussion. The study looked at the basic design of the satellite, a reconfigured design to determine whether any improvements to the accessibility could be made, and if a specialized hangar attached to a space station might offer a solution. It was also possible to gauge the size and handling requirements that would be required in order to undertake such an operation at a space station.

### Why Nimbus?

Foremost, Nimbus was designed as a modular structure with equipment bays housing key components and systems. The equipment bays featured removable panels to provide access for installation and orientation prior to launch. But this configuration was never intended to be serviced in space. Furthermore, the design was compact and dense in order to overcome launch vehicle limitations.

However, being representative of what could be achieved after some redesign, Nimbus offered the best solution at that time to initiating more detailed studies and evaluations. To make the study more realistic the internal volume of the two most densely packed areas, the sensor ring and the attitude control system housing, were expanded by a factor of three. In contrast, the sensor equipment mounting was not studied because it was on the lower outer face of the sensor ring and thus accessible for orbital servicing or replacement. Focusing on the sensor ring brought the bays and internal frame into review, because it was in this area that components for the power supply, electronic sensors, telemetry and data handling, and command system were housed— these being the most likely candidates for replacement or modification during the life of the satellite.

The study also investigated a modified attitude control system housing and an enlarged hexagon internal frame structure that satisfied factory fabrication requirements whilst still offering a general expansion which preserved the geometry of the design. This would then offer greater opportunities for servicing at the expense of volume and mass penalties. The revisions offered access to the solar array drive, support system and associated electronics, control mechanisms, attitude control system flywheels and propellant gas supply, and the wire harness connectors. In addition, it was proposed that the solar array drive mechanisms be separate assemblies in order to make them accessible, and that all 'black box' apparatus should be located on hinged panels. It was apparent that adding access panels with hinged doors would offer greater flexibility for in-space servicing, but the means of that servicing was not revealed.

**Designed for servicing**

The Nimbus study also evaluated an advanced redesign of the satellite to support in-space servicing which had separate systems and supporting electronics, plus a dedicated support system unit that included a range of operational and housekeeping systems. The solar arrays were replaced by radioisotope thermoelectric generators (RTG) that offered more scope for reconfiguration because support equipment to operate the solar arrays was no longer required. The cylindrical support compartment that housed all of the associated systems would also be reached by separate hinged access panels.

The study also considered the use of a "pressurized space-hangar" in conjunction with a service tug or space station. This could offer a shirt-sleeve environment and more extensive servicing support systems (similar to MORL and MOT). What wasn't reviewed was how the Nimbus would arrive at the hangar which, in order to house the satellite, would require to be at least 22 feet (6.70 meters) in diameter.

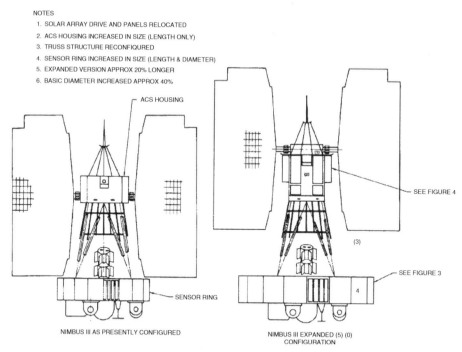

(*Left*) Nimbus III 1969 configuration. (*Right*) Nimbus III 1969 expanded configuration proposed for on-orbit servicing. (Courtesy Bellcomm Inc., AIS Collection)

**Conclusions**

Although restricted in its scope, this study provided some pointers for further work which would have direct links to the Hubble Space Telescope, including:

- Equipment should be configured in modular containers
- A decrease in volume utilization would be favorable
- Access panels either hinged or readily removed were essential
- All fasteners should be of the captive type
- Areas of access should be free of cabling and harnesses
- Cabling and harnesses should be routed in areas where no disassembly is required
- Connectors should be reasonably accessible through equipment access doors
- Insulation should be segmented and fixed to access doors
- Using RTGs instead of solar panels could greatly simplify the configuration for servicing purposes.

This idea of designing hardware for servicing and maintenance in space was not lost on those developing the LST, and as the concept gained support in the early 1970s it directly influenced the design of new generations of other satellites.

## Launch vehicle

Early studies by NASA evaluated the use of existing technology to keep the costs down. One option was for a telescope with a 13.1 foot (4 meter) mirror, which was the largest that could fit inside the aerodynamic shroud on top of the S-IVB upper stage of a Saturn launch vehicle. Then as a further drive to take advantage of current technology and resources, it was decided to use the production facilities that made the optical elements for national security programs. This meant reducing the mirror to a diameter of 10.5 feet (3.2 meters). If the telescope were to be launched by shuttle then its mirror would have to be reduced to 7.87 feet (2.4 meters).[3] As the Phase A study report for the LST was being written, some consideration was given to using a member of the Titan III family (used to launch military satellites having large optical systems) instead of the shuttle in order not to compromise too far on the size of the mirror.

## Warring tribes

NASA is not a single-site agency but a network of individual field centers distributed across the country. Over the years, this had caused frustrations and disagreements in the division of work and responsibility in space projects, programs, and missions. A definition of NASA in such circumstances has been explained as "a loose confederation of warring tribes", and this friction often surfaced during the development of Hubble.

The Marshall Space Flight Center had supported concepts and studies for a large space telescope for many years and its managers were keen for a leading role in the development of the telescope. By 1970 its Program Development Directorate had a "telescope team" that was managed by James A. Downey III and chief engineer Jean Oliver.

This team began by studying a coordinated program involving the space telescope and a space station, but there was little support for a station and so they turned to an independent and untended concept for the telescope. Marshall had a rich history in the engineering and management of developing launch vehicles, but was inexperienced in optical systems and astronomy. On the other hand, Goddard Space Flight Center had a number of professional astronomers on its staff and had accrued vast experience with astronomical satellites. The proximity of Goddard in Maryland to NASA Headquarters in Washington DC was another advantage to be considered, but Marshall had more experience in managing large projects, which such a telescope certainly would become. Marshall had also experience managing a large human space program, having been lead center for the Skylab space station, and so it had experience in systems engineering. Despite the intense rivalry between the two centers over the years, it was also apparent that if they worked together they could match up their individual strengths and weaknesses.

Thrown into this mix was the long history of competition and rivalry between Marshall and the Manned (later Johnson) Space Center for management of manned space programs, which the Houston team deemed to be their birthright.[4] This rivalry had created difficulties during the development of Skylab. Now the Houston team faced the prospect of sharing a manned or man-tended space telescope with not just one but two centers. A further rivalry would emerge between Houston and Goddard when tools had to be created for the Hubble service missions.

Nancy Grace Roman, NASA's first Chief of Astronomy with a model of the Orbiting Solar Observatory (OSO).

was to foster more realistic studies of the feasibility of the telescope that would be acceptable to as many stakeholders as possible." She created study committees to allow astronomers and NASA engineers to work out the most feasible solution for a telescope which would have a primary mirror with a diameter of 10 feet (3 meters).

For over a decade, she wrote Congressional testimonies on the need for the telescope and argued the case to the various committees and groups in Washington DC which would later prepare budget reports for both the President and Congress—although she never personally testified before Congress, preferring instead to provide the material that would enable others to present the case.[2]

## GETTING STARTED

By the early 1970s, various telescope concepts had been evaluated. The result called for an unmanned telescope to be launched on the shuttle, then either regularly serviced in space or returned to Earth for maintenance and upgrading.

Looking back from 2015, it is easier to recall the ups and downs of Hubble after launch, striving to understand why it would not work as designed and then planning to restore that capability, whilst dismissing the equal number of years required to get the telescope off the ground in the first place and create an infrastructure to maintain it during what was hoped would an operational life of 15 years.

**Securing the budget**

Acquiring sufficient funds for a new program is always a challenge, especially in the space business where failure is an element along with the uncertainty of operations in the hostile vacuum environment, often owing to inadequate testing. When the idea of the Large Space Telescope began to be seen as a distinct possibility and gained a strong measure of support from the scientific community, the development of the shuttle had yet to be authorized by Congress. Some thought was given early on to launching the LST on an expendable launch vehicle of the Saturn or Titan class, but the expectation was that it would ride the shuttle and it was designed for that. The reusable shuttle offered the potential for in-space servicing and maintenance, and if necessary returning the facility to Earth for major maintenance and then re-launch. Of course, the hurdle was to secure not only scientific support but the backing of Congress.

Several delays were incurred on the journey to authorization, as each successive NASA budget was proposed, challenged, cut, amended and finally approved, often at a lower level than originally requested. This delayed, canceled, or seriously affected other programs at a time when the priority was to proceed with the shuttle program. Balanced against this were delays in the shuttle program itself, difficulties in manufacturing the LST, and delays in the overall test program.

In her oral history the former NASA Chief of Astronomy, Nancy Roman recalled, "In the early days, funding was pretty flush. Even by the late 1960s, well before the Apollo program ended, funding was becoming a major problem. Most of the big missions had to be de-scoped to save funds. The HST is either a horrific or an excellent example, depending on which way you want to look at the thing. We had to cut costs."[1]

**Nancy Roman: A driving force**

Astronomer Nancy Grace Roman fervently strove to drive forward the proposal for a large telescope in space. After gaining her PhD in astronomy from the University of Chicago in 1949 she worked at the Naval Research Laboratory. Then in 1959 she joined NASA as its first Chief of Astronomy in the Office of Space Science. She remained with the agency for the next 20 years. She was initially only responsible for the management of the ultraviolet programs but after a year she took the whole electromagnetic spectrum; in fact as long as it was outside the solar system she was responsible for heading those studies. Her experience encompassed the development of the Orbiting Solar Observatories, as well as the Orbiting Astronomical Observatories and, much later, the Cosmic Background Explorer. Even after officially retiring in 1979, she continued to work part-time at Goddard for the next 18 years and was a consultant to one of the Hubble contractors.

Roman was very aware of the gap between what the astronomical community required of the telescope and what the aerospace industry perceived it to be. "I decided that my role

## Organizing the LST

Before discussing in detail the servicing of the space telescope, it is worth summarizing the difficult years of the 1970s and 1980s, focusing on a few key factors that were influential in the development of servicing techniques and operations. One of these influential decisions was that of pursing the space shuttle concept, even though that meant losing the large space station whose construction and servicing had been cited as one of the primary tasks for the shuttle.

The US Air Force played an important role in 'selling' the shuttle to Congress. In return, NASA was obliged to design the orbiter to suit the Air Force's requirements, which affected both its shape and the size of its payload bay. By making the bay large enough to carry the largest payloads envisaged by the military, NASA knew that the shuttle would also be large enough to carry a wide variety of commercial and scientific payloads over the next 20 to 30 years, in particular pressurized science laboratories and a space telescope. Furthermore, the shuttle was promoted as being ideal for servicing and maintaining satellites in low orbit, for recovering and returning them to Earth, and for dispatching payloads to high orbits and into deep space. In order to do this, the shuttle would require rendezvous and navigational aids to enable it to track and approach a target, a robotic manipulator arm to reach out and grab the target for stowage in the bay, and apparatus to enable spacewalkers to perform maintenance tasks.

Funding is always vital with a large technological project, especially one as expensive as space exploration and particularly so when it is taxpayers' money. Although support for the LST began to gain momentum in the early 1970s, Congress did not authorize funding until 1977. Even then the struggle was not over; for 6 years, until the hardware arrived at KSC, the launch of the telescope was repeatedly delayed by difficulties in preparing the shuttle for its first flight and then qualifying it to fly 'operational missions' routinely. Then, in January 1986, with the telescope firmly manifested for launch that September, Challenger was lost, grounding the program for over two and a half years and, as it would turn out, delaying the launch of the telescope by 4 years.

## PIECES OF A JIGSAW

The early history of the Space Telescope was a long and at times frustrating round of ideas, proposals, symposiums, and studies. By 1970 the idea that a large telescope was achievable began to gain momentum and over the ensuing decade and a half this led to a wide range of developments, milestones, techniques, skills, and decisions.

Today, looking back at 25 years of Hubble Space Telescope flight operations, it is easy to overlook the intensity of these developments, often conducted out of the limelight, which, in small ways, added those missing pieces to the larger jigsaw puzzle that enabled Hubble to be accepted, funded, built, tested, and launched.

The period 1970 through 1989 was a rollercoaster of ups and downs, gains and losses, dreams and nightmares, and a lot of hard work for countless numbers of workers across the United States and Europe, all of whom were striving to reach the common goal of placing a large telescope in Earth orbit in order to achieve a better understanding of our universe.

## AFTER THE MOON

The American desire to reach the Moon during the 1960s, both with automated and human missions, inevitably diverted funds and resources away from other programs. As frustrating as this was, it was clear that many of the more advanced ideas would take years to develop. To a degree, the apparent ease with which the Moon was reached so quickly belied the truth of how difficult space exploration was. The success of Gemini and Apollo, plus advances in unmanned robotic exploration, helped to establish an overly optimistic and expensive long range plan for venturing far beyond the Moon as well as false confidence in the ability of the hardware that either existed or was likely to become available in time to support such grand plans. The reality check was the lack of public and political will, which manifested itself as financial cutbacks.

During 1969 the public face of NASA focused mainly upon the Apollo program and two unmanned Mariner flybys of the planet Mars. As Apollo 'flew the flag' for America, studies and proposals for the 1970s and 1980s were being developed. It was hoped that a follow-on program would emerge as an extensive space infrastructure that would allow humans to land on Mars early in the 1980s. The anticipated infrastructure included Earth and lunar orbiting space stations and a regular ferry system to transfer crews, hardware and logistics between the Earth, space stations, and a lunar research base. It was also predicted that such a system would see an expansion of human colonization and utilization of space for the benefit of all humanity. That is what NASA envisaged as the Stars and Stripes was planted on the Sea of Tranquility in July 1969, but within 18 months this grand vision had been ditched for more modest goals.

Social unrest at home and abroad, the ongoing and difficult conflict in south-east Asia, a growing awareness of the fragility of the environment (to some extent prompted by pictures of Earth taken by crews journeying to the Moon), the fear of losing a crew on the surface of the Moon without hope of rescue, the decline in American audiences for the TV coverage of astronauts picking up rocks and planting flags, and the evident lack of interest by the Soviet Union for sending cosmonauts to the Moon, have all been cited as having contributed to the decline in support for an adventurous American program of space exploration.

For years, NASA had struggled to gain adequate support for developing programs to be pursued after Apollo. However, with two lunar landings having been achieved in 1969, the new administration of President Nixon was more concerned about sustaining public support for the Vietnam War on the other side of the world than it was in promoting an expedition to Mars. As a result, NASA's budget was reduced. In January 1970 the planned total of lunar landing missions was cut from ten to nine. Three months later, Apollo 13 had to be aborted on the way to the Moon, resulting in a tense few days during which it was far from certain that the three astronauts would make it home safely. Then in September another two Apollo flights were deleted, as were plans for the second and possibly a third Skylab space stations. The prospects for a lunar research base, a 50-person space station, and sending people to Mars slipped back into the realm of dreams.

# 3

# A dream becomes reality

> *Astronomers could hardly wait for the LST, a project that unified the space science community as nothing ever did in the past.*
>
> James J. Hartford, Executive Secretary of the American Institute for Aeronautics and Astronautics (AIAA), May 1974

A study of the early development of what became the Hubble Space Telescope reveals the dedicated effort that drove the concept forward over a 25 year period. It began in the mid-1940s with the onset of the Cold War. By the late 1950s the theory of space flight had been established, and in the following decade it was greatly expanded upon to the point of landing men on the surface of the Moon and sending the first robots to fly close by the planets of the inner solar system. While the financial cost of such an adventure was extremely high, so too were the gains in technology, science, national pride, and political propaganda. Though the expected age of space commercialization was yet to be realized—and indeed is elusive half a century later—the birth of satellite applications suggested there would be significant rewards from studying the Earth's resources, from space sciences, and from technological spin-offs. However, the expansion of space exploration was not as rapid as glossy publications issued by some studies had predicted.

If the 1960s was the decade of great dreams and plans, then the 1970s was a realization that all that was space was not 'gold' and that even winning the race to the Moon was not a free-pass to greater achievements. As the 1970s progressed, NASA's long range plans were significantly revised. In 1972 the agency announced that it was to develop a reusable space shuttle which, amongst other things, would service and maintain satellites on-orbit, such as the Large Space Telescope. However, both programs suffered setbacks and delays, and the shuttle, whose inaugural flight was originally expected in 1979, was repeatedly delayed by technical issues to April 1981. The objective was to provide routine access to low orbit, but the dangers were vividly and tragically revealed by the loss of Challenger on a cold day in January 1986.

6. A Review of Space Research, 1962, National Academy of Sciences—National Research Council, Chapter 2, Astronomy, publication number 1079
7. Precursors of the Hubble Space Telescope, Donald C. Morton, in 400 Years of Astronomical Telescopes, Springer Science, 2010
8. Space Research: Directions for the future, 1966, publication No 1403 of the National Academy of Sciences-National Research Council, Part 2 Astronomy and Physics
9. MORL, Encyclopedia Astronautica, Mark Wade, http://www.atronautix.com/craft/morl.htm last accessed 27 June 2014; also A Manned Orbital Research Laboratory (MORL) Design and Utilization, R. J. Grunkel and C. E. Starns, BIS *Spaceflight,* Vol. 9, No. 3 March 1967, pp81–94
10. Apollo: The Lost and Forgotten Missions, David J. Shayler, Springer-Praxis, 2002, pp18–45
11. Apollo Lost and Forgotten missions pp52–64
12. Preliminary Program Plan and Cost for The Manned Orbiting Telescope (MOT), D2-84042-3, prepared for NASA Langley Research Center, Hampton, Virginia under contract NAS1-3968 by Boeing Company, Seattle, Washington, January 1, 1966
13. MOT report p25
14. MOT report p56
15. MOT report p59
16. Effects of In-Orbit Servicing of Nimbus Configuration—Case 105-3, A. S. Kiersars, Bellcomm Inc., Washington DC, July 18, 1969, NASA CR-106547
17. A Long Range Program in Space Astronomy, Position Paper of the Astronomy Mission Board July 1969 NASA SP-213, 1969, BIS Library Classification 15-04, Access R1525, BIS HQ, London
18. Astronomy Mission Board, 1969, pp272–273
19. NASA's Scientist Astronauts, David J. Shayler and Colin Burgess, Springer-Praxis, 2007
20. The Shuttlenauts 1981–1992: The first 50 missions, Volume 2, STS-Flight Crew Assignments, p57 entry for January 25, 1990, compiled by David J. Shayler, December 1992, AIS Publications

## A role for the future?

The 1969 report closed by suggesting some key roles that crewmembers could be assigned in support of future astronomy missions:

- They could increase the lifetime of an instrument by replacing a failed subsystem or perform "simple" maintenance operations.
- Obsolete instruments could be replaced with newer versions, or subsystems could be added which featured the latest state of the art across different fields.
- Participating in the activation of new major installations either by the assembly of the installation on-orbit or by operations which led to the activation of the instrument or hardware during installation periods.
- Retrieval of film and resupply, where ever larger data storage capacity was required, as long as there were no automated or instrumentation advancements to render that technology obsolete.
- It would be advantageous, both scientifically and psychologically, if any instrument or item of hardware designed for automation could be operated for a short period by an onboard "instrumentalist" [astronaut] in cooperation with the ground.

One area not listed in the above was the capability of servicing consumables (including refueling) to prolong both the orbital and operational life of a facility.

Interestingly, it is exactly these points that were addressed during the Hubble service missions, and instead of direct operation of an instrument in space, the modern "outreach" scope of the Hubble service missions was not lost on NASA or the scientific community.

The 1970s saw major changes in the American space program, with the lunar program being abandoned in favor of more economic operations closer to Earth, with an embryonic space station created using left-over Apollo hardware, a dramatic docking mission with a Soviet spacecraft, new robotic missions to Mars and the outer planets, and a new focus on looking back at Earth. But would this new direction also include the LST?

## REFERENCES

1. Lyman Spitzer Jr, by Denise Applewhite, Princeton University, www.spitzer.caltech.edu/mission/241-Lyman-Spitzer-Jr- last accessed August 2012
2. Dreams, Stars, and Electrons, Lyman Spitzer, Jr, in the 1989 Annual Review of Astronomy and Astrophysics, 27:1–17
3. History of the Space Telescope, Lyman Spitzer Jr. Quarterly Journal of the Royal Astronomical Society (1979) 20, 29–36, an article based upon a paper delivered by Spitzer at the Large Space Telescope Symposium, 12th Aerospace Sciences Meeting of the American Institute of Aeronautics and Astronautics (AIAA), Washington DC January 31, 1974
4. Exploring the Unknown Selected documents of the U.S. Civil Space Program Volume 1 Organizing for Exploration p332
5. Orbiting Astronomical Observatories, pp259–263, NASA History Data Book Volume II, Programs and Projects 1958–1968, NASA SP-4012, 1988

scientist-astronauts would operate instruments in space. Although spending months in Earth orbit or even on the Moon did not appeal to some of the piloting fraternity in the Astronaut Office, it was precisely in order to gain such an opportunity that many of the scientists had applied for the astronaut program.[19]

Despite being warned that they would have a long wait for a flight, if they got to go at all, that they would have to pass a jet pilot course *before* completing astronaut training, and that in the meantime opportunities to conduct their own research would be limited, many of the scientist-astronauts decided to stick with NASA in the hope of securing a mission to one of various planned AAP space stations (which ultimately became a single Skylab) or the new space shuttle and its multi-flight program involving carrying a pressurized laboratory in the payload bay. Would a scientist-astronaut accept the role of an "instrument operator" on an LST mission? The answer would probably be affirmative if it meant getting an early space flight, even though, as the 1969 report spelled out, the tasks would be essentially limited to maintenance because the control of the telescope and its instruments would be done on the ground, not in space. Although professional astronomers, the astronauts would have little if any opportunity to make direct observations using the telescope.

With the introduction of the space shuttle imminent in the late 1970s, the role of scientist-astronaut had morphed into 'Senior' Mission Specialist, then simply Mission Specialist. In time, in addition to carrying out experiments and investigations under the 'science program' of a mission, the role expanded to support operations on the vehicle such as Flight Engineer, RMS operator, and EVA. For dedicated 'science' missions such as the Spacelab series, non-career astronauts flew as Payload Specialists, focusing their skills on the payload or certain experiments.

If the early series of proposed astronomy-dedicated payloads and missions had evolved further under the Apollo Applications Program, or the shuttle, then specialist astronomers might have been able to operate experiments or telescopes in space, if those were designed for man-tended operations. But for astronomy payloads which flew on the shuttle, such as Spacelab 2 and the ASTRO flights, the astronaut's role was fulfilled by Payload Specialists or Mission Specialists, with the role of Payload Commander being added in January 1990 in order "to provide long range leadership in the development and planning of payload crew science activities" and assigned to an experienced Mission Specialist.[20] All that said, in the case of the ASTRO flights the Payload Specialists were astronomers who had been involved in developing the telescopes.

It is reasonable, therefore, to assume that if the payloads had been flown in conjunction with Apollo-type vehicles, then the designations of Commander, CM Pilot and "instrument operator" may have applied; or simply 'Science Pilot' as in the case of the Skylab missions. For the shuttle, especially with payloads such as the early designs of the LST having a man-tended capability, then the designation "instrument operator" would probably have merged into Mission Specialist or Payload Specialist once the hardware was operational.

Of course, at the time that the AMB report was written, the fate of Apollo-type vehicles had been all but sealed, and it would be the shuttle that would have been at the forefront if any LST hardware had actually made it to orbit.

Although the Board was not in a position to definitely answer these points, it did debate the pros and cons of involving astronauts versus an entirely automated program that might include man-tended capabilities. Indeed, it stated that "if the cost of manning astronomical experiments in space isn't regarded as part of the cost of astronomical research, the presence of man could well be an asset". Such issues highlighted the need for further studies of the overall national space program, but such matters were far beyond the scope of the Board's remit. Whilst direct involvement of astronauts in astronomy experiments and observations had been conducted during later Gemini missions, and would continue on Skylab and several shuttle missions (e.g., Spacelab 2, and ASTRO 1 and 2), and such involvement was envisaged for the MOT, there would be no direct involvement in LST/Hubble observations by astronauts in space. Nevertheless, underlining the assertion "the presence of man would be an asset", it can be argued that this was proven during the Hubble service missions as a merger of automated, robotic, and human space activities.

The Board couldn't specify the precise role of man in orbital astronomy but it did give consideration to the role of man "if available". However, instead of directly addressing the question, "Can, or should man in space be involved in space astronomy; and indeed are they really needed at all?" the Board evaluated the type of role that a crewmember might usefully serve in a future space astronomy program.

It was recognized that an imaginative scientist could adjust his instruments in real time, if on hand to improve an observation within the limitations of the hardware. But as the Board noted, "In order to exploit fully a large astronomical telescope in space, a substantial number of astronomers, together with their associates, will have to participate actively in its operation—just as is the case for major ground-based installations. In view of this rather large quantity of scientific manpower required for a major space instrument, we conclude that in general the scientific operation of a large space telescope will most effectively be carried out from the ground."[18]

It was also proposed that a direct link from the scientific research teams to the telescope would be preferable to working indirectly through astronauts in space, "since the later could easily introduce serious psychological barriers between the scientists and the instrument". It is interesting to point out that the single exception to this proposal was in the study of solar science, where phenomena can develop rapidly and an astronaut in space could react more effectively than a scientist on the ground. This was frequently demonstrated aboard Skylab, where crewmembers monitored the ATM console ready to react to the warning systems of pending activities on the Sun and manually target the instruments more effectively when a solar flare was imminent.

**Scientist, astronaut or instrument operator?**

The debate on the role of man in space was interesting at a time when the primary role of an astronaut, at least in NASA, was that of a pilot/explorer first and researcher/scientist second. In 1965 and 1967, two groups of scientist-astronauts were recruited to support an extensive scientific program that was intended to follow the early Apollo landings. These missions fell under the Apollo Applications Program, and in addition to longer flights to the Moon and in Earth orbit they were to undertake research in which the another human program in the face of uncertainty in securing finance for more ambitious Apollo missions and starting up the space shuttle and the proposed large space station.

Once the scientific issues had been formulated, each Panel of the Board considered how best to acquire the knowledge desired, based upon a series of missions involving equipment of increasing size and sophistication. Two proposals were devised. The "minimum rate" was the very least that would be necessary to attract and keep the leading scientists in that field. And the "maximum rate" represented the most that the community would be able to pursue, because if it were more substantial then "there would not be enough good groups of scientists to carry out the program".

The recommendations of each Panel were reviewed by the full Board and a plan was put forward for the period 1971 to 1985 in which the Board allotted each sub-discipline its own annual percentage of an overall budget that it intended to recommend to NASA. It felt that the proposed long range program would require a minimum of $250 million per annum for that period and the maximum program would require $500 million annually. Taking into account their relative priorities, the dates of the various projects were then adjusted to match the two funding rates. Optimistically, work on the LST would start in 1971 and it would be launched in 1980 or 1981. The other plan would involve starting this development in 1976 and launching in 1986. Either way, it would require a 10 year program of development and fabrication from go-ahead to launch. (In reality, this could have been 9 years in Hubble's case had it not been delayed by the loss of Challenger.)

The LST would have a primary mirror of at least 120 inches (3.04 meters) diameter and, with manned maintenance capability, an indefinite life. Covering all options, the Panel also suggested that totally unmanned versions of the LST be launched "every few years" which might offer comparable data to the LST. In addition, it strongly recommended further study of a series of man-tended and automated telescopes that could be used in a variety of fields and research studies.

There was also discussion of the future research and development that would be required in order to create and operate very large telescopes in the future. One area highlighted was the development of cryogenic systems to chill instruments and detectors, and keeping such equipment as small and as light as possible. This also created the basis for investigating the potential for replenishing such structures in space and repairing or replacing instruments on the telescope—a field which was addressed during shuttle development and on early flights in the 1980s.

**What is the role of man in space astronomy?**

This important question was considered by the AMB while compiling its report. The Board investigated the role of man in supporting astronomical research using large instruments in space. In particular, "would there be enough manned space activities in Earth orbit or on the Moon to warrant developing a system which is dependent upon manned support, and would the added costs of a man-rated transportation system be worth the investment?" Such issues extended far beyond the scope of the Board's report and indeed continue to this day, not just in the field of astronomy but in the whole debate concerning the investment in and the return from human space exploration against totally robotic operations.

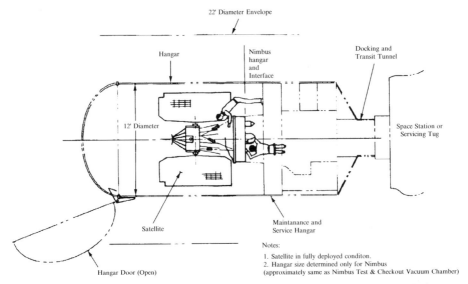

A Nimbus satellite with on-orbit servicing capability inside a satellite support hanger concept. (Courtesy Bellcomm Inc., AIS Collection)

## ASTRONOMY MISSION BOARD REPORT, JULY 1969

In September 1967, NASA created the Astronomy Mission Board (AMB) to offer advice in planning and conducting operational astronomical experiments in space. The wide ranging activities included developing and reviewing the scientific objectives for ground-based and space-based astronomy, for the design of space astronomy missions, and for the experiments and auxiliary equipment to be carried by those missions. Eighteen members of the scientific community, drawn from both academic and national laboratories, formed the Board, and it had seven panels (including the Optical Space Astronomy Panel) and two ad-hoc consulting groups that included a further 31 scientists. It met most months for the next year and a half, and on July 5, 1969 submitted a long range plan.

The Optical Panel was chaired by Lyman Spitzer and included Helmut A. Abt, Arthur D. Code, George H. Herbig, Gerry Naugebauer, C.R. O'Dell, Helen J. Smith, Fred Whipple and, as the NASA representative, Nancy G. Roman.[17] Mindful of tight budgets, the Panel's report offered both a maximum and minimum program from 1971 to 1985. They projected a series of seven further Orbiting Astronomical Observatories (OAO-D through -J) beyond the three already constructed, to be launched one per annum from 1972 to 1978. Then between two to four OAOs of a new Astra-class would be launched every other year between 1976 and 1982. These would have mirrors ranging in diameter between 40 and 60 inches (1 and 1.5 meters), improved instruments, and include a man-tended capacity. Included in these plans, under the heading of an "ultimately desirable instrument" was the Large Space Telescope. Notably, a 'manned' space telescope was *not* mentioned, as presumably the Board was mindful of the challenge of seeking support and funds for

In 1971, NASA Headquarters proposed that Goddard draw up the scientific specifications for the telescope and its scientific instruments, and manage the operation of the telescope in space, and that Marshall manage the overall development program and develop the hardware and optical payload. This was formalized in April 1972 when NASA announced Marshall as lead center for the LST, handling the Optical Telescope Assembly and the Support System Module, with Goddard being responsible for the scientific instruments and flight operations. Over the next two decades this would become a successful, if fraught union that the official NASA history of Marshall likened to "a troubled marriage" due to there being overlapping responsibilities in the division of technical tasks. These difficulties would result in "a lot of wasted effort and dollars."[5] Nancy Roman agrees that many of the problems encountered in the early years derived from the split of management between Marshall and Goddard.[6]

The background to the development of the telescope, its science package, and the funding and political struggles are beyond the scope of this volume but can be explored further in the official NASA History of Marshall Space Flight Center, as well as *The Universe in a Mirror* (Zimmerman, 2008; revised 2010), *The Hubble Wars* (Chaisson, 1994; revised 1998), and *Hubble Legacy* (Eds. Launius and DeVorkin, 2014) listed in the Bibliography of this volume.

**Early servicing plans**

By March 1972 the plan was to construct three telescopes, firstly an engineering model and a 'precursor' flight unit, both of which would be used in developing the necessary engineering and flight systems prior to constructing the actual LST, with a total design and development cost in the range $570–$715 million.[7] This extortionate approach soon came under close scrutiny, and in March 1973 the two test items were canceled. By following the 'prototype' approach that was being successfully applied by the military, a single vehicle would satisfy both a test and flight role. The servicing plan was also changed from the initially envisaged extensive astronaut involvement from inside a pressurized module to returning the telescope to Earth and conducting most of the servicing and repairs on the ground, then re-launching the telescope to continue its science program. Deleting the pressurized module simplified the design of the telescope and reduced costs even further, but not sufficiently to secure the firm support of Congress.

The original cycle for maintenance operations on the telescope was conceived at a time when plans were being drawn up for a fleet of shuttles which would make *25 to 60 flights a year*, operating so routinely as to resemble an airline. The telescope itself would be remotely controlled from Earth but be capable of having its instruments changed, serviced or repaired in space or on the ground. The realities of operating the shuttle were poorly understood and it was suggested that orbital maintenance could be scheduled on an "as required" basis simply by adding a flight into the manifest as an "unplanned shuttle flight"—almost rolling out and launching a mission whenever a problem or requirement came up. This naive approach was the opposite of what was actually required in planning and operating a service mission.

From the late 1970s, the availability of a huge water tank astronaut training facility saw Marshall contribute significantly to the underwater development of the servicing

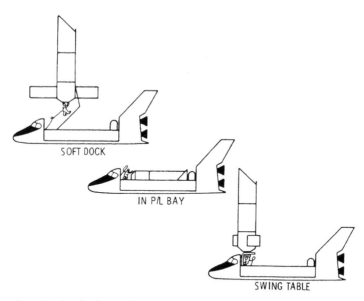

An early sketch of servicing options for the Large Space Telescope (LST).

techniques based upon experiences gained from Skylab, where spacewalking astronauts had released a jammed solar array, deployed thermal shields, improvised repairs to a failing gyroscope and exchanged film packs on the Apollo Telescope Mount. Marshall's support laboratories also assessed the reliability of systems and hardware in order to identify which were more likely to fail, and developed communications and control systems as modular units that astronauts would be able to replace. The scientific instruments were also designed to use a framework that could be slid in or out of the structure of the telescope to allow a rapid exchange of the bulky items. The decision to standardize connectors, fixing bolts and latches also helped in defining the servicing tasks and procedures. In addition, astronauts and JSC engineers who were specialists in EVA got together with colleagues from Goddard to develop special tools and support equipment to aid servicing crews. In the decade before Hubble was launched, a program of dives were carried out in support of planning for contingency EVAs that might become necessary in deploying the telescope, and for establishing the basic guidelines for swapping hardware during in-space servicing.*

In the development of servicing techniques, the Marshall repair and refurbishment team planned to store "orbital replacement units" on the ground and devised a method to extract technical information from them. This information, it was thought, would help in reducing the numerous hardware tests on new updated equipment and hence save development

---

*The water tank at Marshall proved so successful that a larger tank was later built at JSC to assist in the development of space station assembly and maintenance techniques. As soon as this became available, astronauts switched their Hubble training to JSC in order to save time on traveling to Marshall.

funds for other uses. But it was subsequently argued that the concept of designing equipment for repair whilst attempting to reduce the risk of system failure in operation really contributed to pushing up the costs due to the additional requirements to allow the unit to be replaced. This fueled a debate on the merits of adapting hardware and systems on an automated spacecraft for human interaction, instead of producing slightly less reliable components at a fraction of the cost and installing them on a series of spacecraft that would collectively operate over the same period of time. The financing of a project as complex as the space telescope continued to be a challenge.

A 1970s artist's impression of EVA servicing at the Space Telescope (ST).

## Charles Robert O'Dell: A skilled negotiator

On August 30, 1972, Dr. Charles R. O'Dell was appointed NASA's Project Scientist for the LST. Over the next decade he would feature prominently in securing the necessary funding. While a full time professor of astronomy at the University of Chicago in 1971 he had joined the NASA advisory group of elite astronomers and engineers that was asked to evaluate the feasibility of the LST. In his new role, he embarked on a 6 year battle with major research institutes to support the concept and then with Congress for funding. He persevered because he was sure that the LST would become one of the most powerful telescopes of a generation.

In February 1974, NASA put $6.2 million in its FY1975 budget request for LST planning funds. And in May, James J. Hartford, the Executive Secretary of the American Institute for Aeronautics and Astronautics (AIAA), testified before a House Committee that scientists had been skeptical of the likely utility of the space shuttle for space science, but the early results from the Apollo Telescope Mount on Skylab had produced some surprising results. Hartford pointed out that "astronomers could hardly wait for the LST, a project that unified the space science community as nothing ever did in the past."

That may have been the case amongst the scientists, but the politicians were not so easily convinced. In June the $6.2 million was struck from the agency's budget request because the LST was not one of the top four projects selected by the National Academy of Sciences. The same month, the Academy issued its 'Scientific Uses of the Space Shuttle' report, based on the July 1973 conference at Woods Hole, Massachusetts. This document stated that the cost-effective recovery and refurbishment of payloads in space required additional study. It also suggested that a "sophisticated" LST of the type proposed for launch on the shuttle might be worth the cost of a revisit, service, or return to Earth, but cautioned that the likely maximum recoverable payload mass available to the shuttle might restrict such plans. The timing of this report seemed to help the case, because on August 16 the Senate restored the $6.2 million for LST planning that had been deleted from the budget just 2 months earlier.

But in June 1974 the funding was deleted again because Congress, querying the scientific support for the project, said NASA should propose a less ambitious program. This prompted another phase of uncertainty and frantic lobbying. By December 1974 the telescope had to be down-sized to finely balance a performance that would address the scientific objectives with a cost that would satisfy the politicians. The first step was to cut the diameter of the primary mirror from 10 feet (3 meters) to just less than 8 feet (2.4 meters). Not only would this save funds, but by allowing the system to exploit the same technologies that had been developed for making the optical systems of military satellites it would be simpler to build and would require a year less to polish. The plan was also to reduce the number of science experiments from seven to four and to replace the proposed sophisticated steering system with a simpler pointing system. It had also been decided that innovative electronic detectors would replace film in the cameras, as there was a fear from the astronomers that the film might deteriorate and lose data, or might incur damage during changeout during the service missions. In any case, the revised payload delivery estimates for the shuttle showed that the 10 foot telescope would be too large to fit into the payload bay. Reducing the size of the mirror would have a beneficial knock-on effect on the primary support equipment, such as the batteries, reaction wheels, gyroscopes etc., enabling them be arranged around the telescope to make it easier to service. The Support System Module was now moved from the aft of the spacecraft forward about one third of its length, as a doughnut section around the primary mirror. Trimming the mass of the module from 24,000 pounds (10,886 kg) to 17,000 pounds (7711 kg) yielded an immediate saving of 7000 pounds (3175 kg).

Of course not everyone was happy with the changes, because scaling down the mirror put at risk some of the scientific objectives. Above all, astronomers wished to better measure the value of the Hubble constant, which is the ratio between the speed at which a galaxy recedes and its distance, and the only way to achieve this was to view far out into intergalactic space. It had been calculated that anything smaller than a 7.8 foot mirror would be unacceptable for this work.

However, trimming the size of the mirror offered further advantages, such as eliminating the need to build vast new test chambers to accommodate the large mirror. Furthermore, the reduced inertia of the vehicle would simplify the pointing and maneuvering systems. It was evident that the budget would remain tight even with such changes, but the revised proposal was sent to Congress.

There was an ongoing concern that allocating more money for the shuttle meant less for the science programs, particularly the telescope, and vice versa. As the official NASA history of Marshall Space Flight Center noted, this presented a 'Catch 22' situation where the shuttle was justified by its ability to carry scientific payloads and support on-orbit servicing of assets such as telescope, but was also criticized because its development costs were putting at risk the major science payloads whose missions it was supposed to facilitate.[8]

The fine balance between scaling down the costs and expectations of the space telescope in order to get the program authorized, also threatened the reason for launching it in the first place. Delays in developing the shuttle had knock on effects on the envisaged manifests for payloads. The loss of Challenger early in 1986 grounded the fleet. Retaining the completed telescope on the ground for another 4 years inevitably pushed up the costs of the project, not least because this occurred at a time of major inflation in the American economy.

June 1975 saw a funding campaign between the Large Space Telescope and the planned Pioneer-Venus mission. The proposal to impose a 12 month delay on the planetary probe to save funds that could assist the LST project was short-sighted and might result in losing the launch window. Juggling funds like this was inappropriate, as the two projects were not in direct competition. That same month, Dr. George M. Low, NASA's Deputy Administrator, said that if the agency had unlimited funds then he would move quicker to launch the space telescope, which he considered one of the most important projects from a scientific point of view. Funds for both programs were restored in July 1975. It was also Low who decided to delete the 'Large' designation from the project name, reasoning that proposing it simply as the 'Space Telescope' might help in convincing Congress that there was a serious effort to trim extravagance from the project. Although this may well have helped, it would still take over a year of lobbying and refinements before Congress gave its approval in 1976 and the project officially began in 1977.

There was still a long way to go, because the shuttle was still under development and was suffering delays. The original expectation of a maiden launch in 1979 would repeatedly slip, ultimately to the spring of 1981. Operational missions did not begin until late 1982. Clearly, as important as the Space Telescope was, it was not destined for an early ride to orbit.

Table 4 Hubble construction milestones.

| | |
|---|---|
| **1977** | |
| June 17 | NASA issues Project Approval Document for Space Telescope |
| | The Perkin-Elmer Corporation, Danbury, Connecticut, is awarded the contract for optic systems from NASA; Lockheed Missile and Space, Sunnyvale, California, is to develop the housing for the telescope |
| July 25 | NASA announced the selection of Lockheed Missile and Space Co., Sunnyvale, California and Perkin-Elmer Corp., Danbury, Connecticut to negotiate contracts totalling more than $131 million to two major elements of Space Telescope. Lockheed would design, fabricate, and integrate the telescope support-systems module, provide systems engineering and analysis, and support NASA in ground and flight operations. Perkin-Elmer would design, manufacture, and deliver the optical assembly and equipment, including systems engineering, support of the launch, verification of the orbit, and plans for mission operations. Launch now late 1983. MSFC overall project management and would manage contracts for the support module and the optical assembly. Goddard Space Flight Center would manage scientific instrument development |
| October 7 | ESA announced the signing of a memorandum of understanding with NASA on cooperation in the Space Telescope program. ESA would provide a faint-object camera for high-resolution imagery in the ultra-violet, visible and near-infrared portions of the spectrum, with associated photon-counting detector; the solar array to power the scope |
| October 19 | NASA awards the contract for the primary mirror to Perkin-Elmer of Danbury, Connecticut |
| December late | MSFC reports Corning Glass Works, Canton, New York had commenced work on a huge primary mirror blank for NASA's Space Telescope |
| **1978** | |
| January 11 | MSFC announced Corning Glass had begun work on a foot-thick primary mirror blank for Space Telescope. Aperture of mirror blank was about 94 inches, diameter of center hole 2 foot |
| April 25 | MSFC is designated as the lead center for the design, development, and construction of the telescope. GSFC is chosen to lead the development of the scientific instruments and ground control center |
| December | Rough grinding operation begins at Perkin-Elmer in Wilton, Connecticut |
| **1981** | |
| | Perkin-Elmer completes the fabrication of the telescopes mirror using a computer-controlled laser; (a programming glitch that occurred while grinding the mirror would be the cause for the blurred images discovered after launch in 1990) |
| April 29 | Perkin-Elmer completes polishing of the 2.4 meter primary mirror |
| October 23 | ST 'main ring' is delivered to Perkin-Elmer Corp from Exelco Corp., which fabricated the ring over a period of 18 months |
| December 10 | MSFC report a milestone in ST development. Perkin-Elmer Corporation finished putting an aluminium coating 3 millionths of an inch thick on the 94-inch primary mirror, an 18,000-pound polished glass blank. Engineers from the firm and MSFC had ensured the coating had adhered to the mirror and exhibited the correct reflectivity |

## 1982
| | |
|---|---|
| March | During the month the Critical Design Review of the ST support systems module is completed and the design is declared ready for manufacture |

## 1983
| | |
|---|---|
| | Science instruments were delivered for testing at NASA |
| | Faint Object Camera delivered by ESA to NASA |
| February 4 | NASA Administrator James M. Beggs tells House Science and Technology Committee that technical problems developing the electronics and guidance and pointing system will delay launch of telescope, and increase costs |

## 1984
| | |
|---|---|
| May 31 | Five science instruments complete acceptance testing at Goddard SFC |
| July 12 | Technicians at Perkin-Elmer clean the primary mirror NASA states the cleaning has confirmed 'the observatory will have the very best optical system possible' |
| August | Structural fabrication of Optical Telescope Assembly completed at Perkin-Elmer Corp. |
| Fall | Electrical checkout is completed on Optical Telescope Assembly, Perkin – Elmer Corp. |
| November | Optical Telescope Assembly delivered from Perking-Elmer to Lockheed via Super Guppy cargo aircraft |
| December 6 | Goddard's Telescope Operations Control conducts a successful Command and Telemetry Tests with the Hubble at Lockheed Missile and Space corporation, this is the first of seven planned assembly and verification tests |

## 1985
| | |
|---|---|
| July 8 | Lockheed Martin and Space company complete the final assembly of the telescope and commence testing program |
| August 15 | Lockheed Missiles and Space Co. personnel fitted the JPL wide field/planetary camera (WFPC) in Hubble ST. It was then removed and returned to JPL for modifications and testing. All went fine with no reported problems with JPL equipment |

## 1986
| | |
|---|---|
| January 28 | Loss of Challenger delays Shuttle program and grounds fleet until suitable fixes can ensure a safe Return to Flight. Hubble was to have been delivered to from Lockheed in California by ship via Panama Canal to KSC |
| February 27 | Hubble completes acoustic and dynamic and vibrational response tests to ensure telescope will endure launch environment |
| May 2–June 30 | Thermal-vacuum testing is conducted at Lockheed |
| May 21 | Solar arrays arrive at Lockheed from ESA |
| May 27 | Hubble completes thermal-vacuum testing in the Lockheed thermal-vacuum chamber |
| August 8 | Hubble completes 2 months of rigorous testing including temperature and pressure testing; some tests linked to STOCC Goddard |
| November | Solar arrays (rolled up) fitted to Hubble |
| Late | Four of five scientific instruments removed for telemetry servicing |

(continued)

## 96  A dream becomes reality

(continued)

| | |
|---|---|
| **1987** | |
| March 17 | A 3 day ground system test begins at Goddard simulating 39 hours or 28 orbits involving all five instruments, second of three major ground tests planned |
| August 31– September 4 | Goddard STOCC, Marshall Space Flight Center and the Space Flight Telescope Science Institute completed a joint orbital interface test |
| September 9 | Hubble completes the re-evaluation of Failure Mode and Effects analysis (FMEA). This included the Critical List/Hazard Analysis part of the Space Telescope Development Division effort to return the shuttle to flight |
| **1988** | |
| June 20 | Fourth Ground Test (GST-4) begins This is the most comprehensive and longest lasting 5.5 days including activating all six instruments in various operating modes (include Fine Guidance Astrometry) |
| June 24 | GST-4 is completed successfully, except for timing incompatibility between the computer and science instruments corrected by adjusting software |
| August 31 | Launch of Hubble delayed from June 1989 to February 1990, as a result of NASA remanifesting Shuttle missions to meet need of two planetary windows in 1989 |
| **1989** | |
| March | New improved solar arrays delivered from ESA to Lockheed |
| October | The telescope was shipped by modified Air Force Galaxy C-5A from Lockheed in California to the Kennedy Space Center in Florida for launch processing |

Normally it is fairly straightforward to obtain details of the fabrication and testing of space hardware. However due to the fact that the Hubble was built at Lockheed Sunnyvale Facility in California, the same location that fabricates the US Air Force's classified satellites, it has been more difficult to detail the assembly and testing sequence for the physical telescope structure. Additional information courtesy NASA Historical Data Book Volume V, NASA Launch Systems, Space Transportation, Human Spaceflight and Space Science 1979–1988, Judy A. Rumerman, NASA SP-4012, 1999 pp518–524.

Regarding the cost of Hubble, Nancy Roman told Congress that "for the price of a night at the movies, every taxpayer would receive 15 years' of exciting scientific results."[9] Although this prediction proved wide of the mark, it was basically true; for perhaps a little more than one night at the movies the American taxpayer has received over 25 years' of exciting science and stunning images from the real universe as opposed to one created in a studio!

**Gaining a name but losing the Earth return option**

On October 3, 1983 the space telescope was named to honor of American astronomer Edwin Powell Hubble (1889–1953). In 1924 he determined that our galaxy, called the Milky Way, was not the only one in the universe, there were many others. Three years later he found that the majority of galaxies were receding from us at rates which increased the farther they were from us. This was the first evidence that the universe was in a state of expansion rather than being static, as had been presumed. This would form the basis of the theory of the Big Bang. Hubble also devised a classifications system for galaxies that grouped them by their content, distance, shape, size and brightness.

**Getting started** 97

Edwin P. Hubble (1889–1953).

The evolution from the Large Orbital Telescope and the Manned Orbital Telescope to the Large Space Telescope, to the Space Telescope, and finally to the Hubble Space Telescope, had lasted almost 20 years.

During the 1970s plans called for ground refurbishment of the telescope every 2.5 years, based on the expected lifetime of the hardware and the predicated reliability of its onboard systems. The idea to refurbish the telescope in space rather than on Earth was first raised on May 14, 1984. Several studies had indicated that payload contamination and structural loads concerned with bringing the telescope back from orbit, working on its systems, and then re-launching it was far more risky and expensive than simply leaving the telescope in space for its expected 15 year lifetime and sending shuttle crews to service it. As the years rolled by, confidence in the systems and materials extended the time between servicing to 3 years, which would imply at least four service missions. In 1985 it was decided to delete the idea of returning the telescope to Earth for servicing, and instead to service it in space on a 36 month cycle.

In the event of something occurring which could not await the regular service mission, a contingency mission could be added to the shuttle manifest at a convenient time. The same could be done if a scheduled service mission proved unable to finish a crucial task.

## BRINGING THE PIECES TOGETHER

While the political and administrative process progressed through the 1970s, work to define the telescope program continued. At the same time as hardware was being developed, efforts in other areas influential to support for the telescope and its servicing also moved forward.

Included in this development was the crucial decision to develop the space shuttle as the next major investment in American human space exploration. Because this was the primary option to launch the telescope, it would also carry out the service missions. So integral to the shuttle program was the telescope, that in addition to having its own difficulties in securing funding and developing the hardware, anything that delayed the maiden flight of the shuttle and its entry into operational service would also affect the telescope.

One thing that strongly influenced the telescope was selecting the payload bay size of the orbiter. This was essentially dictated by the Air Force. It was also crucial to servicing the telescope that there be a means of capturing the satellite on-orbit and placing it in the bay so that spacewalkers could safely work on it. This required the development of an appropriate rendezvous profile and a safe means of approaching the telescope to enable a robotic arm to grasp it. The choice of a Canadian contractor to develop the robotic arm reflected NASA's desire for international cooperation in what it called the Space Transportation System. The European Space Agency was also involved by developing a pressurized laboratory module for carriage in the shuttle. Europe also supplied several key items of hardware for the space telescope and, as part of its overall involvement in the shuttle program, a critical part of the hardware supporting the servicing program.

Other major decisions and studies supporting the launch of the telescope and its use and servicing involved the structure of program management. It was essential to ensure that the hardware that would be flown on the shuttle was fit for purpose and safe for the astronauts during launch preparations, the ascent, orbital activities, atmospheric entry and post-landing activities. It was also vital to understand the environment in which the telescope would fly and be serviced.

# A SPACE SHUTTLE SYSTEM

In an effort to encourage commercial interest in space flight, and with the support of the US military, the one program that did survive the 'Grand Plan' envisaged in the late 1960s was the shuttle. The motivation to create a reusable spacecraft and launch system that could ferry items to and from orbit was to dramatically reduce the enormous launch and operating costs of one-shot space vehicles. NASA soon realized there would be little support for a broader program involving expanded lunar flights, a proposed 50-person space station, and human Mars expeditions. This did not bode well for a proposal to develop a shuttle whose primary purpose was to assemble and service that space station.[10]

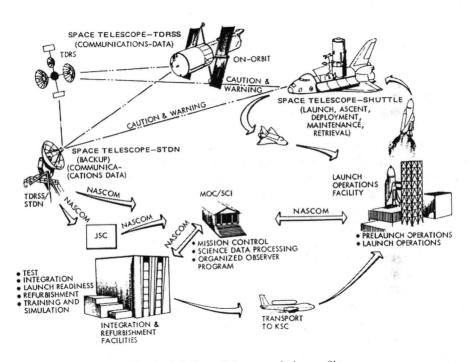

The shuttle's Space Telescope mission profile.

Committed to devising a program to follow on from the very successful Apollo era, the space agency realized that the prospects for the shuttle would improve if the Department of Defense (DOD) could be persuaded to endorse it.

In May 1969 NASA began negotiating with the Air Force on the Terms of Reference for a joint study of 'Space Transportation'. Initially working separately, NASA and the Air Force defined their concept of a vehicle that would meet their individual needs. The next step would be to trade off these specifications to arrive at a joint compromise. It was therefore suggested that the payload bay should be able to accommodate a package of 50,000 pounds (22,700 kg) traveling in either direction. A volume of 10,000 cubic feet (283 cubic meters) would permit a payload to have a diameter in the range 15–22 feet (4.5–6.7 meters). This suited the Air Force, which was developing payloads of similar size for its Titan III-C expendable launch vehicle.

As the studies progressed, several "mission concepts" emerged. One was that the shuttle would not only deploy satellites but also service and maintain them, or possibly repair failed satellites in space and perhaps even return them to Earth for refurbishment and re-launching. In a July 1969 presentation to the Science and Technology Advisory Committee at NASA Headquarters on the plans for the shuttle, the new section of 'Astronaut Tended Spacecraft' was included. This suggested exploiting the ability of the shuttle to revisit large automated satellites and telescopes in order to perform on-orbit maintenance "including the installation of upgraded instruments by the flight crews."[11]

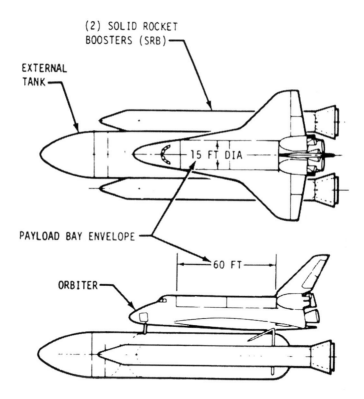

The main components and payload dimensions of the Space Transportation System.

Originally, the shuttle had been proposed as a fully reusable vehicle with a recoverable manned launcher, liquid fueled boosters and manned orbiter, but as the design progressed over the next 3 years a series of budget limitations, changes in the configuration of the shuttle and the cancellation of the space station resulted in a new design being revealed in

1972 that envisaged the manned orbiter being launched affixed to a huge one-shot external fuel tank augmented by a pair of segmented and reusable solid rocket boosters.

The original ambition was for a universal launching and recovery system that would make expendable launchers obsolete. Further, by making access to low orbit "routine", this would significantly reduce the cost of lifting a given mass into orbit. The prospect of having a fleet of vehicles fly dozens of missions per annum was revolutionary. In addition, launches from Vandenberg Air Force Base in California would be able to reach high-inclination orbits that were inaccessible from Florida. It was envisaged that the US military would become one of the largest customers and there was a prospect of the Air Force purchasing its own shuttles, but this plan did not evolve as envisaged.

In return for supporting the shuttle in Congress, the Air Force demanded that it possess a payload bay capable of accommodating the largest of military payloads. Once the size of the bay was specified, this in turn would also define the dimensions and mass of payloads to be carried on the vehicle, including the Space Telescope. This early decision on the size of the bay also ensured there would be sufficient room on service missions to accommodate both the telescope and supporting equipment, thereby eliminating the cost of creating a special 'hangar' to support the servicing of the telescope.

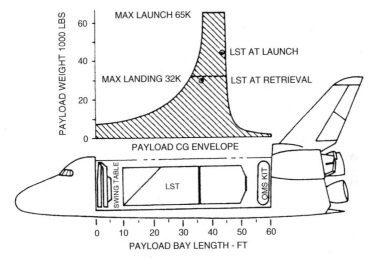

Configuration concept of the Large Space Telescope in the shuttle payload bay.

## Sizing the payload bay

NASA wanted a payload bay that was 40 feet (12.2 meters) long, for compatibility with its idea of a *small* space station module having a diameter of 14 to 15 feet (4.2 to 4.5 meters). But the Air Force wanted a longer bay to accommodate large military payloads. The early suggestion was 22 feet (6.7 meters) wide and 30 feet (9.1 meters) long, but the Air Force, knowing the length of its reconnaissance satellites and their upper stages were increasing with each generation, required it to be 15 feet (4.5 meters) wide and 60 feet (18.3 meters) long.

The payload of the reconnaissance satellite family called *Corona* was integrated into its Agena upper stage. These were initially 19 feet (5.8 meters) in length but had grown to 26 feet (7.9 meters), and were in the process of being superseded by a larger spacecraft called *Big Bird* that was 40 feet (12.2 meters) in length and 10 feet (3.04 meters) in diameter. Its systems were so sophisticated that its introduction had prompted the cancellation of the Air Force's Manned Orbiting Laboratory program. A third generation of satellites was already under development, named *Kennan*. These retained the 10 foot (3.04 meter) diameter, but were 60 feet (18.3 meters) or more in length.

When Dale Myers, NASA's Associate Administrator for Manned Space Flight, wrote to suggest reducing the length of the payload bay, Grant Hansen, the Assistant Secretary of the Air Force for Research and Development replied, "The length of the payload bay is the more critical dimension offering DOD mission needs… [because] the 15 foot diameter by 60 feet length payload bay size previously stated as the DOD requirements is based upon payloads presently in the inventory."[12] Hansen also made it clear that the desired dimension also took into account the plans for a reusable upper stage for "high energy missions" and if a shorter bay were chosen it would "preclude our full use of the potential capability and operational flexibility offered by the shuttle." If NASA were to insist upon a shorter payload bay, then around half of the payloads projected by the DOD for the years 1981–1990 would have to ride Titan III launchers, and if a supply of expendables had to be retained then, according to Hansen, "the potential economical attractiveness and the utility of the shuttle to the DOD is seriously diminished."

NASA was clearly not going to let a disagreement about the size of the payload bay put the development of the shuttle at risk, but the debate on the reasons for various diameters of the payload bay and total lift capability of the shuttle continued between NASA and the Air Force for some time.

In parallel contractor studies into the various concepts put forward for the shuttle system, the "best buy" was with a vehicle that could "carry a payload with a mass of 65,000 pounds (29,250 kg) in a payload bay measuring 15 by 60 feet (4.5 by 18.29 meters)." Anything less, it was suggested, would restrict the future options of launching payloads aboard the shuttle versus expendable launch vehicles. It made little sense to opt for a payload bay that had less capacity than an expendable, on a vehicle which had the added complication and expense of sustaining a human crew.

Of course, designing the shuttle system involved much more than the size of the payload bay, but it was the dimensions of 15 by 60 feet (4.5 by 18.29 meters) and a payload mass of 65,000 pounds (29,250 kg) that appeared in the final design that was approved by President Nixon on January 5, 1972.

One footnote to this saga is that in its 30 year history of 135 flights, the shuttle payload capacity came nowhere near what it was designed for. Following the loss of Challenger in 1986 and the withdrawal of all the planned DOD payloads that were to have been launched into polar orbit out of Vandenberg, the upper payload weight of the shuttle was lowered to 55,000 pounds (24,948 kg).

What is also interesting is that the original push by the Air Force for the 60 foot bay was driven by the size of its reconnaissance satellites. These were orbital telescopes designed to look down at Earth rather than out into space as the Space Telescope would and, ironically, they were all built by Lockheed at the same facility in Sunnyvale, California.

President Richard M. Nixon and NASA Administrator James C. Fletcher discuss the Space Transportation System concept.

## THE REMOTE MANIPULATOR SYSTEM

An essential element in the proposed satellite servicing and maintenance work planned for the shuttle was the development of a robotic manipulator system. The concept was far from new, because futuristic artwork had been depicting their use for years. For example, a 1967 model of the Apollo Applications Orbital Workshop that would become Skylab had featured a "sepentuator" (serpentine actuator) designed to move hardware around the exterior of the station in support of EVA operations in much the same way as robotic arms currently do on the International Space Station.[13]

On August 25, 1972 the Manned Spacecraft Center (as it then was) awarded a $226,256 contract to Martin Marietta Corporation to study systems for handling cargo in the payload bay of the shuttle. Resembling the human arm and electrically powered, these manipulators were to be 9–12 meters (30–40 feet) in length and to be operated from the shirt-sleeve environment of the crew cabin. In addition to moving cargo, the arm was expected to be of assistance to astronauts performing spacewalks.

The following year a 2 day meeting of the NASA Committee on Remote Manipulator Systems and Extravehicular Activity was held on July 10–11 at the Marshall Space

104  A dream becomes reality

Caption: Components of the Shuttle Remote Manipulator System (RMS).

# The remote manipulator system 105

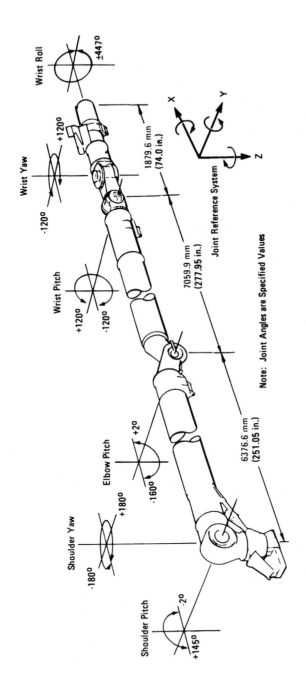

Joint angle motions for the RMS.

## 106 A dream becomes reality

Flight Center. Members of the committee met with industry and military representatives to review current and forthcoming robotic manipulator technologies and consider their use aboard the shuttle, including assisting astronauts servicing the space telescope. It was from these early studies that the specifications for the Remote Manipulator System were drawn up. This was developed through a Memorandum of Understanding signed in 1975 by NASA and Canada, which had started preliminary development work the previous year.

On February 11, 1981 the RMS was officially christened 'Canadarm' and it first saw use on STS-2 in November of that year. At the time, it was reported that the arm was capable of deploying a 65,000 pound (29,484 kg) payload that was no greater than 15 feet (4.5 meters) in diameter and 60 feet (18.29 meters) in length, and that it could retrieve a similarly sized payload of 32,000 pounds (14,515 kg) and put it in the payload bay. The arm formed part of the Payload Deployment Retrieval System (PDRS) that also featured payload retention and deployment devices such as keel fittings, longeron fittings, and a closed-circuit TV system. The prime contractor for the RMS was Spar Aerospace, based in Toronto, Canada.[14]

The jointed manipulator arm was attached to the port longeron of the payload bay, had a length of 50 feet (15.24 meters), had two boom sections, and was capable of six degrees of freedom. Beginning at the orbiter interface, the joints provided shoulder yaw, shoulder pitch, elbow pitch, wrist pitch, wrist yaw, and wrist roll. The upper boom that linked the shoulder and elbow joints was 21 feet (6.4 meters) long. The lower arm boom that linked the elbow to the wrist joint was 23 feet (7.0 meters) long. An end effector on the wrist roll joint served as the hand.

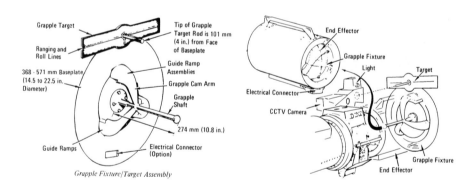

Detail of the RMS end effector and the target grapple fixture.

Working at the aft-right flight deck position, the arm operator employed a combination of direct vision, views from payload bay cameras, and views from the wrist and elbow cameras to align with the target and position the end effector above a grapple fixture that was part of the payload. After a system of three snare wires incorporated into a retractable carrier within the end effector had closed around the projecting shaft of the grapple fixture, the wires were retracted into the end effector to firm up the capture. Guide ramp cams on the grapple fixture mated with compatible key-ways in the end effector to correct any misalignment during the final capture sequence. The movement of the RMS was commanded by the General Purpose Computers (GPC) of the orbiter, whose software interpreted the signals from the operator's hand controller and decided which joint should be moved, in what direction, at what angle, and for how far. The software was capable of monitoring each joint every 80 milliseconds so that in the event of a failure the computer would be able to automatically apply brakes to all joints and inform the operator of the error.

The RMS played an essential role in 91 of the 135 shuttle missions over a 30 year period. Five were built and delivered to NASA between April 1981 and August 1983, one of which was lost on Challenger in 1986. All six Hubble-related shuttle missions carried an RMS. In the wake of the loss of Columbia in 2003, all subsequent missions carried the Orbiter Boom Sensor System (OBSS) that was 50 feet (15.24 meters) long and was operated in conjunction with the RMS to inspect the thermal protection system of the orbiter in search of any breach in the heat shield prior to entry.

**Table 5** Remote manipulator operators for shuttle Hubble missions.

| Flight | Mission | Prime RMS |
|---|---|---|
| Deployment | STS-31 | S. Hawley |
| SM-1 | STS-61 | C. Nicollier |
| SM-2 | STS-82 | S. Hawley |
| SM-3A | STS-103 | J.-F. Clervoy |
| SM-3B | STS-109 | N. Currie |
| SM-4 | STS-125 | M. McArthur |

The mobility and reliability of the RMS were critical to the success of the Hubble Space Telescope service missions. It captured and berthed the telescope in the payload bay, then supported the activities of the astronauts and finally redeployed the telescope. Often praised by the astronauts who controlled or rode on the end effector during their spacewalks (which in itself was an unforgettable experience) the RMS was a key decision early in the program which repaid its investment cost many times over during both the shuttle and space station programs.[15]

## A MEETING IN SPACE

Another major challenge for planning the series of service missions to the space telescope was the development of techniques to position the shuttle so that the robotic arm could reach out and capture the satellite and thereafter release it back into orbit. Developed over several years, these Rendezvous and Proximity Operations skills would also have direct application to the subsequent Shuttle-Mir and ISS assembly missions.[16]

The shuttle's final approach to the telescope. (Courtesy British Interplanetary Society)

## Planning

The design of a shuttle rendezvous and docking system was included in the studies carried out in the early 1970s as a method of physically connecting the orbiter to a second vehicle, ideally a space station. A variety of methods and components were proposed, assessed and planned. Although the shuttle first flew in 1981, docking equipment was not required until 1995—when the space station target was not American but Russian. By that time, the first Hubble service mission had already taken place. The skills of rendezvousing on-orbit and flying alongside a target had been developed over many years and trialed by the shuttle on a number of satellite retrieval missions during the 1980s. But the ability of the robotic arm to capture a payload and precisely install it in the payload bay meant there was no need for the telescope to incorporate a docking system, merely a grapple fixture.

However, when the formal planning for the service missions started, the designs of the telescope and the shuttle had been settled many years earlier and it turned out that some of the experiences of the actual shuttle rendezvous missions had not been fully appreciated or understood.

In the preparation of a flight into space there is a period of planning that encompasses a series of trade studies and simulations, then extensive technical discussions to differentiate essential tasks from contingency operations. The next task is to plan the timeline to achieve the best trajectory to place the orbiter in close proximity with the target, at minimum risk to both vehicles, ready to close in for final contact.

The rendezvous techniques for Hubble featured a series of maneuvering burns that began very shortly after the shuttle entered orbit and continued through to the start of Flight Day 3. On that day, the radar and star tracking navigational sensors gathered relative measurements which were used to update the relative navigational state in the shuttle's onboard computers. The computers then calculated a series of engine burns that would bring the orbiter to within 2000 feet (approximately 610 meters) of the target, at which the close proximity operations would begin. It was during these close proximity operations (known as Prox Ops) that care had to be taken to control relative motion between the two vehicles and to ensure that when the orbiter fired its maneuvering thrusters their efflux would not damage the surfaces of the target. In the case of Hubble, it was essential to protect the optical systems. Throughout the proximity operations, the Hubble controllers at Goddard constantly monitored the thermal and power constraints of the telescope.

The capture of Hubble was achieved by the RMS snaring one of two grapple fixtures on the side of the telescope. Then the telescope was gently lowered onto a support structure in the payload bay. The arm then withdrew from the telescope in order to assist the servicing EVA teams. Care had to be taken to ensure the telescope remained within the constraints of thermal, lighting or communication requirements, and that it was not damaged by firing the orbiter's engines. With the servicing complete, the RMS would once again grapple Hubble, hoist it clear of the payload bay, and release it just as if it were being deployed for the first time. The procedures were amended over the years to take into account flight experiences, but each of the five service missions followed essentially the same profile.

### From pad to orbit

The first challenge was to get the shuttle off the launch pad at a time that would produce the optimum 'catch up' profile to the target. The space ballet of orbital rendezvous and docking is normally handled with an 'active' spacecraft chasing the 'passive' target. The shuttle was the active vehicle, of course. The major constraint in performing a rendezvous with Hubble was its altitude of 380 miles (600 km), since the orbiter's propellant reserves were *only just* sufficient to reach Hubble, boost its orbit to cancel out the orbital decay caused by air drag during the periods between service missions, and finally return to Earth.

### Phases of rendezvous

Ideally, the launch had to be timed so that orbital insertion would occur in the same plane as that of the target vehicle. Any difference in plane between the two orbits would require the chase vehicle to consume more propellant, and potentially render a rendezvous impractical. Even in the ideal case, the chase vehicle cannot simply fire its engines in order to catch up with its target. The rendezvous starts with the two spacecraft in orbits which have

## 110 A dream becomes reality

different maximum and minimum altitudes and orbital periods. If the shuttle were to fire its engine to increase or decrease its velocity, then this would change the parameters of its orbit. Higher circular orbits have a lower velocity than lower ones, so if a vehicle fires its engines in the direction of flight it will increase orbital velocity. Owing to the nature of the gravitational force of the Earth, such a maneuver will convert the circular orbit into an elliptical one that has its high point opposite that at which the maneuver was made. And because a spacecraft flies more slowly at high altitude, raising the orbit lengthens the period of the orbit. On the other hand, firing the engines against the direction of flight in a circular orbit decreases the orbital velocity, lowering the spacecraft and shortening its orbital period.

When a shuttle initiates a rendezvous shortly after orbital insertion, it will be in a lower orbit than its target and so must undertake a 'ballet' in which it varies its orbit in a way that will enable it to 'catch' its target, then it will station-keep alongside in readiness to initiate Proximity Operations. This rendezvous sequence is divided into 'phases'. Drift Orbit A is also known as 'Out of Sight' rendezvous because it occurs when the chase vehicle is still out of contact with the target, separated from it by about 6200 miles (10,000 km). Drift Orbit B occurs when the vehicles are in sight and contact, separated by about 3300 feet (1 km).

For Hubble, the final maneuvers were controlled from the aft flight deck, usually by the mission commander assisted by the pilot. With the RMS operator standing by to reach out and grapple Hubble, other members of the crew would use laser ranging devices to verify closing speeds and take photos as the two vehicles closed in on each other. The Proximity Operations A phase would normally reduce the separation from 3280–330 feet (1000–100 meters) over 1–5 orbits. Prox Ops B would then reduce the range from 328 to 33 feet (100–10 meters) over an interval of 45–90 minutes (i.e. either half or one full orbit). In Hubble's case, the final RMS grapple would take place from a range of 33 feet (10 meters) and usually last about 5 minutes, during which the orbiter would make small translational 'yaw' (left-right) and 'pitch' (up-down) maneuvers and rotate (roll left-right) to refine the relative geometry for a more accurate contact at grapple.

### Approaching the target

There are different methods for bringing two spacecraft together:

- *V-bar* is where the chasing spacecraft is along the velocity vector (flight path) of the target and the final approach occurs from either ahead or behind it.
- *R-bar* is where the active spacecraft approaches the target from either above or below along its radial vector (essentially from the upper or lower sides of the spacecraft as it travels on its orbital path).
- *Z-bar* is where the active spacecraft approaches 'horizontally' from either the left or right of the plane in which the target is flying.

The target can present its docking/grapple fixture towards the approaching spacecraft to aid in the final capture maneuver.

### Developing the shuttle techniques

The shuttle rendezvous technique derived from the Phase B studies conducted during 1970 and 1971. During this 2 year program, four shuttle reference missions were simulated to evolve methods first used by Gemini and Apollo to the needs of the shuttle. It was in 1973 that contamination and efflux from the RCS thrusters in the braking phase became a concern to the various payload customers, and measures were conceived to alleviate this problem by restricting the directions in which the thrusters could be fired during this phase, by installing protective covers, or by adjusting the attitude of the orbiter. By 1977 the 'Low-Z mode' (see below) had been conceived to minimize contamination during the final approach to satellites such as the Long Duration Exposure Facility and Solar Max, both of which were designed to be grappled by the shuttle.[17]

**Flight experience**

Between 1983 and 1985, a number of shuttle missions undertook operations to develop the rendezvous and proximity techniques. This began with STS-7 (SPAS 1 pallet satellite) and included STS-41C (Solar Max retrieval), STS-51A (retrieval of both the Palapa and Westar comsats), STS-51D (Leasat 3 re-rendezvous) and STS 51I (Leasat 3 on-orbit repair). These early successes raised confidence in planning the more ambitious operations that would be required to service Hubble and to assemble the proposed Space Station Freedom.

Bob Crippen commanded STS-7 and STS-41C, which exercised rendezvous techniques during the early years. He and his STS-7 crew were acutely aware of the implications of the rendezvous for the retrieval, repair, and re-deployment of satellites. Their task was to show that the concept worked and develop the basic capability prior to more advanced operations on later missions. It must be remembered that in 1983 no American astronaut had flown a rendezvous since Apollo-Soyuz 8 years before. As Crippen has pointed out, at that time there were no thoughts about the extensive work that would be required to assemble Space Station Freedom since that had not yet been authorized (that would be the following year); the primary focus in the Astronaut Office was on satellite repair missions.[18]

In approaching Solar Max during STS-41C, Crippen's main concern was the Sun angle. He had "just looked at the satellite" and not at the moving background of Earth. This was a situation that Tom Stafford had noted during a rendezvous from above by Gemini 9 back in 1966 as something to avoid. Although the Sun was very bright, Crippen noted that practice and planning, "and a good pair of sunglasses" solved the problem. It was, of course, a much simpler and cheaper option than developing a complicated shade that would probably block the vision out the window.

According to Crippen, there had been an earlier call for rendezvous and Prox Ops to be included in the missions which performed the Orbital Flight Test phase of the program, but this was not pursued. STS-7 was the earliest opportunity to conduct a rendezvous because it was to release and then retrieve the SPAS free flying pallet satellite. STS-41C would make the first direct ascent, a technique that eliminated the immediate post-insertion OMS-1 burn, circularizing instead at the first apogee (OMS-2). This relaxed the pace of the checkouts and allowed the crew more time on the flight deck so they could "fold in" as many test objectives as possible. It was no coincidence that Crippen, one of the most experienced members of the Astronaut Office at that time, was selected to command both important missions tasked with developing shuttle rendezvous techniques.

Reflecting on the lessons learned from STS-41C's rendezvous with Solar Max, Crippen said, "The biggest problem was [that] the drawings of the Solar Max were not exactly right; the chest mounted grapple fixture on 'Pinky' Nelson's MMU would not work, which caused an unplanned tumble of Solar Max." Crippen had trained as backup to Terry Hart for RMS operations, and pointed out that the most important thing in rendezvous and Prox Ops in his experience was "to know what you are going to do and what the satellite would look like." The fact that the drawings were incorrect did not help the operation. Another problem which Crippen found on STS-41C was that the autopilot control on the RCS was so touchy that he inadvertently moved Challenger closer to Solar Max when he really wanted to stop. To solve this, improvements were made to the digital autopilot. Crippen also explained that it was not a simple task to station-keep with either SPAS or Solar Max, though he found that disabling the up-firing jets helped him. As for the drawings that they had reviewed pre-flight for Solar Max, Crippen dismissed this as being "one of those things that people always tell you—that their drawings are right." The veteran astronaut also noted that the person who did the most to get the Solar Max repair going was the recognized pioneer of the on-orbit satellite servicing concept, Frank J. "Cepi" Cepollina of the Goddard Space Flight Center.

**The Hubble experience**

New techniques were devised to aid in Prox Ops with the Hubble service missions. Since the shuttle resumed flying in 1988, only six missions had rendezvoused with independent targets prior to the first service mission. These were STS-32 (LDEF retrieval), STS-39 (SPAS 2-01), STS-49 (Intelsat 603), STS-56 (Spartan 201-1), STS-57 (EURECA), and STS-51 (ORFEUS-SPAS 1). Although LDEF and Intelsat were large vehicles, they were not as complex as the telescope.

During a rendezvous, Hubble had to be correctly managed to ensure that the solar panels stayed aligned with the Sun to recharge the batteries; therefore the attitude of the telescope and its solar panels had to be carefully coordinated. It was also realized during the missions that the flimsy arrays and their support structures and rotation methods were very sensitive to the firing of the orbiter's RCS, particularly over-pressure forces, as well as being vulnerable to contamination from the efflux. So an analysis program was devised by the Flight Design and Dynamics team in Houston to address both nominal (planned) and contingency (backup and emergency) Prox Ops in the approach, grapple, and redeployment phases that would not violate plume constraints on the telescope. Great care had to be taken to ensure that, during the periods of Prox Ops, Hubble could generate electrical power, even if its attitude control system was degraded. Before the shuttle initiated the final phase of its approach, controllers at STOCC would shut the aperture door on the telescope to protect the sensitive components inside. Hubble does not possess thrusters, it employs Reaction Wheel Assemblies (RWA) to control its attitude. There is also a non-redundant backup set of gyros called the Retrievable Model Gyro Assembly (RMGA) which can provide coarse attitude data for a brief period in support of shuttle Prox Ops and grappling activities.

**Contingency approach**

In the event of an anomaly during the approach of a shuttle, Hubble had two attitude control safe modes that would ensure that it remained in a positive power configuration. There were implications for the Prox Ops and RMS operations.

- *Hardware Sun Point* (HSP) relied on RMGA data and would orientate the +V3 axis to the Sun and maintain an inertial attitude hold, whilst aligning the solar arrays with the V1 axis and programming the aperture door to close.
- *Zero Gyro Sun Point* (ZGSP) would point the +V3 axis towards the Sun and impart a slow spin rate around the V3 axis. In addition, the solar arrays would be aligned with the V1 axis and the aperture door commanded to close.

Fortunately, these contingency options were not required during the service missions.

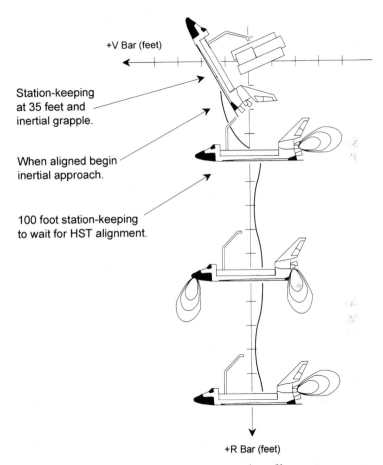

The +R Bar rendezvous approach profile.

**Nominal approach**

Prior to the final approach of the shuttle, Hubble would be commanded by STOCC to stow and lock its V3 high gain antenna and to rotate its solar arrays in parallel with the V1 axis. It would then perform a roll maneuver to locate the RMS grapple fixture on the 'north side' of the orbital plane. (There were two grapple fixtures mounted on the telescope, both of which could have been removed and reinstalled by the servicing crews had this become necessary.) One of the trade-offs in grappling Hubble was that although the preferred attitude simplified the task of approaching and grappling the telescope, it was not optimum for the telescope to generate power using its solar arrays. Therefore, as a safety measure, the completion of the roll maneuver initiated a 180 minute Sun-pointer timer. If the telescope was not grappled by the RMS within this period, Hubble would execute a slow maneuver to an attitude that was better suited to generating electricity.

This contingency function was not called upon during any of the service missions, but it had to be developed, tested, inserted into the system, and trained for by both the ground and servicing crews in case it became necessary in flight. There were many such contingencies, redundancies, and backup procedures factored into each Hubble-related mission under the 'what if' category.

## Low-Z mode

As mentioned above, the design of shuttle Prox Ops was worked out during the early 1970s. These procedures were updated from flight experience. One update was the introduction of the Low-Z flight mode for maneuvering in close proximity to sensitive targets. By the time this issue was considered, the design of the orbiter had been finalized and construction was underway. In normal use, the Z axis thrusters direct their plumes directly towards the target while braking during a final approach. With a spacecraft as sensitive as Hubble, the risks of contamination or over-pressure damage remained high, even with the aperture door closed.

The Low-Z mode was a workaround to remedy this. This used the X-body axis thrusters, whose efflux was directed off to either side of the target instead of toward it. The downside, of course, was that because the thrust was off axis more propellant was consumed to attain a given effect, and Hubble missions were already right on the limit of the shuttle's propellant capacity.

An advantage of stowing Hubble on a cradle in the payload bay of a shuttle for servicing was that the orbiter was able to use its maneuvering engines to raise the orbit and overcome the effects of orbital drag on the telescope, thereby securing its operational usefulness until the next service mission. In order to prevent too much stress on the attached combination during this operation, it was achieved by a series of small engine burns. As there would be years between service missions, this added benefit compensated for Hubble not having its own orbital maneuvering system.

Care had to be taken when employing the Low-Z control mode during Prox Ops and later carrying out the orbital boost maneuver, to ensure that sufficient propellant would remain to enable the shuttle to return to Earth.

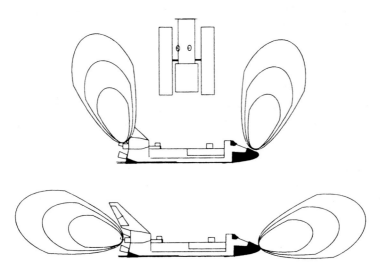

The Low-Z RCS plume firing mode.

**Table 6** Time between shuttle Hubble missions.

| Flight | Launch | Mission | Next mission | Date | Duration |
|---|---|---|---|---|---|
| STS-31 | April 1990 | Deployment | STS-61 | December 1993 | 3 years 8 months |
| STS-61 | December 1993 | Service Mission 1 | STS-82 | February 1997 | 3 years 2 months |
| STS-82 | February 1997 | Service Mission 2 | STS-103 | December 1999 | 2 years 10 months |
| STS-103 | December 1999 | Service Mission 3A | STS-109 | March 2002 | 2 years 3 months |
| STS-109 | March 2002 | Service Mission 3B | STS-125 | May 2009 | 7 years 2 months |

There was roughly 2.5–3 years between missions except for the final flight where the delay was over twice that of earlier missions due to the cancellation and reinstallation of the flight following the loss of Columbia

## MANAGING HUBBLE

The management of an enormous, long term project such as Hubble could not be handled by just one or two NASA field centers. It was decided to divide the task across five centers, plus numerous contractors and universities in the United States and Europe.

Here is a summary of the primary roles that the various facilities played in the program at the time of launch.

- *NASA Headquarters, Washington DC:* The agency operates its entire space science program under the direction of its Office of Space Science and Applications (OSSA). The Hubble program was part of the Astrophysics Division, with the Space Telescope Program Manager having responsibility for policy, goals and the administration of the telescope's resources, while the Program Scientist oversaw the science policy. NASA Headquarters oversaw the field centers, coordinated the

# A dream becomes reality

program within the agency's budget and guidelines, and reported to Congress on progress and results.

- *Marshall Space Flight Center, Huntsville, Alabama:* This was the lead center for the overall management of the program, with responsibility for early development of the telescope and for the orbital verification of the instruments following deployment by the shuttle. It also managed the cost, schedule, and technical performance goals of the telescope. In the early stages of the program there was also a significant amount of simulation and training conducted in the water tank at Marshall to supplement other training facilities at Johnson and at Goddard in developing servicing tasks.

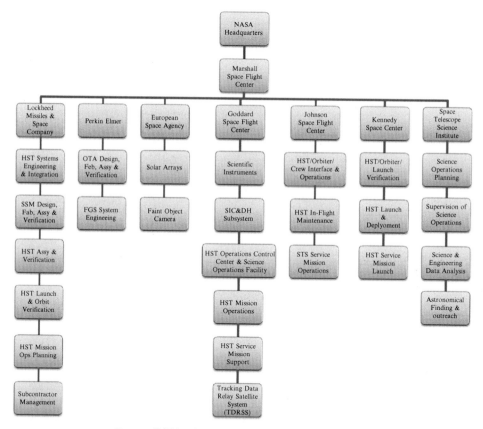

Responsibilities for the Space Telescope (circa 1990).

- *Goddard Space Flight Center, Greenbelt, Maryland:* This held responsibility for the development of the telescope's scientific verification and coordinating its day-to-day operations by way of the Space Telescope Ground System. It worked closely with the Space Telescope Science Institute (STScI) established at Johns Hopkins University in Baltimore, and with the prime contractor Lockheed. It was responsible for providing equipment and instruments for the service missions. Goddard also houses the Space Telescope Operations and Control Center (STOCC). In September 1988 responsibility for what was then referred to as the "Maintenance Mission" (the first servicing flight) was transferred from Marshall to Goddard. From then on, responsibility for developing the subsequent servicing program rested with Goddard.

- *Johnson Space Center, Houston, Texas:* Home of the astronauts, this was responsible for shuttle flight operations from the moment the vehicle cleared the launch tower in Florida to shortly after wheelstop on the runway at the end of the mission. It was also the prime center for astronaut training and was responsible for all interfaces between the orbiter and the telescope. If joint integrated activities were required in servicing Hubble, the shuttle crew communicated through the Mission Control Center in Houston with the STOCC at Goddard.
- *Kennedy Space Center, Florida:* NASA's primary launching facility lies just north of the Cape Canaveral Air Force Station of the Eastern Test Range. Utilizing the former Apollo launch facilities at Launch Complex 39, the center was responsible for the pre-launch processing, launch, and post-mission de-servicing of shuttle flights. It was the primary landing site for the shuttle end of the mission (EOM) activities, supported by secondary landing sites at Dryden Flight Research Center in California, White Sands Test Facility in New Mexico, and a network of emergency sites distributed across the globe. It was also responsible for recovering the twin Solid Rocket Boosters from the ocean and their de-servicing in readiness for return to the primary contractor, Morton Thiokol in Utah. As regards Hubble, KSC prepared the telescope for launch and later the payload package for each service mission.
- *Office of Space Tracking and Data Systems:* This managed the majority of the ground to spacecraft communications for the Space Telescope Project using both the network of Tracking and Data Relay Satellite System (TDRSS) and NASA Communications Network Satellites (NASCOM).
- *Space Telescope Science Institute, Johns Hopkins University, Baltimore, Maryland:* This was responsible for the management and coordination of the science program in between the service missions, in collaboration with the STOCC at Goddard. Planning a science program around in-flight failures and service missions required considerable teamwork. This approach ensured that an effective time management approach was achieved for the optimum observation periods and a maximum of effectiveness during equipment failure. It also managed the establishment of economical downtime during the servicing period, and the test and verification that followed the initial deployment and each service mission to ensure that the science program was resumed as soon and as safely as possible.
- *Lockheed Missile & Space Company, Sunnyvale, California:* As prime contractor for the development of the Hubble Space Telescope, this company managed the design, development, fabrication, and assembly of the telescope, plus the

verification of the Support Systems Module. It was also a co-prime contractor and supervisor for many of the subcontractors. Subsequent to participating in the post-assembly integration testing, it provided support to NASA during the servicing operations. A team from Lockheed assisted the STOCC with control and science operations.
- *Perkin-Elmer Corporation:* This co-prime contractor was responsible for the design, development and testing of the Optical Telescope Assembly prior to its shipment to Lockheed. It also developed the fine guidance sensors.
- *Scientific Instrument Contractors:* For each of the scientific instruments on Hubble, both for launch and subsequent servicing changeouts, NASA contracted with several investigators and subcontractors to develop, build, and test the particular instrument, with a Principal Investigator being responsible for its design and operation. In return for this effort, the PI was awarded primary observing time during the first months of their instrument's operation. In total, between 15 and 20 contractors, subcontractors and PIs worked on developing hardware for the Hubble Space Telescope.

**Tracking and Data Relay Satellite System**

This network of geostationary satellites can receive data from spacecraft in low Earth orbit and retransmit it to the White Sands Complex (WSC) or Guam Remote Ground Terminal (GRGT). In the case of Hubble, the data is then relayed to the Space Telescope Operations Control Center at Goddard for processing and forwarding to the Space Telescope Science Institute in Baltimore.

There are operational (and non-operational) satellites stationed over the Atlantic, Indian and Pacific Oceans. In 2014 there were nine working TDRS satellites available, with three serving in the primary role and the rest acting as spares. At least one satellite needs to be in line of sight with Hubble in order to receive data. Each high gain antenna on the telescope incorporates a two-axis gimbal to allow it to rotate 100 degrees in either direction to track a geostationary satellite. The STOCC can interact with Hubble to issue updated commands or install software 'patches' into its computers. The locations of the satellites in space do not affect the observations that can be undertaken by Hubble, as the necessary commanding is completed in advance of initiating an observation. When none of the satellites are in direct line of sight, the telescope stores data onboard for transmission after it has regained a link.

# Managing Hubble

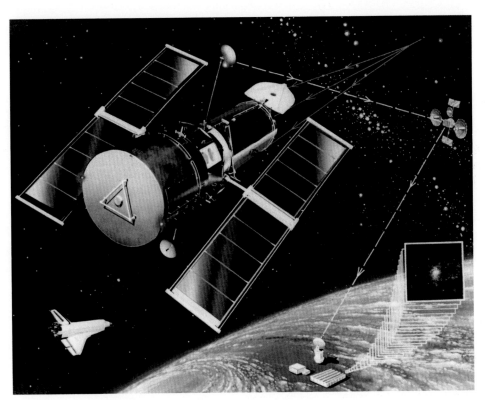

Space Telescope communications links through a Tracking and Data Relay Satellite (TDRS). (Courtesy British Interplanetary Society)

The TDRSS program was created in 1973 and managed by Goddard Space Flight Center to provide NASA with continuous, round-the-clock communications services and improved data reception, with reduced reliance on ground stations in other countries. Three generations have been developed, with each spacecraft receiving a letter designation prior to launch and a numerical designation after passing its on-orbit testing. To date 12 had been launched, of which one was lost aboard Challenger in 1986.[19]

Seven of the nine working satellites are currently operational (two first generation, three second generation and two third generation). In addition two first generation are in storage (TDRS 3 for the Atlantic and TDRS 5 for the Pacific). After 20 years of service, two other first generation (TDRS 1 and 4) were shut off and then super-synced into an orbit 300 miles (482.7 km) above the fleet. Of the operational satellites TDRS 6 and 9 are serving over the Atlantic, together with TDRS 12 after recently passing its testing phase; TDRS 7 and 8 are over the Indian Ocean; and TDRS 10 and 11 are over the Pacific. TDRS M is scheduled for launch in 2016 and there is an option to build an additional spacecraft TDRS N.

# 120 A dream becomes reality

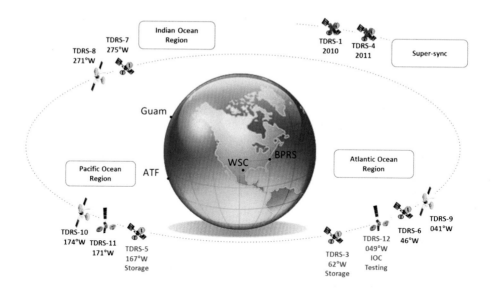

The TDRSS network circa 2014.

**Table 7** NASA tracking and data relay satellite system.

| Satellite | Operational | Launched | Vehicle | Current status |
|---|---|---|---|---|
| *First Generation* | | | | |
| TDRS-A | TDRS-1 | April 4, 1983 | STS-6, Challenger | Retired Fall, 2010, Disposal June 2010 |
| TDRS-B | [TDRS-2] Intended | January 28, 1986 | STS-51L, Challenger | Launch failure, loss of vehicle during ascent |
| TDRS-C | TDRS-3 | September 29, 1988 | STS-26, Discovery | In storage December 2011 |
| TDRS-D | TDRS-4 | March 13, 1989 | STS-29, Discovery | Retired December 2011; Disposal April 2012 |
| TDRS-E | TDRS-5 | August 2, 1991 | STS-43, Atlantis | In storage, 2014 |
| TDRS-F | TDRS-6 | January 13, 1993 | STS-54, Endeavour | Active |
| TDRS-G | TDRS-7 | July 13, 1995 | STS-70, Discovery | Active, replaced TDRS-B |
| *Second Generation* | | | | |
| TDRS-H | TDRS-8 | June 30, 2000 | Atlas IIA | Active |
| TDRS-I | TDRS-9 | March 8, 2002 | Atlas IIA | Active |
| TDRS-J | TDRS-10 | December 5, 2002 | Atlas IIA | Active |
| *Third Generation* | | | | |
| TDRS-K | TDRS-11 | January 13, 2013 | Atlas-401 | Active |
| TDRS-L | TDRS-12 | January 24, 2014 | Atlas-401 | Active |
| *TDRS-M* | *TDRS-13* | *Planned 2016* | – | – |
| *TDRS-N* | *TDRS-14* | *Option* | – | – |

## A EUROPEAN PARTNER

Towards the end of 1974, as the struggle to secure funding for the project continued, NASA sought assistance from Europe. The hope was that foreign participation would not only help to reduce the level of budget request to Congress but also heighten the chances that it would pass the authorization hurdle.

In 1975 the European Space Agency agreed to join the Large Space Telescope program, and in October 1976 it announced that its sixth meeting of member states had unanimously approved contributing €80 million to the project. Twelve months later, in October 1977, the two space agencies signed a Memorandum of Understanding whereby ESA would provide a Faint Object Camera for high resolution imagery in the ultraviolet, visible and near-infrared regions of the spectrum, along with an associated photon detector. The Faint Object Camera would be built by Dornier in Germany and Matra in France. The photon detector supplied by British Aerospace would be capable of detecting a candle from 24,856 miles (40,000 km).[20] "The instrument," the announcement explained, "is to be left on the Space Telescope as long as considered scientifically useful." It was also agreed that ESA would make the solar arrays for the telescope and support the Space Telescope Operations and Control Center at Goddard to manage the observatory and its science program. ESA participation would be managed by the European Space Technology Center (ESTEC) at Nordwijk in the Netherlands. In return for providing 15 percent of the costs, Europe would receive 15 percent of the observing time of the telescope for "the duration of the program." Fifteen staff members from ESA would be assigned to the Space Telescope Science Institute in Baltimore.

According to Lothar Gerlach, who worked on the Hubble program for ESA between 1986 and 2012, the European agency benefited from NASA's experience in both space technology and mission management. Over his long association with the project, Gerlach was impressed with the professionalism of the Americans, ensuring that any small cultural differences were minimized. His only negative recollection was in the wake of the terror attacks of September 11, 2001, when the level of security was heightened in such a way that the situation became very uncomfortable for the European HST team. "We had to be escorted everywhere [even] into the rest rooms."[21] This was a strict rule everywhere the team went. It had to be adhered to, and as the teams were both counterparts and good friends after so many years of working together, no one wanted to cause any problems, so they simply got on with their jobs.

Further to the 1977 agreement, the cooperation between NASA and ESA would see two European astronauts fly on Hubble service missions. Claude Nicollier (Switzerland) was on STS-61, the first service mission, in December 1993 as the prime RMS operator, then as an EVA crewmember on STS-103 in December 1999. The Anglo-American NASA astronaut Mike Foale accompanied Nicollier on his spacewalk, with Jean-François Clervoy (France) serving as the prime RMS operator.

### A British involvement

There were other British links to Hubble and its servicing activities. The Earth Observation and Science Division of British Aerospace Space Systems in Bristol, England, was awarded the contract to develop the solar arrays as part of the ESA contribution. It later supplied the improved arrays that were substituted during the first service mission.

## 122 A dream becomes reality

Although less well publicized, the second British association with the Hubble servicing extended across the five missions. This was the development, fabrication, and supply of a dedicated reusable modular pallet as a cargo carrier. BAe worked under contract to ERNO (Zentralgesellschaft VFW-Fokker MbH) and ESA to produce these units as part of ESA's contribution to the shuttle program. Spacelab pallets were employed on numerous shuttle flights during its 30 year history, carrying a wide variety of payloads and equipment. For Hubble they were the primary carrier for the orbital replacement units used to service the telescope. In addition to transporting scientific instrument to and from orbit and acting as temporary stowage locations during EVA activity, the versatile pallets carried apparatus, tools and spares for use by astronauts whilst working on the telescope.

The Spacelab pallet had a U-shaped design which optimally matched the cross-section of the payload bay and gave the maximum protection of flat surfaces for equipment mounting. Available in single, double or triple pallet configurations to suit the mission requirements, it was designed to maximize flexibility to the user and be rapidly modified for a new payload. This flexibility made it an ideal, although often overlooked, element of the STS system. Its carefully designed attachments were complementary to the orbiter's fixtures, making all of the removable items interchangeable. Approximately 1000 "load cases" (combinations) of mass distribution and configuration could be analyzed to match the miscellaneous payloads and operations of the shuttle. Their low mass made them perfect carriers for the irregularly shaped equipment required to service Hubble, and return older equipment to Earth.

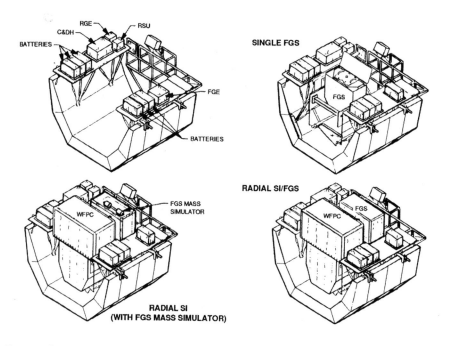

Four configurations of the Hubble Orbital Replacement Unit (ORU) carriers based on the shuttle pallet system, supplied by British Aerospace in England under contract through the European Space Agency (ESA).

Each pallet was covered with aluminum honeycomb panels, and could accommodate a payload envelope 11.92 feet (3.63 meters) in diameter. The pallet was 9.66 feet (3 meters) long, but could accept payloads of greater length. The structural mass of a pallet was 1400 pounds (636 kg) and it could carry a payload of 9086 pounds (4130 kg). It had 24 primary attachment points for either special fittings or direct fixing. In addition, a secondary facility offered a matrix of 24 attachment points on an inner panel for mounting smaller items. Each pallet had a 10 year life or 50 missions, whichever occurred first.

A variety of platform facilities were developed and fixed at either of two levels in order to accommodate different payloads. A platform assembly consisted of two elements, a platform and a link. The platform was rectangular and available in different sizes to offer the greatest flexibility to support smaller items. The elements of a platform were attached using four link elements that could enable attachment at high or low levels. Across the open structure, these platforms could accommodate a load of between 90.7 and 181.5 pounds per square foot (200–400 kg/m$^2$). BAe also offered half-pallet and quarter-pallet designs, as well as options for future space stations. A number of "pallet mission" proposals were evaluated for Spacelab, with autonomous free-flying pallets for a range of Earth, stellar or Sun viewing instruments to be deployed and retrieved by the shuttle.

Although these options were not developed as envisaged, the pallet elements were the unsung heroes of each Hubble service mission, transporting the new hardware into space, serving as temporary or permanent storage locations during EVAs, and then as carriers for returning retrieved hardware to Earth.[22]

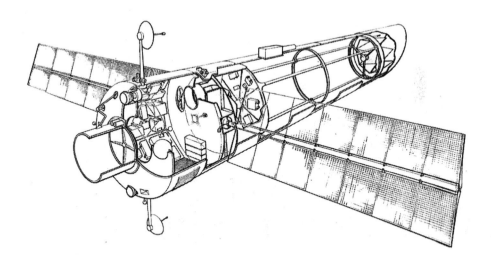

Hawker Siddley Dynamics' concept for a European Astronomy Experiment module circa 1970. (Courtesy British Interplanetary Society)

## A British/European 'Hubble'?

Interest in the development of a large space telescope was not confined to the United States. In the fall of 1970 the European Space Research Organization (ESRO, forerunner of ESA) received a presentation from a team of the Space Division of Hawker Siddley Dynamics in Stevenage, England. Led by Watson Lang, the team included Jim Heaton, John Farrow and John Davis, and was eager to participate in post-Apollo activities.[23]

Their proposal, 'Design Study for an Astronomy Experiment Module', centered upon a space telescope that would have a primary mirror 10 feet (3.05 meters) in diameter and Ritchey-Chretien (modified Cassegrain) optics. Similar to the MOT/LST designs, the HSD facility would orbit Earth at an altitude of 805 miles (500 km) inclined at 55 degrees to the equator. Each 94.48 minute orbit would permit about 37 minutes of observing time. It was expected that its solar arrays would have a lifetime of either 5 years or 10 years and that its batteries would have a 3 year lifetime. Using its own propulsion system it would dock at a space station every 3 months for servicing, then return to its operating orbit. An overall operating life of up to 10 years was intended. It was expected that the station would have a pressurized habitat (similar to that proposed for the earlier MOT) in which to carry out the servicing.

## PROTECTION AGAINST CONTAMINATION

The 1970 Hawker Siddley Dynamics study included an interesting list of potential sources of contamination (still current today, in fact) illustrating the tremendous care that must be taken to ensure that *any* item of hardware is not contaminated when it leaves Earth. In the case of the Hubble Space Telescope this applies in planning shuttle visits to repair or replace failed or outdated items. Anything that comes into contact with the telescope undergoes a rigorous campaign to prevent back contamination. There must be rigorous checks to verify that every item of hardware is launched in as pristine a condition as possible. Contamination can come from many sources and despite stringent procedures, checks and precautions, the hot, humid environment of Florida can challenge this task with natural phenomena such as mosquitoes and hurricanes.

The HSD study addressed the contamination that can occur in manufacturing arising from the materials used, the tools selected to fabricate the equipment, and even the methods used. For example, there could be metal or lubricant contamination or directly from the engineers fabricating the spacecraft. If the personnel are not properly suited, human hair, skin flakes, dandruff, finger oils, and exhaled breath can impart contamination onto the smallest items, unseen by the naked eye. In 1984 this became evident with the aborted launch of STS-41D. Post-flight examination of the failed component concluded that the most likely cause of the abort was minuscule contamination by human fluid, most probably a sneeze or cough, that had trapped liquid in a circuit and later shorted it out. The result was a multi-million dollar delay to the mission.[24]

Even after equipment has been built, it must be stored and handled to form the integrated payload that will be flown. This process creates further risk of contamination. In addition to surviving the stresses and vibrations of launch aboard a shuttle, it was possible for exhaust products to invade the cargo bay. Experience has shown that on-orbit some components do not match their design specifications. A payload can suffer from water dumps, outgassing, thermal variations, physical stresses, and impacts by tiny items of space debris. There are simple accidents such as an inadvertent touch by an astronaut or a slipped tool. And it has been known for apparatus to pass all its checks on the ground and then fail in space for no apparent reason.

Risks can occur during assembly, integration, testing, and any phase of the mission from launch to post-landing. Highly motivated workers are trained and provided with protective suiting, and they rigorously follow the safety procedures under intensive supervision in an effort to ensure that each item of hardware is as clean and functional as possible. This has to be done for *everything*, ranging from the largest item right down to the smallest tool, screw, connector, or circuit breaker.

For Hubble, this attention to detail had to be sustained not only for its launch, but for all five service missions over a period of almost 20 years. The lessons learned from the Hubble contamination program were subsequently applied to the even more complex International Space Station.

**Considering the operational environment**

Creating the eagerly awaited Hubble Space Telescope was a challenge for the designers not only in terms of the reliability of its onboard systems and instruments, and later planning the servicing and maintenance missions, but also in understanding the environment in which the telescope would operate for what was initially hoped would be 15 years and turned out to be over 25 years at the time of writing this book and may ultimately exceed 30 years.

Orbit after orbit, the telescope has been subjected to high levels of thermal and structural loads. NASA maintains a database on conditions in the space environment. This underpins the agency's parameters and guidelines for the development and operation of spacecraft. In the 1970s these research notes were used to define the environment in which the telescope would fly and be operated. These environmental factors hindered (or sometimes assisted in) the development of apparatus and systems, and the mission rules that would make the space telescope a feasible, safe, and reliable instrument for research. They also determined a safe environment to send astronauts out to tend the telescope.

**A natural working environment**

In planning for shuttle operations, the criteria specified in these reports involved radiation characteristics and other natural environmental requirements in the Earth's atmosphere at orbital altitudes. NASA constantly updates a wide range of documents which determine a safe environment to fly, and an example of these used in the early stages of the telescope development were:

- NASA TMX-64587, *Terrestrial Environment (Climatic) Criteria Guidelines for U.S. in Space Vehicle Development, 1971 Revision* (May 10, 1971 edited by Glenn E. Daniels).
- NASA TMX-64627 *Space and Planetary Environment Criteria Guidelines for use in Space Vehicle Development, 1971 Revision* (November 15, 1971 edited by R. E. Smith).[25]

Using the latest versions of these guidelines,* the environment in which the shuttle would have to fly while servicing Hubble became better understood. As a payload of the shuttle, the telescope was subjected to further environmental checks peculiar to the shuttle for assembly, checkout, launch, and orbital flight through to deployment; in short, the telescope was subjected to the standard specifications for "shuttle cargo."

Profiles were made of the atmosphere at orbital altitudes, including the gas properties of its constituents, solar flux, and the geomagnetic index which provides a means of predicting the conditions for 12 to 36 hours during an extremely large magnetic storm. The studies of radiation focused on charged particles, galactic cosmic rays at times of solar minimum and solar maximum, trapped radiation and the radiation belts close to Earth, synchronous orbit environments, and solar particle events.

The Hubble designers reviewed how all these factors would affect the components of the telescope during normal operations, as well as their implications for the frequency and risk during a service mission, thereby defining the protection and shielding requirements for the crew to minimize these effects.[26] Hubble was designed on the assumption that there was at least a 95 percent chance of it sustaining damage by a meteoroid puncture, so it had to be provided with suitable protection against the loss of functional capability. The fact that the telescope would operate at a relatively high altitude meant that it would be at less risk of a collision with man-made orbital debris. Nevertheless, a study was made and the risk to the shuttle during the deployment and service missions and was found to be acceptable.†

---

*Current versions of these documents are NASA/TM-2008-215633, *Terrestrial Environment (Climatic) Criteria, Guidelines for Use in Aerospace Vehicle, Development, 2008 Revision*, D.L. Johnson, Editor, Marshall Space Flight Center, Marshall Space Flight Center, Alabama; and NASA Technical Memorandum 82501 *Space and Planetary Environment Criteria Guidelines for Use in Space Vehicle Development, 1982 Revision (Volume II)*, Robert E. Smith and George S. West, Compilers George C. Marshall Space Flight Center Marshall Space Flight Center, Alabama. These guidelines provide updated information on the natural environment for altitudes between the surface of the Earth and 90 km altitude for the principal areas. These guidelines supersede all editions of NASA Technical Memorandum 4511 *Terrestrial Environment (Climatic) Criteria Guidelines for Use in Aerospace Vehicle Development* dated August 1993. This was recommended for use in the development of design requirements and specifications for aerospace vehicles and associated equipment. The origin of the *Terrestrial Environment (Climatic) Criteria Guidelines* dates to the early 1960s. It was originally conceived by Glenn E. Daniels of the NASA Marshall Space Flight Center's Aerospace Environment Division, and the early editions were prepared under his direction. He continued updating the document until his retirement from MSFC in 1974. After Mr. Daniels passed away in 2004, later editions were dedicated to his memory.

†Following the launch debris strike on Columbia in 2003, further studies were made prior to the final Hubble service mission in 2009. From this research the likelihood of a serious debris impact with the shuttle during SM-4 was a 1:229 chance; somewhat lower than had been thought. For SM-1 in 1993 the value was 1:150, then 1:761 for the 1999 SM-3A and 1:365 for the 2002 SM-3B flight. Information gathered during earlier missions provide valuable data on the study of orbital debris, in particular from natural impacts and their effects on various types of hardware such as solar panels, the metals used in fabrication, and thermal protection materials. Studies of images and returned materials from Solar Max (1984) and LDEF (1990), a number of retrieved and repaired satellites, the shuttle orbiters, and various samples retrieved from spacecraft, together various experiments, have generated a huge database of the potential risks and effects from space debris.

## Ensuring safety and integrity

As Hubble was a payload to be launched on the shuttle and subsequently maintained by astronauts, a number of checks were carried out to ensure that the basic structure and its components conformed to NASA's strict regulations regarding interaction with the shuttle system. In addition to initial processing, launch and deployment, this applied to each of the service missions which would deliver or retrieve science instruments and equipment. For repeated visits to a progressively ageing spacecraft in Earth orbit over a long period of time this was a challenging task. However, this experience was applied in developing the safety policies for the Shuttle-Mir and International Space Station programs.

## Payload safety on the shuttle

As the shuttle required an astronaut crew, all payloads assigned to a mission, for reasons of safety, had to undergo stringent guidelines to qualify them for flight. Particular attention was given to anything for which a potential failure could result in a catastrophic safety hazard. At the onset of the program a set of guidelines were in place that spelled out safety policies and the requirements for the structural development of payloads. This was particularly important because the shuttle was intended not only to carry payloads funded by NASA but a wide variety of hardware developed, constructed, and supplied by domestic and foreign agencies, private companies, and universities.

The guidelines were listed in documents such as

- *General Environmental Verification Specifications for STS and ELV Payloads Components and Subsystems*
- *Mass Acceleration Curve for Spacecraft Structural Designs*
- *Structural Design and Verification Criteria*
- *Payload Verifications Requirements*
- *Implementation Procedures for National Space Transportation System (NSTS) Payload Safety*
- *Fracture Control Requirements for Payloads Using the NSTS*
- *Safety Policy and Requirements for Payloads Using the NSTS.*

Of course, not only would an item of hardware or software intended to be flown aboard the shuttle be *designed to satisfy* these requirements, it would also be *checked and tested* to verify that it did so.

## A BROAD AND INVOLVED DEVELOPMENT

It is amazing how connections and decisions affect the outcome of some actions. From the desire to launch a large optical telescope into space emerged, two decades later, studies and plans to do precisely that. Occasionally such studies go far beyond their basic requirements, and create an overly complicated proposal that can become too expensive to realize.

Debates and arguments for and against the proposal push the idea through the political, financial, and public arenas as the designers and engineers refine their design to satisfy the evolving constraints. Hopefully, at the end of this arduous process a project emerges which will provide useful, if not ground-breaking results, and a success which satisfies all parties.

The Hubble Space Telescope project has arguably achieved this, thanks to its design, the people who have worked so diligently to implement the design, and the planners, engineers and astronauts who deployed and serviced it in space.

Often, decisions made years before can have a variety of consequences decades later. The choice of primary mirror size on the telescope had to match the minimum requirements of the science community but was governed by the need to save money in the budget. The reduction in mirror dimensions also helped to reconfigure the components of the telescope that required servicing, thereby making future servicing tasks much easier. The sizing of the payload bay of the shuttle was defined by the large military payloads that it was originally meant to launch. This enabled the telescope to be carried comfortably. If NASA had held out for its preferred payload bay with a length of 30 feet (9.14 meters) to 40 feet (12.19 meters), then Hubble, at 43.5 feet (13.2 meters), simply would not have been able to be carried. The length of 60 feet (18.29 meters) accommodated not only the telescope but also its support equipment. During each service mission there was room for the telescope's platform and pallets to transport the servicing equipment and support apparatus, with plenty left over for the astronauts to work comfortably and safely. With the exception of a few small experiments, no other hardware was flown on the service missions because the astronauts would be fully occupied with the primary tasks.

The development of the shuttle procedures for rendezvous and proximity operations, the Canadian RMS, and the European Spacelab pallets were all significant events in the overall shuttle program and provided proven hardware, concepts and procedures for use during the Hubble service missions.

The development of the TDRSS network to meet a wide range of objectives for NASA was crucial for the Hubble Space Telescope. Hubble continues to communicate through the TDRSS ground stations, both to receive commands and to return engineering and scientific data. Of course this geostationary network also supported communications with each of the shuttle missions sent to the telescope.

Finally, the myriad of checks, tests, procedures, guidelines, and rules which the agency established during the early years of the space program and constantly updates, provided safety and integrity guidelines for all items of payload for each mission, coupled with an understanding of the environment in which the telescope and shuttle would fly. A strong managerial team ensured that each mission was flown as safely as possible, both for the telescope and the shuttle crew.

## An almost completed jigsaw?

By the late 1970s, the Space Telescope project had been defined and authority to build the hardware was in place. Nevertheless there would still be a struggle to secure the funding to complete the plan. As these budget wrangles continued, other pieces of the jigsaw fell into place. The key was the decision to go for full on-orbit servicing during its planned 15 year lifetime, instead of returning the telescope for ground maintenance every few years (which varied between every 2.5 or 5 years). To do this required the authorization of the shuttle and the sizing of its payload bay, as only then would it be possible to establish that there would be sufficient room to carry the telescope into space and to service it there. The reduction of mirror size not only helped in the tight funding parameters of the day, it also made it easier to fit the telescope in the payload bay, and that in turn required a rethink of component layout with the benefit of simplifying the orbital servicing.

In support of that servicing, the components of the shuttle had to be developed, built and tested, including the Remote Manipulator System that would play such a critical role on all telescope missions. Then the method of shuttle rendezvous and proximity operations had to be developed, software prepared, and techniques decided both for nominal and contingency approaches. Throughout the 1970s and into the 1980s, key supporters of the program argued the case for continued scientific, financial, and political support. Meanwhile astronauts and engineers commenced developing and rehearsing telescope-related EVA activities in huge water tanks. While the scientific instruments were developed, international cooperation was sought from Europe. This yielded not only a science instrument but also the solar arrays and the oft-overlooked pallet support systems to carry orbital replacement units to and from the telescope. Also, systems and procedures were put in place to prevent contamination of the hardware, and an understanding was achieved of the environment in which the telescope would conduct its normal operations and be serviced by shuttles. In parallel the safety and integrity of the shuttle, its systems, and crew were evaluated.

While all of this was going on, the detailed planning for the service missions was begun. The fact that astronauts were often able to make a shuttle flight look straightforward was a testament to the enormous amount of planning by an army of people on the ground prior to launch. And in terms of sheer complexity, the Hubble Space Telescope was near the top of the list.

## REFERENCES

1. Roman Oral History, 2000
2. NASA Oral History Project, Nancy Grace Roman, September 15, 2000
3. Hubble's Legacy, page xii
4. Skylab America's Space Station, D. Shayler, 2001, pp16–17
5. The Hubble Space Telescope, Chapter 12, Power to Explore: History of MSFC, Andrew Dunar and Stephen P. Waring, The NASA History Series, SP-4313, 1999, pp473–425
6. Power to Explorer p488
7. Power to Explore p479
8. Power to Explore: p484
9. Roman, Oral History, 2000
10. The Space Shuttle Decision, NASA's Search for a Reusable Space Vehicle, T.A. Heppenheimer, NASA SP-4221, 1999

# 130 A dream becomes reality

11. Heppenheimer p139
12. Letter Hansen to Myers, June 21, 1971, in 1999 Heppenheimer, p226
13. *Skylab America's Space Station*, David J. Shayler, Springer-Praxis 2001, p40
14. RMS Press Briefing transcript, Part 1, November 2, 1981, AIS archives, STS-2 Box File
15. SRMS History, Evolution and Lessons Learned, Glenn Jorgensen and Elizabeth Bains, NASA JSC, AIAA Paper; Canadarm STS-3 Press Kit, 1982, AIS Archive STS-3 Box File
16. *Rendezvous and Proximity Operations of the Space Shuttle*, John L. Goodman, United Space Alliance, Houston, Texas, AIAA, 2005; *Hubble Servicing Challenges Drive Innovation of Shuttle Rendezvous Techniques*, John L. Goodman, USA, and Stephen R. Walker, NASA JSC, AAS, AAS-09-013, 2009; also *To Orbit and Back Again: How the Space Shuttle Flew into Space* Davide Sivolella, Springer-Praxis, 2014, pp411–417
17. *Rendezvous and Primary Operations of the Space Shuttle,* John L. Goodman, USA, 2005 AIAA; forthcoming *Space Shuttle Space Station Assembly Missions*, David J. Shayler, Springer-Praxis, publication planned for 2016
18. AIS interview with Robert L. Crippen, January 23, 2013
19. Tracking and Data Relay Satellite (TDRS) NASA, http://www.nasa.gov/directorates/heo/scan/services/network/text_tdrs.html last accessed December 18, 2014
20. Europe's Space Program To Ariane and Beyond, Brian Harvey, Springer-Praxis, 2003, pp211–213
21. AIS interview and subsequent emails, Lothar Gerlach February 8, 2013
22. Pallet and Half Pallet Shuttle Dedicated, Re-Usable Carriers, 3rd Edition, 1981, British Aerospace Dynamic's Group
23. Presentation Copy (photocopy) of the Astronomy Experiment Module, presentation to ESRO, Paris, September 30, October 1, 1970, (Contract 1035/70/HP), Hawker Siddley Dynamics, Space Division, Stevenage, UK, archived in the library of the British Interplanetary Society, London.
24. Presentation made to the British Interplanetary Society, by David R. Woods, June 7, 2014
25. Natural Environment Design Requirements for the Space Telescope, by George S. West and Jerry J. Wright, Space Sciences Laboratory, Science and Engineering, NASA MSFC, Alabama, July 1976, NASA TM X-73316
26. The estimation of galactic cosmic ray penetration and dose rates, March 1972, NASA TN D-6600, M.O. Burrell and J.J. Wright

# 4

# LST becomes ST, becomes HST

> *The development phase is always the serious part of the program and is always expensive. In general the customer wants the technology of a Formula 1 car for the price of an economy car, while the industry wants [to] sell the technology of a small uncomfortable car for the price of a Ferrari.*
>
> Lothar Gerlach

Before discussing how the Hubble servicing techniques were developed, it is appropriate to review of the 1972 LST Phase A design and its servicing plan and then briefly describe the main Hubble Space Telescope components and systems as they were flown and which were relevant to the service missions.

## THE 1972 LARGE SPACE TELESCOPE PHASE A DESIGN STUDY

After decades of planning and proposals, by the early 1970s the development of the Large Space Telescope had progressed to the design stage. Although many hurdles remained to be addressed, steady progress was being made.

In 1972, a Phase A study was completed which both refined the current thinking of the planned telescope and directed future efforts towards the final program. This study defined the LST concept based on very broad mission guidelines provided by the Office of Space Science (OSS), the scientific requirements developed by that Office in conjunction with the scientific community, and an understanding of NASA's long range planning at that time.[1]

A low-cost design approach was followed during Phase A, which resulted in the use of standard spacecraft hardware, the provision of maintenance at the black-box level, growth potential in systems design, and offsetting the cost of servicing activities. At that time, the shuttle was being marketed as a way to reduce launch costs. One way was literally to sell payload space. Originally the idea to fill the payload bay volume that was not occupied by servicing hardware with commercial payloads that would be deployed before rendezvousing with the telescope, was deemed to be a means of offsetting the cost of the service mission which, in itself, would have no commercial gain, but was not pursued for the servicing missions.

Table 8  The ABC phases of a space project.

| Phase | Activities |
|---|---|
| Pre-Phase A | Covers the *Conceptual Studies* where the original ideas are first proposed, studied and evaluated. At this point a Science Working Group is established to develop the science goals and requirements and request proposals for experiments. |
| Phase A | This features the *Preliminary Analysis* period from which an early design and project plan is created. |
| Phase B | The *Definition* phase includes a number of reviews emerging from Phase A including a system requirement, a system design and a non-advocate reviews. This phase takes the preliminary plan evolved during Phase A and converts it into a baseline technical document from which the defined design and development programs can be established. |
| Phase C and Phase D | These are the *Design and Development* phases which include the Critical Design Review, the Test Readiness Review and the Flight Readiness Review, which focuses upon the assembly, testing and launch of the flight hardware. |
| Phase E | The final period is the *Operations Phase* of the program and includes the *Mission Operations and Data Analysis (MO&DA)* which comprises of the Primary Mission and [more frequently on unmanned satellite or space probe mission] the option of an Extended Mission. |

The journey of major space hardware, such as the Hubble Space Telescope, is normally a long and at times turbulent path, but all follow this Mission Concept flow path.

## LST mission description

Originally, the LST was to be a telescope with an aperture of 10 feet (3 meters) providing near-diffraction-limited performance. The plan was for a smaller precursor telescope to be operated for 5 years prior to launching the LST, which would operate for 10 years. But by 1972 the precursor had been deleted, and periodic maintenance and refurbishment would enable the single LST to operate for 15 years. The preliminary guidelines also included the requirement to make the LST compatible with both the shuttle and Titan III launch vehicle, but by 1972 only the shuttle was under consideration. The plan was to undertake on-orbit maintenance as required in order to replace failed or degraded components and to upgrade individual scientific instruments. An alternative under consideration at the time, was after "several years" to return the facility to Earth for a complete refurbishment and instrument replacement, but only if the new instruments could not be accommodated on-orbit by the existing support structures on the telescope.

The LST Preliminary Study report prepared by Marshall Space Flight Center and dated February 25, 1972 was the starting point for the Phase A studies. This design envisaged on-orbit maintenance, subsystem replacement, and instrument updates, and was adopted as the reference design configuration for the Phase B activities that would refine the design. There was also flexibility to use the Phase A design to explore alternative concepts, systems, and subsystems prior to progressing to the next stage.

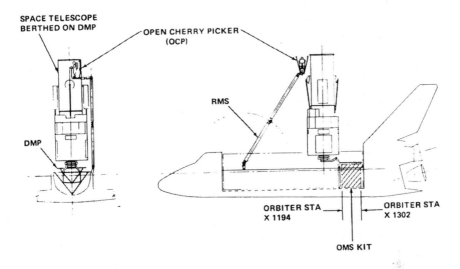

The Large Space Telescope on-orbit maintenance concept.

**Orbit selection**

An essential task in planning a satellite is to select the orbit that will optimize the mission performance within the operational constraints. The main constraints on the LST were that the shuttle or its alternative Titan IIIE/Orbit Adjust Stage (OAS) would be launched from Florida and that because the telescope would have no propulsion of its own, its operating orbit would have to be accessible to the shuttle for the initial deployment (if appropriate), servicing, and possible retrieval for return to Earth. And of course, that orbit had to offer a minimum lifetime of 5 years. The next step was to consider how that orbital environment would influence the design of the hardware and systems.

The mission performance parameters, as a function of orbital operations were listed as:

1. Payload capability
2. Orbital decay rate
3. Ground station contact time
4. Target visibility
5. Target viewing time.

The orbital environments which affected system design were:

1. Trapped particle radiation
2. Magnetic fields
3. External disturbances
4. Micrometeoroid flux contamination
5. Stray light.

A parametric constant-orbit selection analysis was undertaken for the LST design study to determine the minimum altitude requirements for the nominal mission conditions; calculate lifetime and decay histories for the reference orbit and assess the sensitivity of predictions to the configuration of the spacecraft; match the spacecraft to the performance of the shuttle; determine the tracking network coverage; and draw up preliminary mission timelines. After considering these parameters, it was decided the LST would use a circular orbit at an initial altitude of 611 km (380 statute miles) which was inclined at 28.5 degrees to the equator. The circular orbit was selected because the analysis did not identify any significant benefits from any elliptical orbit which was within the capabilities of the shuttle (or indeed a Titan launch vehicle).

The inclination at 28.5 degrees with a due-east launch from KSC was chosen because:

1. A lower orbital inclination required yaw steering and a significant loss in payload capability
2. A higher inclination quickly reduced the prime Earth shadow viewing time
3. The majority of shuttle missions during which another satellite could be deployed prior to performing a rendezvous for LST maintenance were at that inclination
4. The sensitivity of other performance parameters to inclinations between 28.5 and 40 degrees was negligible.

Other parameters investigated in the selection of orbital altitude were:

1. Titan payload capability
2. Orbital decay rates
3. Trapped particle radiation.

Launch vehicle payload capability and the known radiation environments both argued in favor of lower orbits but a higher altitude was preferable to minimize the decay rates. Based upon a projected launch in 1978 a compromise occurred at about 380 miles (611 km), so this was the altitude originally chosen for the telescope. For a launch near the peak of the 'solar cycle' in 1980 it was predicted that the altitude would be able to be reduced to approximately 354 miles (556 km). At the time of the report, LST performance degradation due to residual atmospheric conditions was a factor in final orbital selection. This launch date was an option in case either a 30 percent reduction in radiation environment exposure was required or there was an increase of 400 pounds (181.5 kg) in the Titan payload capability.

**Launch vehicle analysis**

The 1972 report analyzed the capabilities of the shuttle as the primary launch vehicle, not only to deploy the LST into its planned orbit but also for emergency visits and end-of-life retrieval. The study evaluated the performance of the shuttle based upon an orbiter/parallel burn SRBs with OMS reserves of 50 feet per second (15 meters per second). Its capacity to deliver approximately 23,587 kg (52,000 pounds) to the 380 mile (611 km) design reference orbit included adding an additional OMS tank set to the payload bay on a framework which would occupy 5 feet (1.52 meters) of the 60 foot (18.29 meter) long bay.*

---

*The concept was similar to the subsequent upgrades introduced to the shuttle during planned maintenance periods and the additional cryogenic storage tank sets carried in the payload bay during Extended Duration Orbiter missions from 1992.

At this early stage of the LST, a portion of the structure was expected to be capable of being pressurized. All missions would therefore have to carry a docking module of 2000 to 3000 pounds (900 to 1400 kg) that would occupy 7 feet (2.13 meters) of the bay. Although carrying extra tanks and a docking module would challenge both the launch performance of the shuttle and the available space in its payload bay, the report stated, "The shuttle offers ample performance capacity and payload bay volume." It reasoned that adding the second OMS bay kit would enable the shuttle to achieve an orbit of approximately 520 miles (880 km), although at the cost of reducing the payload to approximately 40,000 pounds (18,000 kg). A third such kit would enable the shuttle to achieve an orbit of 683.5 miles (1100 km) with 26,500 pounds (12,000 kg). Of course, having up to three OMS kits and their support structures inside the payload bay would significantly reduce the options to offset the launch costs of a telescope service mission by deploying additional payloads.

When the report was compiled, the shuttle had yet to be formally authorized and wasn't guaranteed to be available to launch the proposed telescope. To cover all eventualities, the Phase A study also analyzed expendable vehicles as potential alternatives. These included: Titan IIIC; IIIC-IA; IIIE/Centaur; IID/Agena; IIIE/OAS; IIID/Burner II, and IIIE/Integral. The Titan IIID was labeled Titan IIIE when flown out of the Eastern Test Range with the Orbit Adjust Stage (OAS), Integral or Centaur upper stages. The Titan IIIC-IA and Titan IIIE-OAS were considered to be the most feasible contenders, as the others were either too expensive or too complex in technical compatibility. At the time of the study, the planned HEAO launch vehicle was the Titan IIE-OAS and this was selected as the backup launch vehicle for the LST.

**LST configuration and system design**

In the Phase A design, the LST was 41.66 feet (12.7 meters) in length and comprised three main assemblies.

The *Optical Telescope Assembly* (OTA) was 25.25 feet (7.7 meters) long and 12.33 feet (3.68 meters) in diameter. It consisted of the primary and secondary mirrors, the metering truss, the primary or main ring, meteoroid shields for the telescope, a light shield, the fine guidance sensors and associated equipment, telescope-peculiar sensors, and the primary structure for the scientific instruments. The *Scientific Instrument Package* (SIP) was 10.16 feet (3.1 meters) long and contained all the instrument payload, the secondary structure for the instruments, and additional instrument-support equipment. The *Support System Module* (SSM) could be pressurized to allow internal crew access. It was 16.41 feet (5 meters) long, had a forward diameter of 12.33 feet (3.68 meters) which was compatible with the Optical Telescope Assembly and, at the opposite end, it had a maximum diameter of 14 feet (4.27 meters) where there was a 3.3 foot (1.02 meter) diameter docking facility and transfer tunnel assembly. It held all the load-bearing structures aft of the primary ring, the attitude control equipment, electrical power and distribution equipment, communications and data handling equipment, thermal control equipment, and the contamination control equipment.

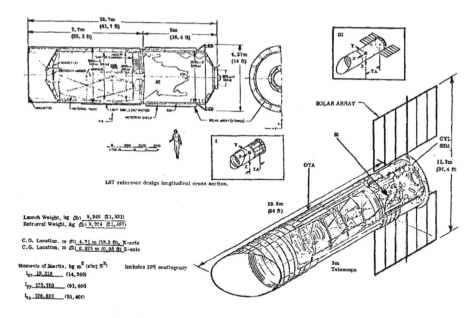

Components of the Large Space Telescope and dimensions (original of poor quality).

## Scientific Instrument Package

An open truss structure allowed access to all areas, and the instruments were systematically arranged to facilitate the removal of items without disturbing other instruments. In order to minimize the need for astronaut dexterity and special maintenance skills when replacing an instrument, the design incorporated self-aligning devices and guide rails. All of the imaging sensors would be able to be removed and either replaced or repositioned without having to remove the associated optical elements or the subassemblies of other instruments. Modular design would allow periodic on-orbit maintenance and repair, as well as the replacement of instruments with upgraded or improved ones.

In 1972 the selection of instruments was still under development, but taking into account the scientific objectives and technological capabilities at that time, a tentative payload was identified as:

1. High spatial resolution camera (f/96)
2. Two high resolution spectrographs
3. Three faint object spectrographs
4. Fourier interferometer
5. Wide field camera (f/12).

## Support System Module

The Support System Module interfaced with the OTA to provide both that unit and the SIP with electrical power, communications and data handling, environmental control, coarse attitude sensing and control, launch vehicle structural and electrical interfaces, and

# The 1972 Large Space Telescope Phase A design study

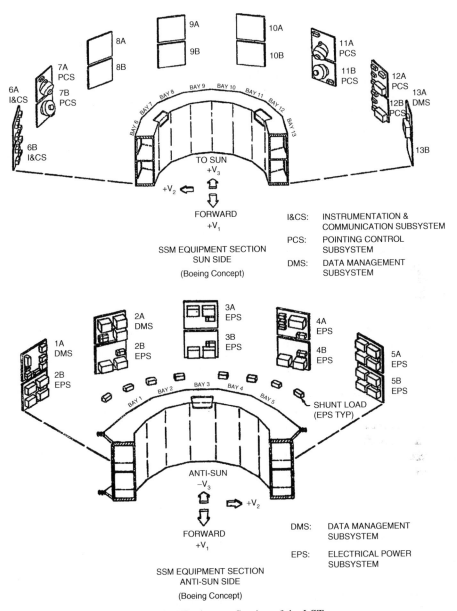

The SSM Equipment Section of the LST.

a shuttle docking structure for on-orbit servicing or retrieval. It was essentially a cylindrical structure with a shallow cone at its aft end that supported a standard androgynous docking assembly. The components of its internal systems were arranged to provide ease of astronaut servicing whilst also maintaining adequate thermal control of both its own systems and the scientific instruments.

## Contamination control

This hardware consisted primarily of ducting and filters to feed clean air from the shuttle to the forward end of the SSM, and through highly efficient particulate air (HEPA) filters into the SIP. On the shuttle side, the equipment included fans, trace contamination absorption beds, oxidizers, and filters. A trace contamination loop was located on the shuttle, but for added crew protection it would be physically isolated from the remainder of the orbiter's habitable environment by a fabric curtain. This design of the telescope featured a direct androgynous docking unit and crew access tunnel to enter the SSM, therefore the contamination control equipment aboard the shuttle would support the contamination control environment in the SSM while the crew was inside.

## System reliability

During the Phase A studies, the systems intended for the LST were evaluated for individual 'failure' rates which would either result in the loss of the telescope or would require a shuttle flight to undertake maintenance. Wherever possible, a reasonable degree of redundancy was desired throughout, including "use of existing equipment, or equipment common with [other programs]". This further reflected the need to cut costs in the design. The report also stated that it was preferable to incorporate greater redundancy into an established system than to start again with a completely new design that would have a lower likelihood of failure, and that non-critical elements had been excluded from the analysis; in fact, some items had to be excluded due to the lack of credible data. The most significant exclusion from the reliability analysis were the nickel-cadmium batteries and their charging units, which had been already identified as a possible redundancy problem in the design. This is an interesting observation, because the service life of the nickel-hydrogen batteries that were flown aboard Hubble was 19 years. Another omission from this reliability analysis was the science instruments, since any failure in these units would be a justification for maintenance actions to repair or replace them.

## Maintenance of the LST

It was recognized that a reasonable level of maintenance on the telescope would ensure and sustain the high level of operational performance that would be necessary to:

1. Provide a means for instrument update whenever warranted by either advances in technology or the evolving nature of scientific interest
2. Assure system performance for long term operations
3. Minimize the total cost of the LST program.

The 1972 study stated that the first two objectives could be met either by using a number of expendable LSTs or a smaller number of maintainable ones. By the expendable approach, a new telescope would be launched whenever the performance of an existing one fell below optimum levels or it was considered necessary to supersede an instrument package. With the maintainable approach, degraded or non-operational components would either be replaced in space to swiftly restore the LST to full operational status or the

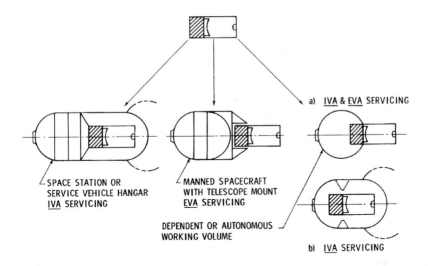

Servicing options for the Large Space Telescope. The center option of a manned spacecraft with EVA servicing capability would evolve into the Hubble servicing mission concept.

telescope would be returned to Earth for refurbishment and instrument replacement, and then be re-launched.

Over the planned 15 year program, it was estimated that either two maintainable or eight expendable LSTs would be required. The maintainable telescopes would include a precursor or test unit that would operate for the first 5 years and then the advanced model for the final decade. It was estimated that each of the expendable units would have a life of 2 years based upon component lifetime. During the 15 year program, a new LST would be launched every 2 years. The total cost of the expendable telescopes was estimated at 1.8 times the total cost of the two maintainable units, presumably owing to the high launch costs. Hence the report recommended that the maintainable approach be adopted as the design reference mode, not just because it was likely to be cheaper but also because it would impose less disruption on the orbital science program.

To determine the feasibility of the maintainable capability, four modes of maintenance were analyzed:

1 On-orbit manned maintenance, pressurized
2 On-orbit manned maintenance, unpressurized
3 On-orbit manipulator maintenance
4 Earth return maintenance.

These were compared in terms of a number of factors, such as the level of maintenance that could be performed, the number and complexity of the likely tasks, the time required, and the cost, growth potential and flexibility of design. In 1972 it was difficult to identify which would be the most effective alternatives because in theory they were all technically possible and each had attractive and detrimental features. It was surmised that because the on-orbit maintenance mode did not utilize the full mass and volume capacity of the shuttle

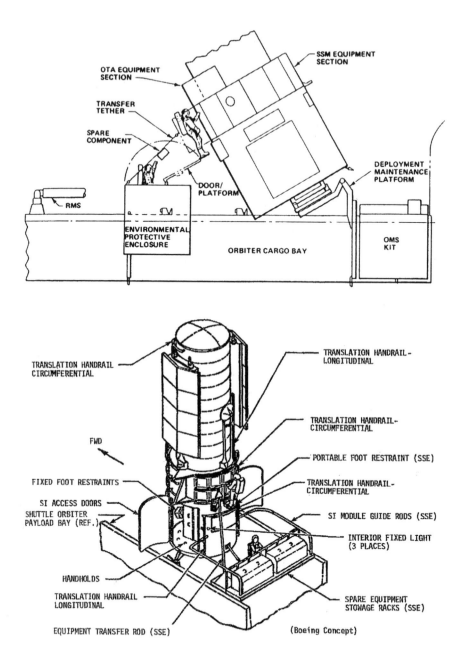

Early concepts for astronaut servicing the Large Space Telescope.

during launch, the costs of this mode could be offset by sharing space in the bay with other equipment. The likely commercial satellites and scientific payloads on small carriers were studied to identify potential payloads that could share a maintenance mission.

Preliminary analysis revealed that of the four options the pressurized maintenance mode offered the greatest potential for mission success, mission flexibility, and maintainability. It would provide the astronauts direct access to the apparatus and eliminate the complexity of manipulators, and also the environmental difficulties of repeated Earth return and re-launch. Although overall cost comparisons were not available at the time of the report, preliminary figures indicated that the pressurized maintenance mode gave a slight cost advantage. As a result, it was chosen for the Phase A feasibility analysis. Subsequent comprehensive analysis found the pressurized maintenance mode to be too complex to pursue further.

**Summary**

The results of the Phase A study recommended a telescope with a primary mirror 10 feet (3 meters) in diameter with f/12 Cassegrain optics. It was to be launched by the shuttle (in the primary mode) or Titan IIIE/OAS (in the alternative mode) and operate for at least 5 years. Whilst there remained considerable work to do on the systems, it appeared that much of the required technology could be transferred from other programs, with the use of proven equipment both reducing costs and improving reliability.

The 1972 report also observed, "The use of man in such unmanned satellite programs as the LST, no matter what its configuration, can be more effective and can be accomplished with much less impact on the design of the hardware than was experienced on the Apollo or Skylab programs, due to the difference in the nature of the missions and the experience from those programs." It also suggested that the manned maintenance option was "a cost-effective approach" because the frequency of shuttle visits would permit the lifetime requirement for apparatus to be reduced to just 2.5 years within the context of a much longer mission.

Sharing payload space on maintenance flights offered a means of cutting costs, and a high degree of borrowing from other programs could reduce overheads even further. An analysis had revealed that approximately 42 percent of the components in the SSM were identical to those of the HEAO satellites, with a further 13 percent from "other programs". It was stated that 13 percent of existing hardware could be adapted if this was found to be a cost-effective approach, and that approximately 23 percent of the hardware intended for LST required new designs but not new technology.

**LST PHASE A DESIGN STUDY UPDATE**

Eight months after the Phase A report was issued, an update was released based on changes in the guidelines and new data. Ground-return maintenance remained the preferred mode but there was an increase of limited on-orbit EVA as another option. This additional data came too late to be included in the Phase A document but was considered sufficiently important to be provided as a supplement for consideration by the Phase B studies.[2] There were several options for maintenance using the Earth return mode as the primary reference. The emphasis was placed on concepts which maintained a small degree of on-orbit maintenance capability, and less on those that offered a more extensive on-orbit maintenance program; although this was not ruled out it was more limited in scope.

Instrument priorities had been defined in separate OTA/SI work statement guidelines and so this study assumed the same complement of instruments as defined in Phase A. However, minor changes could be introduced if they would be beneficial to maintenance activities. For example, moving radial instruments closer to the external wall of the SSM in order to permit radial extraction through a hatch without requiring a long reach inside the telescope might be advantageous.

**Level of maintenance**

The levels of maintenance were identified as:

- Minor maintenance (EVA supported) would replace failed or life-limited equipment and instrument sensors.
- Emergency maintenance (EVA supported) tasks would be less extensive than minor maintenance and allow more relaxed constraints (e.g. lower contamination risks) than would occur under routine maintenance.
- Moderate (Earth return) would replace failed, life-limited or obsolete equipment and instrument sensors and/or entire instruments (scheduled for every 2.5 years).
- Major overhaul (Earth return) would include the replacement of failed, life-limited and obsolete equipment and instrument sensors, entire instruments, resurfacing the thermal control surfaces, repairing micrometeoroid punctures, recoating the optics, replacing equipment that could not be easily replaced on-orbit (e.g. solar arrays, and secondary mirror mechanisms), cleaning to remove contamination, and an extensive testing/verification program (scheduled every 5 years).

The updated document for Phase A continued to see ground turnaround/maintenance as the primary maintenance mode, but included a fairly high degree of on-orbit EVA activity that would have minimum impact on the LST design.

**Key changes from the Phase A study**

In the 8 months after the publication of the original Phase A report, there were significant changes in the development of the LST (and indeed in other programs and supporting areas) that required the implementation of several key changes to the document.

- Although compatibility with HEAO systems was desirable, a re-evaluation of the comparison was required owing to recent changes in that program.
- Recent funding reductions meant deleting a backup or precursor telescope and that the single flight model be capable of an orbital lifetime of 15 years.
- In order to achieve this goal, the facility would have to be refurbished, modified and maintained.
- Earth return maintenance was still the primary mode for the first maintenance flight, scheduled 2.5 years into the mission, but "further studies would determine the impact of performing very limited on-orbit maintenance in addition to the Earth return mode. The study would also determine the impact of deployment after a more extensive on-orbit maintenance program after the initial Earth return."
- The reduction in the number of flight articles made identifying low-cost approaches even more urgent.

- The option of using a Titan launch vehicle as a backup to the shuttle for deploying the telescope was terminated. This tied the LST to the Space Transportation System, and whatever delays, limits, and setback that might suffer.

## Ground maintenance

Ground maintenance operations of the Earth return option for the LST would have begun with the shuttle rendezvousing with the telescope, which would have been commanded to adopt a standby mode with its aperture door closed and its solar arrays retracted. Once the orbiter was in position, it would either dock with the SSM or use its manipulator to grapple the telescope; either way, the telescope would be stowed in the payload bay for a return to Earth.

The report estimated that from the time that any shared component of the shuttle mission was completed and LST observations terminated, the rendezvous, safing of the telescope, and its stowage in the payload bay would take approximately 3 to 4 hours. The entry and landing would follow some 6 hours later.

Back on Earth, the LST would be de-mated, inspected and prepared for transportation to a "special integration-to-maintenance facility"—possibly off-site to KSC—for refurbishment. The activities would obviously depend upon the servicing plans and severity of any repairs or maintenance, but it was envisaged that they would normally take about a fortnight. Then the telescope would be "flown in the Super Guppy back to the Cape" for processing and return to space, hopefully just a few weeks after being removed from orbit.

## Minor on-orbit EVA maintenance

Under this option, approximately 38 hours of on-orbit maintenance time would be available, probably during a shared mission in which the first 34 hours would be reserved for the "other payload". However, the flight, possibly with a minimum crew of 4 or 5 astronauts, would not exceed a 7 day nominal mission.

## Minor on-orbit RMS maintenance

A total of 17 specially adapted equipment trays or modules could be replaced employing the manipulator. Various subsystem elements would be able to be replaced, but not the optical or science instruments. The complete exchange would require about 12 hours (including a crew rest period). This would be followed by a science instrument changeout EVA during another 24 hour period (including rest). The mission would therefore perform 36 hours of telescope servicing.

## Key changes after Phase A

The key changes after Phase A still centered upon Earth return maintenance as the primary option with only limited EVA on-orbit maintenance activities. However, it was beginning to look like there might soon be other possible options because there was a growing interest in developing the potential for robotics for on-orbit maintenance operations on a limited scale and in a way that would not impact the design of the telescope or require it to be returned to Earth. The updated study also included evaluating other options and concepts, such as more extensive on-orbit maintenance at some point in the future.

The option of a tray-mounted packaging scheme could be developed for either manual or robotic operations. Development of this system, it was suggested, would be a valuable tool for a suited astronaut to quickly changeout a large quantity of equipment. It would also keep costs down, because the trays would be standardized for use either by EVA astronauts or by the RMS and would not require the addition of any specialized devices.

**Limits of the arm**

At this time (1973) the idea of an RMS was still under development and its total load forces were not fully determined. With forces of about 10 pounds (4.5 kg) the packaging of payload items on the trays had to be kept relatively small and compact in order to ensure that the arm could perform an assigned function. However, the report did state that, should an open truss configuration be adopted for the LST it would permit the manipulator maximum access, and by carefully positioning of subsystems they could be mounted on tray-type modules for ease of servicing.

**Variables**

A list of variables to be considered in the development of potential maintenance operations was included in the supplementary report. This included:

- *Structure*: There were over 17 structure configurations categorized into three basic types: shell, truss or truss combined with a non-load bearing shell, and the Support System Module.
- *Maintenance mode*: These were either Earth return or on-orbit. Further research was required into pressurized module options. The on-orbit options also included EVA, shuttle manipulator, internal robot, external robot, and hybrid configurations.
- *Degree of on-orbit maintenance*: Either none, minor, or extensive.
- *Modularity/package level*: This could be achieved using components only, a tray, a saddle-bag design, or a box.
- *Location of components*: External, or internal siting.
- *Docking and handling techniques*: This could be done by latching the telescope into the bay using launch locks, or by the attachment of a docking module. An option not considered in this study, because it seemed too risky, was having the shuttle's RMS hold the telescope throughout the maintenance activity.
- *Access methods*: This could be via an axial hatch, hinged rear plates, side hatches, or side accesses without hatches.
- *Solar Array*: Rigid or roll-up designs.

**Unitized configuration**

One option considered was to use the SSM in either a truss or shell configuration, but with interfaces between the scientific instruments and the SSM that could be isolated to facilitate easier removal or replacement. In this concept, all SSM equipment was packaged aft of the scientific instruments and none was installed around them. Using a longer adapter, the SSM could have been adapted to be physically separated at a point immediately to the rear of the scientific instruments and directly in front of the SSM equipment. At this time there was no intention to try to remove the SSM from the rest of the OTA/SI on-orbit, although it was indicated that a subsequent version might permit this in order to cleanly separate the SSM from the OTA/SI in both physical and functional terms. The possible use of this capability on-orbit would require further analysis and development, but separating the SSM from the rest of the assembly would be very useful during refurbishment, service and testing on the ground.

**Accessibility**

Good accessibility remained one of the difficulties for accurately evaluating the capability to maintain the telescope either on the ground or on-orbit by RMS and/or EVA. In addition, the shuttle was also going through its development, and changes were constantly being made as its configuration matured. The report mentioned that "this configuration [of the LST] could be modified to match the latest shuttle attach points" once the design of the shuttle had been finalized.

On-orbit maintenance became even further limited when it was realized that the Control Moment Gyros (CMG) and battery module would be inaccessible to the RMS in the SSM cylindrical shell mode, and therefore least desirable for on-orbit maintenance by the robotic arm. To support EVA astronauts, removable micrometeoroid protection cover plates would allow access, and because such units were unpressurized the access doors wouldn't require complicated pressure seals.

Commonality was always a strong design feature of the LST program. The primary frame assembly for the telescope was chosen as one potential common structure that could be used as a "chassis-bus" for a planned domestic communication satellite series. As the shuttle was now the only launch vehicle under consideration for deploying the telescope and with the frame assembly ideal for on-orbit servicing or Earth return maintenance, this design commonality with other satellites suggests it was likely that these were originally intended for the shuttle with a maintenance capability.

In summary, the report said, "The truss was originally proposed because of advantages to access to equipment which it was felt to have over the shell for a man in a pressure suit or a manipulator. Upon more detailed analysis, however, the access advanced over a suited man appeared more doubtful." The advantageous access that an open truss was initially expected to offer was (at the time of the report) thought not to lend itself to suited operations because the shell design offered ample scope for expansion. As a result, the truss was dropped from EVA maintenance considerations but retained as a consideration for RMS maintenance.

Interestingly, the report identified one area in which a suited astronaut would be the most favored option: that of the scientific instruments. Unfortunately, their location and nature did not tend to make manipulator maintenance easy. As a result, early in the design phase it was determined that exchanging the instruments would have to be done by spacewalkers in all of the proposed configurations.

It was expected that any item of equipment could fail during the planned 15 year lifetime, and the RADC Reliability Notebook suggested that failures might occur at an average of 15 per million hours maximum, with most of the instrument sensors (cameras) early candidates for replacement under the on-orbit minor maintenance option. As such activities would have to be conducted by EVA, and there would likely not be much other equipment needing to be replaced, this could be done by EVAs rather than by RMS. Consequently the truss structure was dropped from further evaluations in the LST program.

## MAINTAINING MAINTAINABILITY

In addition to developing the LST concept and its capacity for maintenance, the study also considered how the servicing would be sustained through the life of the telescope. By the time of the first planned service mission, nominally set 2.5 years after launch, some of the equipment would be nearing the end of its design life. The scientific instruments would be due for replacement by upgraded ones. In addition, there would have been random failures and physical deterioration.

Excluding the OTA and SI, the accompanying table of spare parts indicates the type of equipment that the study thought would require maintenance, listed in deceasing order of expected maintenance and also categorized by the probability of lowest, middle range, or upper band of renewal.

Table 9 LST minor on-orbit maintenance spares and logistics in decreasing order of expected replacement need (determined by reliability, operating hours, and ranking of importance).

| Item | Qty | Total weight kg | lb | Total Volume $m^3$ | $ft^3$ | Total EVA Maint. time Req'd (hr) |
|---|---|---|---|---|---|---|
| Batteries | 7 | 235 | (518) | 0.397 | (14.03) | 3.5 |
| Tape recorders | 3 | 19.6 | (43.2) | 0.036 | (1.26) | 1.0 |
| Fine guidance assembly | 1 | 139 | (308) | 0.145 | (5.12) | 1.0 |
| f/96 camera | 3 | 241.7 | (532.8) | 0.012 | (3.6) | 1.0 |
| Faint object spectr (FOS1) | 1 | 67.0 | (147.6) | 0.15 | (5.25) | 1.0 |
| High res Echelle spectr (FOS1) | 1 | 62.0 | (136.8) | 0.242 | (8.55) | 1.0 |
| High res Echelle spectr (FOS2) | 1 | 62.0 | (136.8) | 0.22 | (7.8) | 1.0 |
| Faint object spectr (FOS2) | 1 | 54.4 | (120) | 0.276 | (9.75) | 1.0 |
| Slit jaw camera | 1 | 50.6 | (111.6) | 0.41 | (14.55) | 1.0 |
| Faint object spectr (FOS3) | 1 | 53.3 | (117.6) | 0.111 | (3.9) | 1.0 |
| f/12 camera | 1 | 81.6 | (180) | 0.048 | (1.71) | 1.0 |
| Faint object spectr (FOS4) | 1 | 21.7 | (48) | 0.15 | (5.25) | 1.0 |
| RGA | 1 | 12.5 | (27.6) | 0.017 | (0.6) | 0.7 |
| DPA | 1 | 7.6 | (16.8) | 0.009 | (0.3) | 0.7 |
| CMG | 1 | 96.8 | (214) | 0.69 | (24.5) | 1.0 |
| Regulator | 1 | 4.3 | (9.6) | 0.012 | (0.45) | 1.0 |
| Remote decoder | 1 | 0.6 | (1.2) | 0.005 | (0.15) | 0.8 |
| DAU | 1 | 0.6 | (1.2) | 0.0006 | (0.0075) | 1.0 |
| Spares subtotal | | 1210 | (2670) | 3.02 | (106.8) | 19.7 |
| EVA tool kit | | 16.3 | (36) | 0.13 | (4.5) | |
| Cargo bay subtotal | | 1226.3 | (2706) | 3.15 | (111.3) | |
| Two suits | | 90.7 | (200) | 0.33 | (11.8) | |
| EVA consumables | | 170 | (375) | 2.0 | (72.0) | |
| Cabin subtotal | | 260.7 | (575) | 2.33 | (83.8) | |
| Totals | | 1487 | (3281) | 5.48 | (195.1) | |

A major overhaul of the telescope would require a return to Earth. This type of servicing would include the replacement of thermal control surfaces, the repair of any damage caused by micrometeoroids or space debris, the recoating of optical elements, replacement of solar cells, and a tear down and cleaning of the entire LST and subsequent a re-verification of all its onboard systems in a program of tests and simulations. Ground maintenance would also allow any concerns or issues involving the main structure to be addressed. The reliability of major subsystems would be examined in depth, and the control of any contamination issues implemented.

**Earth return issues**

Bringing the telescope back to Earth was considered to be a major overhaul. The functional lifetime of some of the hardware would be limited by the time between ground servicings, and items such as the solar panels and thermal coating would normally be replaced on the ground.

One major issue for the Earth return mode was the degree of contamination the telescope might suffer during its flight home. Returning the telescope to the payload bay and closing the doors for atmospheric entry would not guarantee a sealed pristine condition inside. The influx of contaminating hot gases to the bay during entry, however minute, would add to the work on the telescope. There was also concern about the stresses that would be imposed on the telescope at landing and re-launch by returning it home every 2.5 years. It was also quite likely that once the telescope was back on the ground, there would be a tendency to attempt more work than was originally scheduled by adding get-ahead tasks, thereby increasing the likelihood of accidental damage that would prolong the time on Earth, delaying re-launch, and undermining the carefully planned science program.

**On-orbit issues**

It was recognized that the list of spares suggested for the LST was more of a "shopping list" than a specific maintenance payload, as this would change as the needs of the telescope were addressed. Except for the fine guidance assemblies, scientific instruments were not included in the list—although several of their sensors were classified as high-failure items. Over time, upgrades and improvements to instrumentation would provide state-of-the-art replacements that were expected to be lighter, more compact, smarter and to have a longer operational life expectancy.

More extensive EVA operations would become necessary if the number of trays or pallets were increased. If this occurred, then the category of such activities would be increased to the moderate level and sealed containers would be located in the payload bay with the necessary spares and tools for the expanded EVA operations.

For the limited on-orbit maintenance, the LST was to have been locked into the payload bay in a horizontal state with the SSM facing forward. Should maintenance be necessary on the other side of the telescope, the report said, "the manipulator will be used to lift it out of the bay, roll it 180 degrees and place it back into the bay". (This would overcomplicate a case of "limited" maintenance.) The report also stated that if the manipulator maintenance mode was selected, then the tray packaging concept with astronauts changing out the more bulky items of equipment would be the preferred option.

## Design issues

On Earth or on-orbit, ease of access was essential to successful and timely completion of a planned maintenance program. On the ground, a stripped down LST would benefit from a controlled environment, a communications network, access platforms, lights, and various local resources. On-orbit this work would be supported only by the shuttle, which offered more restrictive access. The design requirement for on-orbit servicing were:

- Suited astronauts must have physical access to a failed item.
- Any electrical or mechanical connectors must be designed to facilitate operation either by pressure-suited hands or by tools at the disposal of a crewmember.
- Handholds and footholds must be able to absorb the torque and loads generated by the act of removing and replacing items.
- Adequate lighting must be provided.
- Crew time and energy usage is dependent upon proper design of the fasteners and connections assigned to the task.

For the study, it was assumed that life-limited items such as tape recorders and batteries would be designed specifically to optimize EVA access and that their connectors and tools would be fabricated to permit access with pressure-suited hands, and that wherever feasible the operation would be able to be achieved using only one hand.

For many items that were *not* expected to fail but would require replacement if they did, the report encouraged using a simpler design, even if that would involve supplying special on-orbit tools. All operations were to be evaluated for one or both EVA crewmembers, in order to determine access and torque limits. Adopting a standard design would reduce crew time and physical energy. For major EVA operations, a stronger design philosophy would help to reduce the risk of failure, and it would be advantageous to design to enable efficient removal and replacement of bulky items.

The report scheduled the first on-orbit maintenance for 2.5 years into the orbital mission, with the first Earth return no earlier than the 5 year milestone. This decision would have an impact on the design of the solar arrays, thermal coatings, protection from micrometeoroids, and optical coatings; all of which would require at least a 5 year life expectancy. Under this plan, three ground maintenance sessions were scheduled over the 15 year operational life of the telescope.

One of the problems with the precursor telescope, was it would have offered no orbital maintenance opportunities and this reduction in flexibility in its design would have made it more expensive; that was why it was cancelled. The report acknowledged that while Earth return maintenance was the preferred option at the time, if the final design of the telescope offered at least a partial or more extensive EVA maintenance capability, this would boost flexibility with minimum impact on the overall design. This was an interesting observation that questioned the supposition that returning the telescope to Earth was the most effective option and in turn lent support to those who argued for increasing the EVA capability.

## Structures

In designing the LST for either ground or orbital maintenance, there had to be consideration given to the structural integrity of the SSM configuration. In addition to the opportunities for providing the most optimum access to hardware during maintenance periods, designers also had to consider structural loads during assembly and testing, launch processing, and the ride into space. The Earth return phases would impose additional stresses on the structure during re-entry and landing. Furthermore, in the case of a landing, consideration had to be given to both nominal and potentially much rougher emergency landings.

Apart from the "crash" landing scenario, the dynamic forces on the telescope inside the shuttle were considered to be "nominal" and within the guidelines of +3.25 to −1.30 g for launch and −0.75 to −3.50 g for landing; with a factor of safety of 1.4 g applied. However, loads of 9.0 g could be expected in a crash landing. In this scenario, the telescope was not expected to survive the initial impact and potentially damage the orbiter. It was imperative that large pieces of payload debris must not penetrate the crew compartment bulkhead. By allowing a 10 percent contingency on all masses for the calculation, the maximum mass of the LST was 17,740 pounds (8047 kg). Data from the design loads were considered in the design of the telescope and materials used to lower the deflection of the main frame (made from aluminum alloy), warping, and bending of the structure during a hypothetical crash landing.

**Thermal control**

The design of the telescope had to be one of fine balance. The long term plan was to allow a program of maintenance and servicing to prolong its operational life, yet the designers had to minimize maintenance activities as much as possible in order to minimize both overall costs and the time that was not devoted to the science program. The longer that the telescope could survive without intervention, then the greater would be the scientific return from the capital investment.

One of the challenges in leaving the telescope on-orbit for long periods of time was in the management of the thermal barrier. Each 90 minute orbit imposed regular cycles of extreme heat from the Sun and intense cold while in the Earth's shadow. Studies were carried out of various coatings, radiator plates, louvers, insulation, and polished aluminum foil. The use of thermostatically controlled heaters, component grouping, isolating the battery compartment, and the way the vehicle was oriented in space all contributed to controlling the temperature inside and outside of the structure. Once the telescope was in space, the hardware, scientific instruments, and support systems would all be affected by the balance between the heating and cooling rates of the various components, and a design that addressed these issues would directly impact on the successful operation of the telescope and any future maintenance and servicing.

**Electrical power**

The development of the electrical systems of the LST benefited from experience with the observatories such as OAO and HEAO, from the orbital workshop and telescope mount of Skylab, and probably from classified Air Force programs. The decision to use roll-up solar arrays instead of rigid fold-out designs was interesting. The report stated the roll-up design would offer a lower mass, a more compact storage design, and be better suited to on-orbit maintenance because of its "ease of retraction". (Interestingly, on the first Hubble service mission 20 years later, the retraction of the solar arrays would not be so easy to achieve as expected for the LST.)

### Control and data handling

The update to the report stated that the new guidelines for revised maintenance modes had no functional impact on the Control & Data Handling (C&DH) system.

### Attitude control

The report said "on-orbit unpressurized maintenance of the Reaction Control System would be performed by EVA". Maintenance on-orbit would have featured quick-disconnects and flex hoses for ease of replacement, but during ground maintenance hard fittings and brazed connections would be used. The report also said that the "LST maintenance mode selection had very little impact on the ACS configuration" and so no mention was made of re-fuelling the Attitude Control System on-orbit, either automatically or as an EVA maintenance task.

### Subsystem reliability

The percentage reliabilities for the subsystems of the LST were given as:

- *Attitude control*: 0.98783 for the first year and 0.88555 at 2.5 years
- *Electrical power*: 0.94769 (1 year) and 0.85113 (2.5 years)
- *C&DH*: 0.99807 (1 year) and 0.98674 (2.5 years)
- *SSM*: 0.93435 (1 year) and 0.74328 (2.5 years).

These reliability expectations and their predicted rate of decrease were well within the guidelines and plans to service the telescope after it had been in space for 2.5 years and to return it to Earth 5 years into its mission.

### Contamination control

Experiences from the Skylab program were addressed in the revised planning for the LST, prior to commencing the Phase B studies.

In launching the unmanned Skylab orbital workshop in May 1973, one of the stowed main solar arrays and part of the micrometeoroid and thermal protection covering were ripped off.[3] The way in which these problems were addressed and resolved directly influenced the design of the LST by the selection of materials to fabricate the telescope, the control of internal and exterior environments, the scheduling of events, the cleaning of components, the verification of software/hardware, configuration management, and indoctrination of personnel.

Contamination could seriously compromise an experiment or item of hardware, and once again lessons learned from Skylab were applied to the LST. Deposits of particles deep within the hardware were of particular concern, because these would be harder to access and could seriously impair the optical quality of the telescope. This was a major issue not only during the construction of the telescope but also in its transport, testing, processing, launch, orbital operations, EVA, and return to Earth for ground maintenance and re-launch. The fact that the LST would have to go through the process not once but several times during its operational life added to the concern that at some point serious contamination would put the mission in jeopardy.

## Key conclusions from the updated study

Following the completion of the Phase A study and the subsequent design update, a number of conclusions were drawn as the program progressed towards Phase B.

It was clear that the proposed ground maintenance would not impact the configuration of the telescope, and that although a reasonable level of access was a major consideration, any limitations could be overcome either by disassembling the telescope or by inserting access hatches in the sidewalls.

As regards astronauts servicing the telescope, the report said "a fairly great degree of on-orbit EVA maintenance is possible with minimum configuration impact". But this would require there to be sufficient spacing between items of equipment to accommodate either a gloved hand or special tool. In addition, the design of electrical connectors and mechanical fasteners must be compatible with gloved hands and tools. There must be suitable lighting and foot and hand holds to aid the astronauts and minimize their work load and exertions.

The report noted that the proposed docking system for the LST might not be included in future designs. If this were eliminated, it would save 437 pounds (198 kg) on the telescope structure. The deletion of the associated docking module aboard the orbiter would eliminate 3000 pounds (1361 kg) and release 7.5 feet (2.2 meters) in the payload bay. Furthermore, the cost savings from not developing, fabricating, integrating, and testing a docking system would not adversely influence either the retrieval or ground maintenance of the telescope.

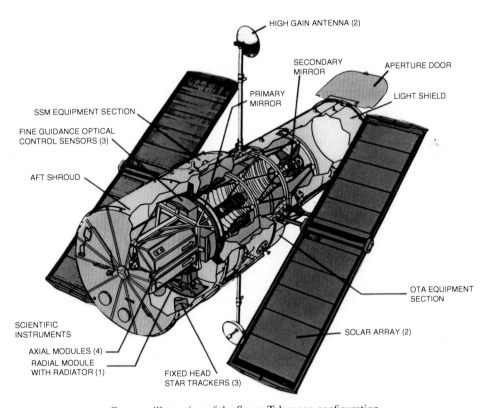

Cutaway illustration of the Space Telescope configuration.

## EVA AT LST

During the 1970s and early 1980s numerous studies were commissioned into the potential of EVA from the shuttle, including whether the new spacewalking hardware, techniques, and procedures could be applied to intended payloads such as the Large Space Telescope. One such study by the McDonnel Douglas Corporation, published as two volumes in June 1976, detailed potential EVA activities at the LST in support of on-orbit maintenance and servicing to supplement the Earth return option.[4] The Foreword to the report explained that "manned extravehicular activity (EVA) is a qualified, prime candidate for economically conducting on-orbit payload support functions" and that "servicing, repairing, and refurbishing payloads are some of the more significant economic measures that can be applied either through ground based or orbital operations." Though Earth-based servicing was considered paramount for the LST, there was growing evidence that orbital servicing could offer benefits that were as good as (and in some aspects, far greater than) returning a payload to Earth and then re-launching it. However, with the LST it would be almost a decade before the decision was made to opt for orbital servicing instead of Earth return, and this was primarily due to the costs involved.

This study was sponsored by the Bioengineering Division of the Life Sciences Office of NASA Headquarters, and monitored under the technical direction of the Crew Training and Procedures Division, Flight Operations Directorate at JSC. The objectives of the 12 month study, which commenced in June 1975, was to develop a comprehensive baseline description of shuttle EVA systems, to identify and select candidate payloads for EVA applications, and complete EVA planning guidelines for operational procedures and timelines. Four candidate payloads were used to develop these guidelines. One was the LST, and the others were the Advanced Technology Laboratory (ATL) based on the Spacelab laboratory module, the Low-Cost Modular Spacecraft (LCMS), and the Shuttle Infrared Telescope Facility (SIRTF). The LST study featured a 98 page summary in the second volume, almost one-third the size of the document.

Because of potential contamination of the Optical Telescope Assembly and the Scientific Instruments, and the risk to the quality of data collected in those areas, no EVA tasks were planned for either the Forward Shell of the SSM or the interior structure of the OTA. But EVA inspection, cleaning, and replacement tasks were considered as possible future potential targets of opportunity such as those with life limited parameters but these were dependent on changes in the EMU and associated apparatus in order to prevent cross contamination. One area that required further study was the resupply of cryogenic fluid dewars for the infrared instrument rather than changing out the entire instrument. In the study the infrared instrument was designed as a two-stage cryogenic cooling system with a minimum 12 month supply to cool the detectors. The report suggested that replenishment could be more cost effective than instrument replacement.

For the LST, the study listed planned, unscheduled, and contingency EVAs. Planned and unscheduled EVAs were "fully scheduled and unscheduled planned maintenance of expected wear out items accessible externally on the SSM and on the focal plane assembly, including the SI modules". Activities that fell outside the contractors' defined EVA capabilities would require additional equipment which, when utilized on EVA, could return the failed subsystem to operational status. The contingency (unscheduled) maintenance would be "for deployment or retrieval failures (i.e. to provide manual override)" to correct a problem encountered by the LST on-orbit. The study envisaged two orbiter crewmembers being assigned to make an EVA while a third crewmember would be available "to perform Payload Station EVA supporting functions and crew activities monitoring". Essentially, this third role evolved into the EVA choreographer/intravehicular crewmember assigned on actual shuttle EVA operations. When the report was carried out, exact design details were not available for the physical/functional interfaces of LST equipment, and "EVA support equipment conceptual designs were assumed or developed to implement procedures/timeline development."

Clearly a huge amount of work was being conducted into the potential of EVA to support the LST (and other shuttle payload) operations at the same time as the baseline shuttle EVA guidelines and hardware were being developed. As work on the telescope continued into the 1980s, concepts to support its servicing by EVA gained momentum and ultimately became the most cost effective means of keeping the telescope operational for the longest period of time.

**The LST becomes the ST**

The baseline telescope was defined from these early LST studies. But as recounted in the previous chapter, difficulties in securing funding and the inability of the shuttle to launch the telescope with the 10 foot (3 meter) primary mirror caused the design to be revised during 1974–1976. This reduced the size of the mirror, altered the layout of the support systems, and eventually made the option of ground servicing more expensive than on-orbit servicing. It was deemed too costly (and indeed too risky) to return the telescope to Earth and then re-launch it again not once but two or three times. Instead it was decided to leave the telescope in space during its planned 15 year operational lifetime and fly a series of Maintenance and Refurbishment (M&R) missions to it every two or three years, with the option to retrieve it at the end of its life. Additional orbital replacement units and EVA aids were incorporated into the design to aid the astronauts on-orbit. At the launch of the HST in 1990, NASA noted that 70 items on the spacecraft could be replaced on-orbit, including components in the guidance and control system, a computer, solar arrays, and the scientific instruments.

By the early 1980s the now Space Telescope (which would soon be named the Hubble Space Telescope) was under construction as the early shuttle missions were developing the techniques that would be needed for the telescope service missions. By 1985, all hopes of bringing the telescope home for maintenance were abandoned on the basis that it would be too risky to land the heavy payload without incurring damage that would delay its return to orbit and disrupt its important science program.

**THE HUBBLE SPACE TELESCOPE: A BRIEF DESCRIPTION**

The following pages offer a brief description of the Hubble Space Telescope and its major components as a guide to the subsequent chapters covering the shuttle service missions. It focuses on servicing and maintenance, rather than its operation during the scientific program in between the servicings.[5]

| Equipment | Contractor | Equipment | Contractor |
|---|---|---|---|
| Actuator Control Electronics | P-E | Off Load Device | ESA |
| Aft Latch, Solar Array | LMSC | Optical Telescope Assembly | P-E |
| Antenna Pointing System | Sperry | Optical Control Electronics | P-E |
| | | Oscillator | Frequency Elect. |
| Battery | Eagle Picher/GE | | |
| | | Photomultiplier Tube Electronics | P-E |
| Charge Current Controller | LMSC | Pointing Safemode Electronics | |
| Circulator Switch | Electromagnetic | Assembly | Bendix |
| Coarse Sun Sensor | LMSC | Power Control Unit | LMSC |
| Computer | Rockwell Autonetics | Power Distribution Unit | LMSC |
| | | Primary Deployment Mechanism | ESA |
| Data Interface Unit | LMSC | Primary Mirror Assembly | P-E |
| Data Management Unit | LMSC | | |
| Deployment Control Electronics | ESA | RF Multiplexer | Wavecom |
| Dish and Feed for HGA | GE | RF Switch | Transco |
| | | RF Transfer Switch | Transco |
| Elec. Power/Thermal Control | | Rate Gyro Assembly | Bendix |
| Elect. | P-E | Reaction Wheel Assembly | Sperry |
| | | Retrieval Mode Assembly | Northrop/Bendix |
| FHST Light Shade | Bendix | Rotary Drive | Schaeffer |
| Faint Object Camera | Dornier | | |
| Faint Object Spectrograph | MMC | SAD Adapter | ESA |
| Fine Guidance Electronics | Harris | SI C&DH | Fairchild/IBM |
| Fine Guidance Sensor | P-E | SSA Transmitter | Cubic |
| Fixed Head Star Tracker | Ball/Bendix | Science/Engineering Tape Recorder | Odetics |
| Focal Plane Assembly | P-E | Secondary Deployment Mechanism | ESA |
| Forward Latch, Solar Array | LMSC | Secondary Mirror Assembly | P-E |
| | | Sensor Electronics Assembly | P-E |
| Goddard High Resolution Spectrograph | Ball Aerospace | Solar Array Blanket | ESA |
| High Speed Photometer | Univ. Of Wis. | Solar Array Drive | ESA |
| Hinge, Aperture Door | LMSC | Solar Array Drive Electronics | ESA |
| Hinge, High Gain Antenna | LMSC | Star Selector Servo | BEI |
| Image Dissector Camera | | Temperature Sensor | LMSC/P-E |
| Assembly | P-E | Thermostat/Heater | LMSC/P-E |
| Instrument Control Unit | LMSC | | |
| Interconnect Cables | LMSC/P-E et al. | Umbilical Drive Unit | Sperry |
| Latch, Aperture Door | LMSC | Waveguide | LMSC |
| Latch, High Gain Antenna | LMSC | Wide Field/Planetary Camera | JPL |
| Low Gain Antenna | LMSC | | |
| MA Transponder | Motorola | | |
| Magnetic Torquer | Ithaco/Bendix | | |
| Magnetic Sensing System | Schoenstadt/Bendix | | |
| Mechanism Control Unit | LMSC | | |
| Metal Matrix Mast | DWA/LMSC | | |
| Multilayer Insulation | LMSC/P-E | | |

Hubble Space Telescope main contractors.

## The Great Observatories

NASA's Great Observatory program comprised four large space-based telescopes, each of which was equipped to observe in a different region of the electromagnetic spectrum. It had been intended that they would all be launched by the shuttle and their designers considered the prospects of on-orbit servicing, but in the end only the Hubble was able to fully support that capability. This program consisted of:

- *Hubble Space Telescope (HST)*: Launched in 1990 on STS-31 (Discovery). By the spring of 2015, after a number of service missions had replaced its original science instruments, it was surveying the near-infrared, visible and near-ultraviolet ranges.

- *Compton Gamma Ray Observatory (CGRO)*: Launched in 1991 on STS-37 (Atlantis) it was able to observe gamma rays and X-rays. The option of on-orbit servicing had been designed out, so when a gyro failed and impaired attitude control, the satellite was commanded to de-orbit in 2000.
- *Chandra X-ray Observatory (CXO)*: Prior to being named in honor of a pioneering astronomer this was called the Advanced X-ray Astronomical Facility (AXAF). It was launched in 1999 on STS-93 (Columbia). Astronomer-astronaut Steve Hawley, who had previously flown on two Hubble missions, was also a member of this crew. As of 2015 it was still operating.
- *Spitzer Space Telescope (SST)*: Because this was intended to be launched by shuttle and use a Centaur stage to reach its operating orbit, it was initially called the Shuttle Infrared Telescope Facility (SIRTF), but following the loss of Challenger in 1986 the Centaur was canceled for carriage in the payload bay. Planned for launch on a Titan, then Atlas, both versions were canceled due to costs and the delayed telescope was renamed the Space Infrared Telescope Facility (retaining the acronym SIRTF). In a slimmed down format, it was launched in August 2003 on a Delta II rocket. Several months later it was renamed the Spitzer Space Telescope in honor of Lyman Spitzer, the "father" of the Hubble Space Telescope concept. When its supply of coolant was exhausted in 2009 most of its instruments were retired.

**Record setters**

Hubble was classed as a "large payload" for the shuttle because it was about 24,500 pounds (11,110 kg), 43.5 feet (13.1 meters) in length and had a maximum diameter of 14 feet (4.27 meters) at the Equipment Section diminishing to 10 feet (3.1 meters) at the Light Shield and Forward Shell location. Stowage inside the payload bay was simplified by the use of furled solar arrays that would be deployed once the shuttle was on-orbit. Each of the two original arrays measured 7.8 by 39.4 feet (2.3 by 11.8 meters) and was stowed in a cylinder that was no larger than 15 inches (38 cm) in diameter and pivoted alongside of body of the telescope.

Although an impressive payload, Hubble was not the largest to be carried into orbit by the shuttle. That record was set on July 23, 1999 by the Chandra X-ray Observatory which had a mass of 50,161 pounds (22,753 kg); over twice that of Hubble. However, their diameters and lengths were similar due to the dimensions of the payload bay. To reach its operational orbit Chandra was propelled by a two-stage Inertial Upper Stage (IUS) which, combined with the observatory and mounted on its supporting table, occupied 57.1 feet (17.4 meters) of the 60 foot (18.3 meter) bay. As a payload without a propulsive stage, Hubble remains one of the largest launched by the shuttle.

**Hubble in space**

In April 1990 the shuttle Discovery deployed Hubble in a 353 mile (569 km) circular orbit inclined at 28.5 degrees to the equator that enabled it to spend 28 to 36 minutes of each 97 minute revolution in the Earth's shadow. This altitude was sufficient to enable the telescope to conduct its planned science investigations without the atmosphere clouding its images. It was also just within the range of the shuttle for maintenance and service missions.

## 156  LST becomes ST, becomes HST

The Hubble frame of reference has three axes. The primary axis, V1, coincides with the optical axis of the telescope. The V2 axis runs parallel to the solar array masts and V3 runs parallel to the high gain antenna masts. When it rotates, the pointing system uses external references for the three axes that allow it to aim towards intended astronomical targets, to face its solar arrays towards the Sun, or to otherwise alter its orientation.

**The main elements**

The Hubble Space Telescope incorporates three interactive elements that enable it to make observations similar to those made by optical observatories and support facilities on Earth. These elements are the Support System Module (SSM), the Optical Telescope Assembly (OTA), and the Scientific Experiment Package (SEP).

*Support System Module (SSM)*: The SSM is the workhorse of the spacecraft. It houses the servicing systems, including electrical power distribution, data communications, pointing control and maneuvering. In addition, a pair of solar arrays generate electrical power and recharge the storage batteries, and the antenna systems receive commands from Earth and send back data. In particular, it implements communications with the OTA, the Science Instrument Control and Data Handling (SIC&DH) unit, and the scientific instruments.

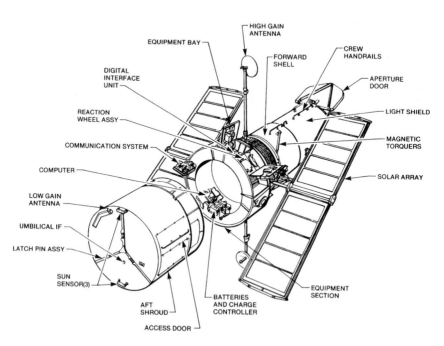

Exploded view of the main components of the Hubble Space Telescope.

# The Hubble Space Telescope: a brief description

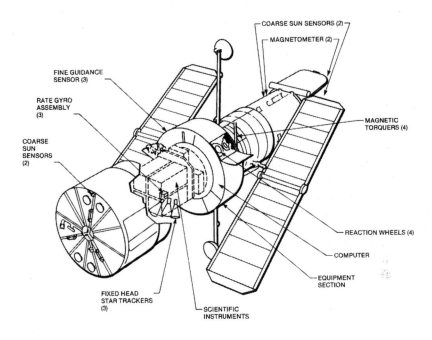

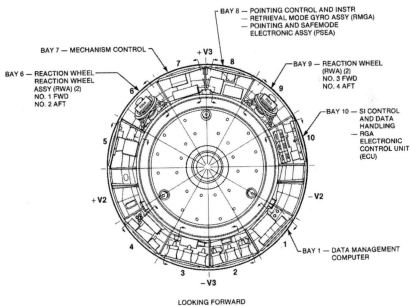

Detail of the main Orbital Replacement Units and Equipment Bay locations.

## 158  LST becomes ST, becomes HST

In detail, the SSM design features include:

- The outer structure is comprised of interlocking shells.
- Four Reaction Wheel Assemblies (RWA) and magnetic torquers are employed to maneuver, orientate, and stabilize the attitude of the telescope.
- Two Solar Arrays (SA) generate electrical power.
- The communications system includes a pair of High-Gain Antenna (HGA) and a pair of S-Band Single Access Transmitters (SSAT).
- Arranged in a ring around the telescope tube are a dozen Equipment Section Bays, ten of which contain miscellaneous support equipment:

    - Bay 1, Data Management Hardware
    - Bay 2, Electrical Power Equipment
    - Bay 3, Electrical Power Equipment
    - Bay 4, Electrical Power Equipment
    - Unnumbered trunnion (support) bay
    - Bay 5, Communications Hardware
    - Bay 6, Reaction Wheel Assembly (RWA)
    - Bay 7, Mechanism Control Hardware
    - Bay 8, Pointing Control Hardware
    - Bay 9, Reaction Wheel Assembly (RWA)
    - Bay 10, SIC&DH Unit
    - Unnumbered trunnion (support) bay.

- Computers for operating the systems and onboard data handling.
- Thermal protection by reflective surfaces and heaters.
- Outer doors, latches, handrails and footholds designed to assist astronauts during on-orbit maintenance tasks. In addition to over 225 feet (68.5 meters) of handrails there are 75 attachment points for foot restraints.

The seven major subsystems of the SSM are: structural and mechanical, instrumentation and communications, data management, pointing control, electrical power, thermal control and safing/contingency.

**Structures and mechanisms**

The Support System Module takes the form of a stack of cylinders with the aft bulkhead on the base with the Aft Shroud above, then the Equipment Section, the Forward Shell, and the Light Shield with the Aperture Door uppermost.

The *Aperture Door* measures about 10 feet (3 meters) in diameter and its single task is to cover the opening of the light shield. It is fabricated from sheets of honeycombed aluminum and the exterior surface is covered with a solar-reflecting material while the inside surface is painted black to absorb any stray light. From the fully closed position, the door can open to a maximum of 105 degrees, which is ample for the telescope's aperture of a 50-degree field of view centered on the +V1 axis. The door assembly includes a protective driving mechanism that can automatically close the door within 60 seconds. When sunlight approaches to within 35 degrees of the +V1 axis, Sun-avoidance sensors can automatically trigger the closing of the door to prevent damage to the internal optics.

# The Hubble Space Telescope: a brief description

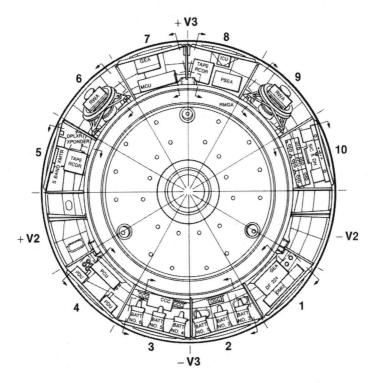

Enlarged view of Hubble's Equipment Bays.

This results in full closure by the time the Sun has reached 20 degrees of +V1. The controllers at the Space Telescope Operations and Control Center can override the protective door-closing mechanism if planned observations fall within the 20-degree limit; for example when using the dark limb (or edge) of the Moon to partially block out light when observing a bright object.

The *Light Shield* provides a connection between the Aperture Door and the Forward Shell and is designed to prevent stray light from entering the telescope. It is 13 feet (4 meters) long with an internal diameter of 10 feet (3 meters), and takes the form of a stiffened corrugated skin barrel of machined magnesium, protected by a thermal barrier. Ten light baffles inside the shield are coated with flat black paint to suppress the propagation of stray light.

The facilities to support servicing work include the placement of an exterior latch on each side for securing the stowed Solar Arrays (SA) and High Gain Antennas (HGA). Near these latches are 30 inch (76.2 cm) metal protective scuff plates used to secure the SA. When the telescope was locked in the payload bay of the shuttle, the Light Shield latches also supported the forward Low Gain Antenna (LGA) and its waveguide. There are also two magnetometers and two Sun sensors in this area. A series of handrails encircle the shield and there are a number of built-in foot restraints which were used to assist astronauts during an EVA.

Content detail of each Equipment Bay.

The *Forward Shell* is the central section of the spacecraft. It houses the main baffle and the secondary mirror of the Optical Telescope Assembly (OTA). It is 13 feet (4 meters) long and 10 feet (3 meters) in diameter, and is made from machined aluminum plating. For added strength, it has external reinforcing rings, ensuring clearance for the OTA inside, as well as internal stiffened panels. Its exterior is covered with thermal blankets.

In the stowage position, both the SAs and HGAs are latched flat against the sides of the Forward Shell and Light Shield. Four magnetic torquers are positioned at 90 degree intervals around the circumference of the Forward Shell. Two grapple fixtures next to the HGA drives would allow the shuttle's manipulator arm to retrieve the telescope out of orbit and re-deploy it at the end of a service mission. A trunnion system locked the telescope in the payload bay during periods of servicing. This section also features a number of EVA handrails and foot restraints.

The doughnut-shaped *Equipment Section* features a ring of 10 stowage bays and 2 support bays. The equipment bays contain approximately 90 percent of the electronics that both ran the telescope during scientific surveys and during the service missions. Located between the Forward Shell and the Aft Shroud, this section is made of machined and stiffened aluminum frame panels which are attached to an inner aluminum barrel. Each bay is trapezoid in shape, with the outer diameter (the access door) measuring 3.6 feet (1 meter) greater than the inner diameter which is 2.6 feet (0.78 meter). Each bay is 4 feet (1.2 meters) wide and 5 feet (1.5 meters) deep. Bays 6 and 9 also have thermal-stiffened panel doors that protect the Reaction Wheel Assemblies. The other eight stowage bays feature flat honeycombed aluminum doors mounted with equipment. The Equipment Section is enclosed by a forward frame panel and aft bulkhead, and the OTA is secured to the inside of the structure by six mounts.

The *Aft Shroud and Bulkhead* completes the main structure. It is the location of the Focal Plane Structure (FPS) that contains the axial scientific instruments and was also the location of the COSTAR unit between 1993 and 2009. The three Fine Guidance Sensors (FGS) and the Wide Field/Planetary Camera (WFPC) are mounted radially, near the connection points of the Aft Shroud and the SSM Equipment Section. It is 11.5 feet (3.5 meters) long and 14 feet (4.3 meters) in diameter. Constructed from aluminum with a stiffened skin, the design features 16 external and internal longeron bars for support, internal panels, and reinforced rings. The Bulkhead is fabricated from honeycombed aluminum panels 2 inches (5.08 cm) thick with three radial aluminum support beams. The Aft Shroud and Bulkhead includes a gas-purge system which was used prior to launch to prevent contamination of the scientific instruments, and the gas-exit vents are "light-tight" to prevent stray light from entering the focal plane of the telescope in space.

Hinged doors on the exterior of the Aft Shroud allowed spacewalking astronauts, assisted by handrails and foot restraints, to remove and exchange apparatus and scientific instruments. During maintenance, the interiors of the compartments were illuminated by internal lamps. The umbilical connections to support Hubble while it rested in the payload bay during launch and servicing were located in the Aft Bulkhead area.

The external SSM structure also carries a number of mechanisms, such as the latches to restrain the solar arrays and antennas; hinge drives to operate the aperture door and deploy arrays and antennas; motors to power the hinges and latches of the arrays and antennas; and gimbals to move the HGA dishes. There are four latches for the antennas, four for the solar arrays and one for the aperture door. They are driven by stepper motors called Rotary Drive Actuators (RDA) and work using a four-bar linkage system. A trio of hinge drives are used on each HGA and one for the aperture door, all with RDAs. In the event of an RDA failure, an astronaut could have used a hex-wrench to manually operate the hinge or latch.

Close up view of the EVA handrails across the Hubble Telescope structure.

Close up view of the base of Hubble showing one of the low gain antennas and circumferential EVA hand rails.

## Instrumentation and communications

The main communication system between Hubble and the ground is through the network of Tracking and Data Relay Satellites (TDRS) in geostationary orbit. All data is processed by the Data Management Subsystem (DMS). The high data rate flow passes via the High Gain Antenna and is reserved for the important scientific data. The HGA is mounted on a mast with a two-axis gimbal and electronics to rotate 100 degrees in either direction and track a relay satellite. The two Low Gain Antennas, located 180 degrees apart on the Light Shield and Aft Bulkhead, handle the low data rates used to receive commands from Earth and to transmit engineering data. They were also used during retrieval and deployment, and in safe mode operations.

## Data management

The *Data Management System* receives data from the various systems of the SSM, the OTA and the science instruments, and communications from the Space Telescope Operations and Control Center relayed via the TDRSS. It allows for processing, storage, and transmission of information by request. Apart from one Data Interface Unit (DIU) located in the Equipment Section, its subsystems are located in the SSM Equipment Section.

- The computer in Equipment Section Bay 1 is used for onboard digital computations. It also stores commands, formats telemetry, computes all processing to maneuver, point and attitude-hold the telescope; generates onboard commands for facing the SA to the Sun and steering the HGA; and monitors the health of the spacecraft. The computer installed in Hubble at launch featured a DF-224 unit with three central processors, two of which serve as backups, and six memory units, two of which serve as backups. The unit measures $1.5 \times 1.5 \times 1$ feet ($0.4 \times 0.4 \times 0.3$ meters) and has a mass of 110 pounds (50 kg). A new computer 20 times faster and with 6 times the memory of the original was installed in 1999.
- The *Data Management Unit* (DMU) is located on the inner door of Equipment Bay 1. Linked to the computer, it encodes data and transmits messages to selected telescope units and all DMS units, powers the oscillators, and is the main timing system for the telescope. It also receives and decodes all incoming commands, then forwards them for execution. Scientific data from the SIC&DH unit and engineering data from each component is either stored on the recorders or transmitted to Earth. It measures $26 \times 30 \times 7$ inches ($60 \times 70 \times 17$ cm) and has a mass of 83 pounds (37.7 kg).
- Four *Data Interface Units* (DIU) provide the link for commands and data between the DMS and various electronic boxes. The OTA DIU is on the OTA Equipment Section, and the others are in Equipment Section Bays 3, 7, and 10. Each DIU is actually two complete units, either of which is capable of handling the unit's function. Such a unit measures $15 \times 16 \times 7$ inches ($38 \times 41 \times 18$ cm) and has a mass of 35 pounds (16 kg).

- Three *Engineering/Science Tape Recorder* (E/STR) units were originally located in Equipment Section Bays 5 and 8. Two were used in normal operation, with the third serving as a backup. They could store up to 1 billion bits of information, including engineering and science data that could not be transmitted to the ground in real time. Each E/STR was $12 \times 9 \times 7$ inches ($30 \times 23 \times 18$ cm) and weighed 20 pounds (9 kg). They were later replaced during shuttle servicing by solid-state digital storage units.
- Two oscillators (timing clocks) can provide very stable central timing pulses for the telescope systems. The prime unit and its backup are located in Equipment Section Bay 2. They are cylindrical in shape, 9 inches (23 cm) long and 4 inches (10 cm) in diameter, with a mass of 3 pounds (1.4 kg).

Hubble's Scientific Instrument Control and Data Handling (SIC&DU) unit.

## Pointing control

A unique system enables Hubble to point within 0.01 seconds of arc at any target and remain in that position with an accuracy 0.0007 seconds of arc during a period of observation lasting up to 24 hours, whilst also orbiting Earth at 17,500 miles (28,157 km) per hour. By selecting guide stars, two Fine Guidance Sensors control the telescope to maintain it stable relative to those stars. When another target is required, the pointing system selects the most appropriate stars and maneuvers the telescope to the desired attitude. The system includes the computer, attitude sensors, actuators, and elements of the spacecraft safe mode (see below).

Five types of sensors form part of the Pointing Control Subsystem (PCS):

- Five *Coarse Sun Sensors* (CSS) are located on the Light Shield and Aft Shroud to measure the orientation of the telescope to the Sun.
- The *Magnetic Sensing System* (MSS) includes magnetometers and electronics to measure the orientation of the telescope relative to the Earth's magnetic field. Two units are located on the front end of the Light Shield.
- There are three *Fixed Head Star Trackers* (FHST). These electro-optic detectors are used in conjunction with star trackers to both locate and track a specific star within the field of view. They are in the Aft Shroud, behind the Focal Plane Structure and next to the RSUs.

One of three Rate Sensor Units, each of which contains a pair of gyros.

- There are three *Rate Gyro Assemblies* (RGA). Each consists of a Rate Sensing Unit (RSU) together with two rate sensors and an Electronics Control Unit (ECU) which measures the motion rates around its sensitive axis. The RSUs are in the Aft Shroud but the ECUs are in Bay 10 of the SSM Equipment Section. The PCS uses the RGAs to control the orientation of the telescope and to provide the attitude reference during maneuvering.
- The three *Fine Guidance Sensors* (FGS) provide angular position in respect to target stars. Using fine-pointing adjustments, the system is accurate to within a fraction of a second of arc. Two of these units are used by the pointing system, the third is used to make positional measurements of specific stars for astrometry research.

Two types of actuators were designed to move the spacecraft into a desired attitude and then provide control torques to stabilize the line of sight once in position:

- *Reaction Wheel Assemblies* function by rotating large flywheels, speeding them up to 3000 rpm or slowing them down and reversing them as necessary to control the momentum of the telescope. It has four RWAs, and the orientation of their axes is such that only three wheels are needed to move the telescope. They are located in Bays 6 and 9 of the SSM Equipment Section. Each wheel is 23 inches (59 cm) in diameter and has a mass of approximately 100 pounds (45 kg).
- *Magnetic Torquers* modify the speed of a reaction wheel by reacting against the magnetic field of Earth to create a torque on the spacecraft. The reaction occurs in the direction that reduces the speed of the wheel. They supply torque in directions perpendicular to Earth's magnetic field lines. This system also serves as a backup to orbital attitude stabilization in contingency modes. They are located on the Forward Shell of the SSM, and each unit is 8.3 feet (2.5 meters) long and has a mass of 100 pounds (45 kg).

**Electrical power**

The *Electrical Power Subsystem* (EPS) supplies Hubble with energy for its other systems and scientific instruments. While the telescope was in the payload bay of the shuttle for servicing, power was routed from the orbiter via an umbilical. During periods of orbital flight between service missions, the extended solar arrays transformed sunlight into electricity to be stored in batteries and subsequently fed out by the Power Control and Power Distribution Units. After servicing, the telescope was not released by the shuttle until its batteries were fully charged.

The EPS consists of the two Solar Arrays (SA) and associated electronics; six nickel-hydrogen ($NiH_2$) batteries; six Charge Current Controllers (CCC), one Power Control Unit (PCU) and four Power Distribution Units (PDU). With the exception of the two arrays that are mounted on opposite sides of the telescope, all the other components of this subsystem are housed around the SSM Equipment Section.

The two solar arrays are the prime source of electrical power. Each array is supported by its own electronics system. This includes the SA Drive Electronics (SADE) Unit that sends pointing commands to the wing assembly; a Deployment Controls Electronics Unit for the drive motor that extends and retracts the wing; and Diode Networks to direct

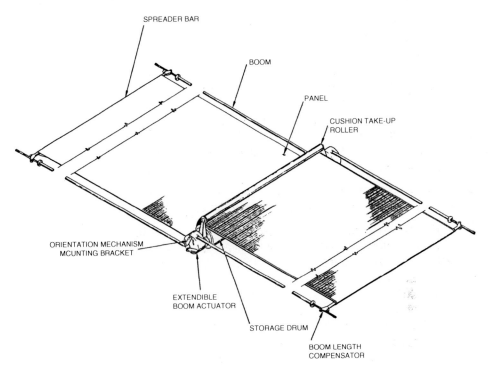

Detail of the first generation of solar arrays.

current flow. The original design featured a pair of double rollout array wings 39.38 feet in length, each consisting of two flexible blankets with five power generating panels. They used a silicon transducer cell that measured 0.78 × 1.57 inches (2 × 4 cm) and was rated at 0.39 ohm-inch (1 ohm-cm) with a relatively high efficiency of 12.7 percent at 25 degrees Centigrade. They were interconnected with 0.0013 inch (0.035 mm) silver mesh wire strips which were used to compensate for the varying coefficient of thermal expansion. Power transfer was by flexible silver mesh strips that were placed between bare Kapton and reinforced by glass-fiber cloth. When the arrays were rolled up for launch, embossed Kapton foil was inserted between solar cells for additional protection. The estimated lifetime of the array structures was 5 years, equivalent to 30,000 cycles at +/–100 degrees Centigrade with initial power requirements of 4 kW after 2 years.

Post-flight inspections of Columbia after the first shuttle mission in April 1981 gave the first indication that atomic oxygen at orbital altitude significantly eroded exposed surfaces, especially Kapton. From STS-3 there was a 35 percent loss in Kapton thickness and STS-8 confirmed the high Kapton erosion rates and used a flight experiment to determine erosion rates on different materials; showing a 0.0045 micron silver thickness reduction in just one week on-orbit. This was 30 percent of the total thickness of the interconnectors between the solar cells of the arrays for Hubble. This meant that the solar arrays, as originally designed and manufactured, were not suitable for the planned orbit at 372.84 miles (600 km). There could be no quick solution that would also be reliable, therefore the substrate and solar cell interconnectors had to be redesigned and re-qualified.[6]

First generation solar array deployment tests in a water bath at British Aerospace Dynamics Group, Bristol, England, in July 1981. (Courtesy British Interplanetary Society)

Lothar Gerlach recounted these early days on the Hubble project in Europe.[7] "From 1978 to 1986 I was working at AEG Telefunken in Wedel in Germany. I worked with teams from the development, manufacturing and testing departments involved with solar generators for space flight, and became involved in solar array design, engineering and management. AEG was responsible for the development and manufacturing of flexible blankets for Hubble. The initial design of the blankets was already completed and a large portion of the flight hardware built. But then from the first shuttle flight came the problem with atomic oxygen (ATOX). It was very quickly concluded that the existing HST blankets were not suitable for the Hubble mission because ATOX would destroy the blankets within a few weeks. There was no way to upgrade them. The message was, back to the drawing board as a completely new design was required. This is where I joined the Hubble 'family'. I was part of the team that designed and developed a solar generator which was ATOX friendly and had a lifetime of over 10 years. Luckily the development of the next generation of solar cells was very successful, allowing us to increase the performance of the solar generator by more than 10 percent. This allowed NASA to increase the power allocations to onboard instruments."

Although initially planned for launch in October 1983, the series of financial and technical difficulties experienced by NASA required rescheduling no fewer than five times, resulting in a three year delay to October 1986. Then, when the telescope was finally being processed for launch, the Challenger disaster occurred. This caused a further three and a half year delay due to the accident investigation and acting on its recommendations, as well as requalifying each surviving shuttle orbiter for flight.[8] For ESA, this had several repercussions. The solar arrays had been delivered to the Cape and installed on Hubble early in 1986. With the grounding of the shuttle fleet after the loss of Challenger, there was concern that the further delayed launch of the telescope could coincide with a period of maximum solar activity, including potentially damaging solar flares. Concerns were raised that at the planned deployment altitude the silver interconnects between the solar cells could be especially vulnerable. Another concern was the increase in NASA's planned power usage,

Fitting the folded solar arrays to the main telescope structure. (Courtesy British Interplanetary Society)

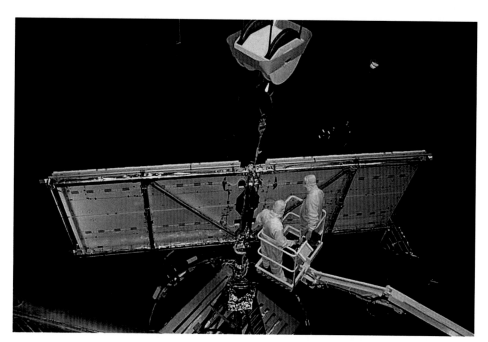

Third generation solar arrays being tested at ESTEC. (Courtesy ESA)

which would render the power budget marginal at least. The decision was made to ship the solar arrays back to Europe for upgrading. Over the next 2 years, the arrays were reworked with high-performance and atomic oxygen-resistant cells and interconnects. In 1988 the improved arrays were once again shipped to the Cape to be fitted on the telescope.

Hubble has six nickel-hydrogen ($NiH_2$) batteries to support the electrical power needs of the vehicle when demand exceeds the capacity of the solar arrays, when the spacecraft is in the shadow of Earth conducting science observations, and while it is in safe mode. Three 22 cell batteries form a module that measures $36 \times 36 \times 10$ inches ($90 \times 90 \times 25$ cm) and has a mass of 475 pounds (214 kg). There are two such modules. The batteries are recharged from the arrays by a dedicated Charged Current Controller (CCC). When Hubble was deployed in April 1990 a fully charged battery held 75 amp hours. That was sufficient to sustain normal science operations for about 7.5 hours (5 orbits) with a reserve for all safe and contingency modes, as well as coping with all enhancements added to telescope operations since launch. Placed in Bays 2 and 3 of the SSM Equipment Section, these units were designed to be safely handled during the service missions. Each module had two large yellow handrails to enable astronauts to easily and safely extract it from its housing and insert the replacement. After the originals had powered Hubble for 19 years, they were replaced by the final servicing mission in 2009.

Three *Charge Current Converters* (CCC) control the recharging of each battery unit by monitoring the voltage-temperature and measuring the progress of the recharging operation. The *Power Control Unit* (PCU) serves as the main power bus, interconnecting and switching current flow between the SA, batteries, and CCCs. It is located in Equipment Section Bay 4, measures $43 \times 12 \times 18$ inches ($109 \times 30 \times 2$ cm) and has a mass of 120 pounds (55 kg). In 2002 the original PCU was replaced with an upgraded unit that required

Hubble's main battery units.

Replacement batteries being installed on a SLIC (Super Lightweight Interchangeable Carrier) for flight aboard SM-4.

A Power Control Unit with its multiple connections. Note the handrails marked 'Astronaut Use Only'.

the telescope to be powered down for the first time since launch. The mood was very tense while the changeout was underway and the relief was palpable when the telescope was powered up again. There are four *Power Distribution Units* (PDU) containing local power buses, switches, fuses and monitoring devices for electrical power distribution around the vehicle. Two buses were to support the OTA, science instruments, and SIC&DH; the other two were to supply the SSM. The PDUs are in Bay 4, along with the PCU. Each unit is $10 \times 5 \times 18$ inches ($25 \times 12.5 \times 45$ cm) and has a mass of 25 pounds (11 kg).

**Thermal control**

To protect the telescope from the space environment, over 80 percent of its exterior surfaces are covered with multi-layer insulation. Additional electric heaters maintain a temperature level within prescribed safety limits. The *Thermal Control Subsystem* (TCS) on the SSM is designed to maintain temperatures within the prescribed limits in the Equipment Section, the science instruments, and the interfacing structures of the OTA. Even in the event of a worst case scenario of temperature variations, when the telescope emerges from the "cold" of the Earth's shadow into the "heat" of sunlight, the TCS maintains a safe operating environment. It also dissipates the excess heat that is generated by onboard equipment.

The features of the thermal control system include:

- The Multi-Layer Insulation (MLI) thermal blankets on the Light Shield and Forward Shell are made from 15 layers of aluminized Kapton, with an outer layer of Flexible Optical Solar Reflector (FOSR) made of aluminized Teflon.

# The Hubble Space Telescope: a brief description

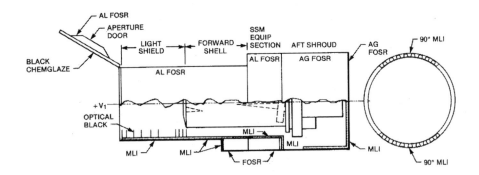

Original thermal coverings across the exterior of Hubble.

- Aluminum (silvered) FOSR tape is on the Sun-facing side of the Aperture Door. For the exterior door panels of the Equipment Section Bay there are specific patterns of FOSR and MLI blankets. There are also MLI blankets on the internal bulkheads to maintain a thermal balance between bays.
- In planning where to locate each item of equipment in the bays, it was essential to match temperature requirements. The designers made the best use of the available volume within each bay; e.g. by installing heat-dissipating apparatus on the side of the Equipment Section that would spend most of its time facing away from the Sun during normal operation.
- To provide added protection for the delicate scientific instruments, silvered FOSR tape is located on the Aft Shroud and Bulkhead exteriors, MLI blankets are on the interiors, and there are radiation shields on the inside of the Aft Shroud doors.
- Over 200 temperature sensors and thermostats are installed both internally and on the exterior of the spacecraft to constantly monitor each component and to control the operation of the heaters.

These coverings, supplemented by reflective or absorptive paints, protect the telescope against the intense cold of space and reflect solar heat so that it is able to operate within its design norms.

## Safing (contingency) systems

Wherever possible, duplication, overlapping, or redundancy in systems ensures a safeguard against potential breakdown or failure. There can never be a fully complete system backup unless an exact duplicate spacecraft is constructed, therefore wherever possible the addition of a contingency or backup option adds to the safety element of a mission.

The *Safing System* enables Hubble to operate in an emergency regime, in the event of a serious failure that threatens its life. This system makes use of most of the pointing control and data management components in addition to the dedicated *Pointing Safing Electronics Assembly* (PSEA). It can operate "indefinitely", without a communications link to ground control. The safing capability includes stabilizing the attitude of the telescope, rotating the solar arrays to achieve maximum solar exposure, and minimizing power usage to conserve electrical power.

During normal operations the Safing System automatically monitors the various systems using "keep alive" signals generated by the onboard computer indicating that all is well. As soon as a failure is detected, the Safe Mode is automatically activated. There are a range of such modes available, depending on the severity of the failure. If a malfunction occurs that does not actually threaten the survival of the facility, then the Safing System places it into a *Software Inertial Hold Mode* (SIHM) in which the telescope is "held" in its most recently commanded position. Should this occur when a maneuver is underway, the system is smart enough to allow that maneuver to complete, and then it holds the telescope in the resulting position and at the same time suspends all science operations. This is a situation which can only be reversed by commands from ground control.

In the event of a marginal Electrical Power System malfunction or a failed safety check from the Pointing Control Subsystem, then the spacecraft will automatically enter what is termed the *Software Sun Point Mode* (SSPM) in which the telescope is maneuvered so that the solar arrays face the Sun to constantly absorb solar energy and thus generate power. The Thermal Control Subsystem is designed to ensure that all onboard equipment is maintained within its operating temperature range and above survival temperatures, pending a return to normal operations. Ground controllers must identify and correct the malfunction before the science program can be resumed.

Several computer software improvements and upgrades to the Safing System have been installed on Hubble since its launch in 1990 to make the responses more robust, but should the situation worsen the system can hand control to the Pointing Safe Electronics Assembly in the *Hardware Sun Point Mode* in which the computer conserves power by turning off all nonessential equipment. The sequence starts by safing the science instruments and facing the solar arrays towards the Sun (if this has not already been achieved). Operating power is then removed from equipment that is not essential to the telescope's survival. The sequence ends with shutting down the computer and, 2 hours later, shutting down the SIC&DH subsystem. Such a serious problem alerts the ground controllers (who may or may not already be aware of a serious degradation of systems or performance). At this point, NASA management will form a failure analysis team to study the problem and explore the options. This team will be led by a senior managerial representative from Goddard Space Flight Center, which hosts the Space Telescope Operations Control Center, and have authority to call upon experts not only from within NASA and the telescope's contractors but from any organization that is, or has been, involved in the wider Hubble project. Once the team has identified the problem it will recommend corrective actions to a higher managerial level at Goddard, where any hardware changes or amendments to the configuration of computer software will require NASA Level 1 Authorization. With the retirement of the shuttle in 2011, it is no longer possible to plan a servicing rescue or repair mission. Over the next few years, as Hubble gradually deteriorates and its systems and components fail or are turned off, the decision to end science operations and dispose of the telescope will have to be addressed.

The Pointing Safing Electronics Assembly consists of 40 electronic printed circuit boards with redundant functions capable of running Hubble even in the event of an internal circuit failure. Located in Equipment Section Bay 8, it has a mass of 86 pounds (39 kg). A dedicated backup gyro, designated the *Retrieval Mode Gyro Assembly* (RMGA), is also located in Bay 8. This has three gyroscopes of a lower rate quality than the RGAs because they are not used during observations.

## Optical facility

The Optical Telescope Assembly features the primary mirror and the main support ring. The mirror collects and concentrates the incoming light onto a focal plane for use by the science instruments. The OTA features the primary mirror, which is supported inside the main ring, reaction plates and actuators, and both the main and control baffles. The Secondary Mirror Assembly is cantilevered off the front face of the main ring.

It was the distortion on the primary mirror that caused the spherical aberration which was discovered during the commissioning process that followed deployment in April 1990. This was corrected by the COSTAR package installed by SM-1 in December 1993. The astronauts had no direct access to the mirrors but they could access the OTA Equipment Section and the axial scientific instruments (see below) in the Focal Plane Structure.

***OTA Equipment Section***: This is a large semicircular set of compartments containing various systems. Located outside the spacecraft on the Forward Shell of the OTA, it holds the OTA electrical power and thermal control electronics systems, fine guidance electronics, actuator control electronics, optical control electronics, and the fourth DIU of the DMS.

Seven of the nine bays in the OTA Equipment Section are used for storing equipment and the other two are for support. All nine bays feature outward opening doors for ease of access during the servicing missions. Each also has electronic connectors, heaters and insulation for thermal control.

Primary mirror construction.

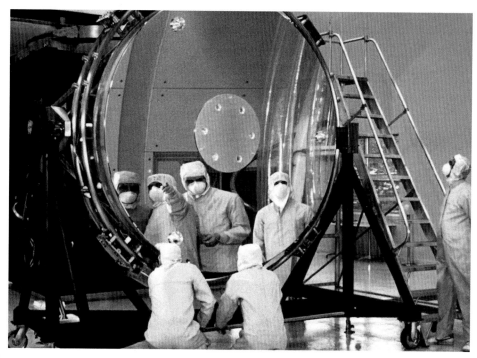

Primary mirror visual inspection.

- Electrical Power/Thermal Control Electronics (EP/TCE) distributes power from the SSM electrical power subsystem to the OTA systems, using thermostat controllers to regulate mirror temperatures as appropriate. The EP/TCE also collects thermal sensor data for transmission to the ground.
- Fine Guidance Electronics (FGE) supplies power, commands and telemetry to each of the Fine Guidance Sensors.
- Actuator Control Electronics (ACE) supplies the command and telemetry interface directly to the 24 actuators affixed to the primary mirror and the 6 actuators on the secondary mirror.
- Optical Control Electronics (OCE) controls the Optical Control Sensors (OCS), with one OCE assigned to each FGS. They use white light to measure the optical quality of the OTA and send this data to the ground.
- Data Interface Unit (DIU) is the electrical interface between the electronics in the OTA and the main telescope command and telemetry system.

**Scientific Instruments**: Four of the main scientific instruments on Hubble are located in an aft section termed the Focal Plane Structure (FPS), at the circumference of the spacecraft in close proximity to the three Fine Guidance Sensors. The instruments (but not the FGS) are controlled by the SIC&DH subsystem. Digital data from each instrument is transferred to onboard computers for processing (encoding), and is either temporarily stored on board or sent to Earth in real time.

**Table 10** Science instrument changeouts

| Exp. Bay | 1990 deployment | 1993 Service Mission 1 | 1997 Service Mission 2 | 1999 Service Mission 3A | 2002 Service Mission 3B | 2009 Service Mission 4 |
|---|---|---|---|---|---|---|
| A1 | GHRS | GHRS | STIS | STIS | STIS | STIS |
| A2 | FOS | FOS | NICMOS | NICMOS | NICMOS | NICMOS |
| A3 | FOC | FOC | FOC | FOC | ACS | ACS |
| A4 | HSP | COSTAR | COSTAR | COSTAR | COSTAR | COS |
| R+V3 | FGS-2 | FGS-2 | FGS-2 | FGS-2R | FGS-2R | FGS-2R |
| R−V2 | FGS-1 | FGS-1 | FGS-1R | FGS-1R | FGS-1R | FGS-1R |
| R−V3 | WFPC-1 | WFPC-2 | WFPC-2 | WFPC-2 | WFPC-2 | WFPC-3 |
| R−V2 | FGS-3 | FGS-3 | FGS-3 | FGS-3 | FGS-3 | FGS-3R |

KEY

| | | | |
|---|---|---|---|
| A1 | Axial Bay 1 | R+V3 | Radial Bay Upper |
| A2 | Axial Bay 2 | R−V2 | Radial Bay Right |
| A3 | Axial Bay 3 | R−V3 | Radial Bay Lower |
| A4 | Axial Bay 4 | R+V2 | Radial Bay Left |

| | |
|---|---|
| GHRS | Goddard High Resolution Spectrograph |
| FOS | Faint Object Spectrometer |
| FOC | Faint Object Camera |
| HSP | High Speed Photometer |
| FGS# | Fine Guidance Sensor (1, 2 and 3) |
| FGS #R | Fine Guidance Sensor Refurbished (1, 2 and 3) |
| WFPC# | Wide Field/Planetary Camera (1, 2 and 3) |
| COSTAR | Corrective Optics Space Telescope Axial Replacement |
| STIS | Space Telescope Imaging Spectrometer |
| NICMOS | Near-Infrared Camera/Multi-Object Spectrometer |
| ACS | Advanced Camera for Surveys |
| COS | Cosmic Origins Spectrograph |

SCIENCE INSTRUMENT CHANGEOUTS

| | |
|---|---|
| GHRS | At launch; removed during SM-2 |
| FOS | At launch; removed during SM-2 |
| FOC | At launch; removed during SM-3B |
| HSP | At launch; removed during SM-1 |
| FGS# | Three original units at launch; #1 removed during SM-2; #2 removed during SM-3A; #3 removed during SM-4 |
| FGS #R | #1 installed on SM-2; #2 installed on SM-3A; #3 installed on SM-4; ALL STILL ABOARD |
| WFPC# | Unit 1 at launch; replaced with unit 2 during SM-1; replaced with unit 3 during SM-4, STILL ON BOARD |
| COSTAR | Installed during SM-1; removed during SM-4 |
| STIS | Installed SM-2; STILL ON BOARD |
| NICMOS | Installed SM-2; STILL ON BOARD |
| ACS | Installed SM-3B; STILL ON BOARD |
| COS | Installed SM-4; STILL ON BOARD |

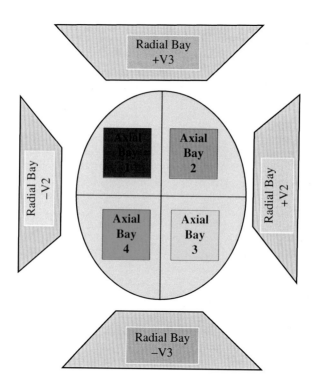

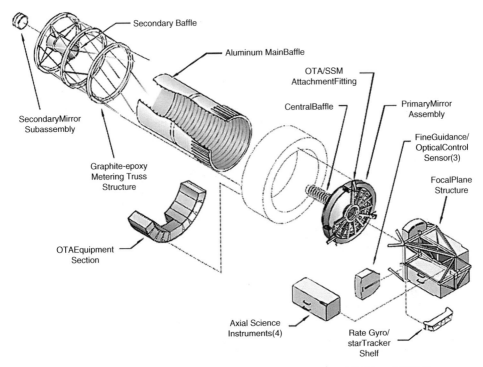

An exploded view of the Optical Telescope Mount. (Courtesy NASA and STScI)

The scientific instruments installed on Hubble have been changed over the years, and the science package on the telescope in 2015 bears no resemblance to that at the time of launch 25 years ago. For reference, the science instruments and their fate are summarized below:

- *Advanced Camera for Surveys* (ACS): Installed in 2002 by STS-109 in place of FOC, then repaired by STS-125 in 2009. It is used to study the 'weather' on planets in other solar systems, study the nature and distribution of galaxies, and conduct surveys of the universe. It is currently active.
- *Cosmic Origins Spectrograph* (COS): Installed in 2009 by STS-125 in place of COSTAR. It is currently active.
- *Faint Object Camera* (FOC): Installed at launch, then removed in 2002 by STS-109 (replaced by ACS). It used the optical resolution of the telescope to record objects in deep space with greater clarity than previously obtainable. The cameras studied the evolution of stellar formations, galaxies and objects such as quasars. It was also used to hunt for extra-solar planets. It was designed by ESA and built by Dornier Systems in West Germany, Matra Corporation in France, and British Aerospace, England. Its Photon Detector Systems had two identical detectors. Photons from the target source entered the image intensifier, were converted to electrons and accelerated by a high voltage and reconverted to photons at a production rate of

100,000 photons for each electron. An intensified image was produced and stored in the video processing unit for later transmission.
- *Faint Object Spectrograph* (FOS): Installed at launch, then removed by STS-82 in 1997 (replaced by NICMOS). Originally intended as a companion for the Goddard High Resolution Spectrograph, the FOS was built by Martin Marietta. It studied the light from very faint objects while GHRS studied bright light in greater detail, thus overlapping their research.
- *Goddard High Resolution Spectrograph* (GHRS): Installed at launch. During STS-61 an electrical connection box was fitted (the Goddard HRS power supply redundancy kit). The instrument was removed by STS-82 in 1997 (replaced by STIS). By being sensitive to ultraviolet radiation, it studied the composition, temperature, and density of stellar objects and giant gas clouds.
- *High Speed Photometer* (HSP): Installed at launch, then removed by SM-1 to make room for COSTAR. It measured the intensity and color of light across the spectrum from ultraviolet to infrared to make precise measurements of the brightness of stars, search for variable pulsars, and test theories about black holes by searching for their surrounding disks of gas.
- *Near Infrared Camera and Multi-Object Spectrometer* (NICMOS): Installed in 1997 by STS-82 in place of FOS. Extra coolant was installed in 1999 by STS-103. Then in 2002, STS-109 installed a cryogenic cooler and radiator to restore its operation. This instrument uses infrared light to view objects that are normally obscured by cosmic dust and gas. This second generation image/spectrograph has been dubbed Hubble's heat sensor. It is currently active.
- *Space Telescope Imaging Spectrograph* (STIS): Installed by STS-82 in 1997 in place of GHRS. In 2009 STS-125 helped to revive it by replacing a failed power supply. It provides information on the temperature, density, motion and chemical composition of an object. It is currently active.
- *Wide Field/Planetary Camera* (WFPC): Installed at launch, then removed by STS-61 in 1993 and replaced with WFPC-2, which was in turn replaced by STS-125 in 2009 with WFPC-3. The main camera system on the telescope, it is used to observe almost everything by using 48 filters allowing studies across wavelengths from ultraviolet to near-infrared. The first unit could produce images and spectrographic, photometric, and polarimetric measurements, making it the most versatile instrument onboard the telescope. The second unit had improved Charge-Coupled Devices (CCD) detectors that were far more sensitive than modern CCDs used in digital cameras. The second unit was originally a spare but after the problem with the spherical aberration which affected the WFPC-1 (necessitating COSTAR) this unit was prepared for flight with its own optical correction. The third unit used elements retrieved from the first (after SM-1) and spares from the second. WFPC-3 is currently active.
- *Fine Guidance Sensors* (FGS): The third sensor can serve as a 'scientific instrument' by measuring the positions of stars in relation to other stars for astrometry to measure stellar masses and distances.
- *Corrective Optics Space Telescope Axial Replacement* (COSTAR): Installed by SM-1; rendered superfluous by SM-3B in 2003, and removed by SM-4 in 2009

(replaced by COS). COSTAR wasn't a scientific instrument but a unit with five optical mirrors in pairs designed to refocus light from the flawed mirror to enable the first-generation instruments to work correctly. After those instruments ceased to be used, COSTAR was removed.

***Science Instrument Control and Data Handling Unit (SIC&DH):*** This system oversees the workings and coordination of all the scientific instruments. The science and engineering data is processed by the data management unit for transmission to the ground, reporting the status of the telescope, its various systems, and scientific data. It consists of electronic components attached to an ORU which is mounted on the door of SSM Equipment Section Bay 10, and a small remote module installed on each individual scientific instrument.

The components of the SIC&DH, which are duplicated for redundancy, at launch were:

- One NASA Standard Spacecraft Computer, Model I (NSSC-I)
- Two Standard INTerface (STINT) circuit board units for the computer
- Two control unit/science data formatter (CU/SCF) units
- Two central processor unit modules (CPM)
- One Power Control Unit (PCU)
- Two Remote Interface Units (RIU)
- Various memory, data and command communications lines (buses)
- Bus coupler unit (BCU).

## THE JOURNEY FROM LOT TO HST

NASA loves acronyms, and in the world of technology over the years it has almost become a language of its own—a NASA'olgy. In a way, the description of how the Hubble evolved is told by means of these acronyms.

Conceived from an idea in the 1940s, Spitzer's concept emerged in the mid-1960s as the Large Orbital Telescope (LOT), which itself morphed into the short-lived Manned Orbital Telescope (MOT). Then came the Large Space Telescope (LST), which by the mid-1970s was deemed too extravagant and was trimmed to Space Telescope (ST). In 1983 the name Hubble was added to improve its public image, making what we know today as the Hubble Space Telescope (HST).

So in NASA'olgy terms, the LOT became the MOT then the LST which became the ST and ultimately the HST!

Away from the terminology, 40 years after Spitzer proposed placing an optical telescope in space to escape the restrictive effects of the Earth's atmosphere, the concept had emerged as a viable instrument. It was a long struggle, fraught with difficult decisions and sacrifices based upon the choices and constraints of the technology available at the time, politics and rivalry, delays, and even tragedy, and of course the battle to convince those empowered to fund the project to commit to sign the checks. Despite all the problems (and still with many more to come in the future) on April 25, 1990 Hubble was deployed on-orbit. The time had come to prove that the concept was worth all the effort.

Under the skin of a telescope. Hubble, yet to receive its protective outer layering, is seen in a special clean room at Lockheed. (Courtesy British Interplanetary Society)

Part of the long process of going through the sequence of construction, testing, thermal vacuum evaluation and systems compatibility. (Courtesy British Interplanetary Society)

The Optical Telescope Assembly is lowered into the Aft Shroud. (Courtesy British Interplanetary Society)

Functional testing of the HST included special acoustic tests to replicate the stress levels likely to be encountered during ascent to orbit. (Courtesy British Interplanetary Society)

The shipping container used to support the transfer of the telescope from Lockheed to KSC (top), and thereafter being moved by hand (bottom). (Courtesy British Interplanetary Society)

Completed telescope undergoing tests at KSC. (Courtesy British Interplanetary Society)

Shrouded in a protective cocoon, Hubble was moved from the east test cell of the Vertical Processing Facility at KSC to the west cell prior to transfer to the launch pad. (Courtesy British Interplanetary Society)

## REFERENCES

1. Large Space Telescope, NASA Technical Memorandum NASA TMX-64726; LST Phase A Final Report; Volume 1—Executive Summary; December 15, 1972; NASA MSFC
2. LST Phase A Design Study Update, NAA Technical Memorandum TM-X-64763, August 1, 1973, NASA MSFC
3. *Skylab, America's Space Station*, D. Shayler, Springer-Praxis 2001 pp163–184
4. Application of Shuttle EVA Systems to Payloads, Volume I EVA Systems and Operational Modes Description; Volume II Payload EVA Task Completion Plans, MDC W0014, June 1976, NASA CR-147733
5. Hubble Space Telescope Media Reference Guide, Lockheed Missile & Space company, Inc., February 1990
6. Information supplied by Lothar Gerlach, by email February 15, 2013, and Power Point presentation on ESA's involvement in the Hubble project: *Good Luck Hubble, Thirty Years of Challenge and Excitement,* Lothar Gerlach, ESA Hubble Space Telescope Project Manager, Head of Solar Generator Section, ESA, dated June 9, 2011
7. Email from Lothar Gerlach February 12, 2013
8. The History of the Hubble Space Telescope and ESA Involvement, R. J. Laurance, ESA Bulletin No, 61 February 1990, pp8–11

# 5

# Simulating servicing

> *The approach to Hubble or any short duration flight where there's a lot to do, has to be massively over trained, way more than is efficient in respect to time management, but never enough if you make a mistake and that's the conundrum you are caught in.*
>
> Mike Foale, Mission Specialist STS-103

The suggestion to service and maintain the spacecraft by astronauts on-orbit was a feature in the earliest studies of the LST/ST/HST project. Nevertheless, initially it was planned that the telescope would either be returned to Earth for major refurbishment and upgrading or feature pioneering robotic servicing. However, astronauts would always be able to conduct EVAs if necessary. When Earth-based servicing was dropped, the emphasis to prepare astronauts for extensive maintenance and servicing tasks on-orbit became even more critical as the program evolved. From its inception, the shuttle was to have the capability to support spacewalks via an airlock attached to the crew compartment. There would be a range of equipment to aid the astronauts go about their EVA tasks, most notably the Remote Manipulator System (RMS), support fixtures, handholds, foot restraints, and an extensive 'tool kit'. For a while, even the Manned Maneuvering Unit (MMU) was expected to be a useful servicing aid.

Preparing for Space Telescope servicing activities became a long and involved program. It had its origins in the 1960s, when satellites were not so reliable. To overcome this expensive hurdle the quality control of satellite manufacture was greatly improved, common designs of components and fixtures were adopted, and tools and equipment were developed to support in-flight maintenance and repair. This encouraged the belief that satellite servicing programs already under development for the next decade would be able to be enhanced and expanded.

Fifty years later, regular spacewalks on the International Space Station demonstrate how familiar working outside a spacecraft appears, and although it could never really be routine, the operations involving the ISS and Hubble clearly showed how things have changed since the pioneering steps of the mid-1960s.

Several things have contributed to this success, including the recognition of the value and importance of space servicing and maintenance, the development of tools and procedures to support such activities, the design of compatible items of space hardware that will simplify servicing, and the vision of one man in particular who drove forward the idea that if an item of hardware were to fail in space, it was not always necessary to replace it with a completely new spacecraft. This concept evolved into the capabilities to service satellites and maintain the Hubble Space Telescope. The skills developed from these operations were also applied to the assembly and maintenance of the ISS, and have led to current developments in the fields of robotic and human space servicing concepts. The application of these developments over the coming decades will bear fruit not only in Earth orbit, but at the Moon and the asteroids or Mars.

## CREATING A CONCEPT

In the mid-1960s placing objects in space was still relatively new, as were the skills required to ensure that satellites could endure the extreme vibrations of the ascent to orbit, prolonged exposure to the harsh environment of space, and return useful data. If systems failed, it was not yet possible to visit the satellite in order either to repair it in-situ or to return it to Earth. Instead it was recognized that the improving manufacturing standards and a high degree of commonality between designs would alleviate many failures, and that when a system for orbital servicing was eventually introduced, it would be possible to exchange faulty items should a failure be non-recoverable.

At that time, approximately one-quarter of NASA's spacecraft would fail within days or months of being launched. This was expensive as well as immensely frustrating. The answer was to make the hardware more reliable. One problem was that it was not clear whether the contractors were performing adequate checks prior to shipment to NASA. To compound the problem, the space agency did not have in place suitable checks on their contractors. It was around this time that Frank Cepollina was a young engineer at Goddard Space Flight Center working on the Orbiting Astronomical Observatory (OAO) program. As leading field center and manager for most of these programs, it was Goddard's responsibility to determine what was going awry with satellites and to recommend remedial action. Cepollina was given the task. Some of the failures that he investigated were quite embarrassing for the agency, like that of an Orbiting Geophysical Observatory (OGO) which successfully reached orbit, then failed when commanded to start its science program. Post-flight analysis revealed the gyro had been connected in reverse, thereby causing it to spin backwards and fail. Such an error should have been spotted during assembly and testing on the ground.

### Frank J. "Cepi" Cepollina

Frank Cepollina, known across the aerospace industry as "Cepi", joined NASA Goddard in 1963 as a spacecraft engineer. He worked on a variety of projects before being assigned by his boss, Joseph Purcell, to find out what was going wrong with satellites, and how things could be improved in order to prevent so many failures.[1]

Frank Cepollina. (Courtesy Goddard Space Flight Center)

By the early 1970s, the Apollo program was winding down to be replaced by the Skylab space station that used surplus Apollo hardware. At the time, Deputy Administrator George Low was urgently seeking to cut significant costs in NASA and make things more reliable. The Skylab program was considered an excellent example, reusing equipment intended for one program to perform another. When Skylab suffered serious damage during its launch in 1973, a remedy was quickly devised. The astronauts went up and installed a protective solar "parasol" on the exterior of the workshop in order to protect its damaged skin and manually deployed a fouled solar array. This human ingenuity and resourcefulness was an impressive demonstration of how astronauts, given the tools and opportunity, could recover a mission that would otherwise have had to be written off at launch. The rescued station enabled three crews to perform highly successful missions of 28, 59 and 84 days.[2]

In his NASA oral history Cepollina pointed out, "That was the message of the late '60s and early '70s. Find a way to do things better. Find a way to take advantage of humans in orbit [in order] to be able to repair, maintain, and prolong the life of valuable space assets. That was our philosophy."

During the late 1960s the rise in the cost of space exploration, coupled with a restricted budget, obliged NASA to rethink its policies and find ways to reduce the cost of achieving low Earth orbit. In 1972 Congress authorized the development of a reusable shuttle for what was grandly called the National Space Transportation System. This would replace the one-shot Apollo missions. One of the mission profiles envisaged for the shuttle was a program known as Maintenance and Refurbishment (M&R) which would service satellites on-orbit. Cepollina realized that for this to be practicable, satellites would need to be designed to aid such servicing.

**On-orbit servicing**

In 1975 Cepollina wrote an article entitled 'On-Orbit Servicing' in which he explained the concept of what became the Multi-mission Modular Spacecraft (MMS). First, spacecraft components should be standardized wherever possible. Second, it would be logical to use the same basic spacecraft to support several different missions. Third, it was important to fully develop the capability offered by the shuttle to either replace systems or install new instruments on a satellite, without having to return it to Earth, repair it and then re-launch it.

Around this time, NASA requested three aerospace companies to review current options and identify the most cost-effective way of using the shuttle to extend the lives of satellites. Grumman Aerospace, General Electric, and TRW Systems suggested four different ways to address this task:

- A replacement satellite could be launched on an expendable booster. [This was the traditional way and was considered an expensive approach to the problem.]
- Launch the replacement aboard the shuttle. [This might be more expensive and risky owing to the added complication of a human crew.]
- Launch the satellite on either an expendable booster or the shuttle, but use the shuttle to retrieve the failed satellite for return to Earth for refurbishment. [This would add the cost of bringing home a defunct satellite, repairing it, and perhaps launching it a second time.]
- Have astronauts on the shuttle replace the failed systems on the satellite in order to restore its operational capability. [This was Cepollina's preferred method, and was deemed to be the most cost-effective.]

In addressing this problem, Cepollina assembled an ad-hoc team of satellite builders and engineers at Goddard. They reasoned that by splitting into "modules" the main components which enable a satellite to function, these could be plugged in from the outside of the main spacecraft framework or "bus". This design could then be standardized in order that similar instrumentation, subsystems and components could be replaced simply by plugging them in as required, and in space if necessary. Therefore the modules for attitude control, command and data handling, and power generation and distribution could all be plug-in units. Even the unique science instruments could be plugged in from the outside if the framework on which they were mounted was compatible with the bus. The intended mission of the bus could be changed as required, without redesigning the whole spacecraft.

"We called that the Multi-mission Modular Spacecraft," said Cepollina, "designed to be easily serviced and repaired by the shuttle, and [which] could be launched by the shuttle, or could be launched by conventional [expendable launch] vehicle, it didn't really matter. As long as the shuttle could reach it and grab it, then astronauts could fix it and upgrade it, and do whatever else was necessary to do."

**Multi-mission Modular Spacecraft**

The idea of "plug and play" units to be snapped into place from outside the main spacecraft framework was attractive to the designers of the Large Space Telescope, in which Goddard was involved at that time. Accessing work areas from the outside of the spacecraft reduced the threat of space-suited astronauts trying to squeeze inside and risk tearing their pressure suits, and eliminated the need to devise complicated robotic mechanisms to operate on the inside. It also saved the ever present challenges to spacecraft designers of having a limited volume, mass, and of course cost, because a generic bus could be used for a wide variety of missions that could be sustained in space either by servicing isolated subsystem modules or by removing an old experiment and replacing it with a new one.

"We took the ownership of that particular MMS concept," Cepollina explained, a concept that evolved into the orbital replacement units (ORU) flown on the Hubble Space Telescope. His team applied the concept to the Landsat 4 and Landsat 5 satellites, Solar Max, the Upper Atmosphere Research Satellite (UARS) and the Compton Gamma Ray Observatory, as well as several smaller satellites such as the Extreme Ultra-Violet Explorer (EUVE). Cepollina estimates that 16 different spacecraft employed the MMS approach during 1980–1994. The idea also became very attractive to the Department of Defense, which used it for several of their classified satellites.

Having solved the original problem of multiple satellite failures, and despite the obvious advantages of the system, it didn't appeal to some levels of management at NASA. Making a satellite bus capable of performing a variety of missions, and using systems that had greater reliability and modularity so that they could be repaired or replaced in space and extend the life of a satellite, severely restricted the role of some managers in developing *new* satellite projects featuring the latest technology, thus stealing some of their thunder.[3] In Cepollina's concept, the latest technology was not always required and often it caused friction between the two sides. Reliability was gained by using proven systems. New technology was risky. But Cepi was a determined, hardworking hands-on engineer, and very eager to explain and promote his concepts. His outspoken manner could cause difficulties but in the end he was usually proved right. Hence to many in NASA, and on the outside, he is a hero in the space program. As a former colleague told writer Robert Zimmerman, "everybody who works for NASA should work for him for a while".[4] In my research for this book, Frank Cepollina was often praised for his work to establish the concept of the Hubble service missions as one of the highlights of NASA in recent years. Even following the final Hubble mission in 2009, Cepi had no plans to retire, despite being in his late 70s, telling Zimmerman, "my plan is to develop a national capability to repair and maintain satellites, anywhere and in any location in space." As he continues to work on plans to have robotics assist in refueling spacecraft in space, who can doubt him.

With the modular approach decided upon for the Space Telescope, work could begin in defining the methods and procedures to achieve that promise of servicing by the crew of a shuttle.

## ESTABLISHING THE GUIDELINES

Between 1965 and 1967, as part of NASA/industry studies into optical technology systems and experiments, the prospect and limits of astronaut interaction with large space telescopes was analyzed in the context of "human factors engineering". These studies helped establish guidelines for early crew involvement.[5]

By the middle of 1972, the Lockheed Company had an ongoing and major Crew Systems Activity program that included the development of full-size hard mockups of the telescope and extensive man-in-the-loop simulations. It was during this time that, in cooperation with NASA, the LST program objectives were developed, interfaces identified, and on-orbit crew system requirements (as well as potential ground servicing options) established. The overall objectives of these early studies were to develop a far reaching capability for on-orbit EVA maintenance; to minimize program cost; to reduce design/operations complexity; to improve operational reliability; and to reduce development uncertainties. It was also recognized that the development of a flexible approach for planned or likely maintenance tasks—using as much off-the-shelf equipment as practical whilst also ensuring maximum flight safety for both the telescope and the shuttle crew—was crucial in developing a smooth and flexible EVA/maintenance system to blend with existing and planned crew systems for the shuttle.

### Key design and operational issues

Several key issues emerged from an analysis of LST systems and planned crew-supported EVA maintenance modes and these prompted several operational and development design points that would require further study under potential EVA operations:

1. Safety to crew, shuttle, and space telescope (in that order).
2. Compatibility with Perkin-Elmer, Itek, and the NASA scientific instrument layout.
3. Telescope/shuttle docking and/or berthing operations and hardware.
4. Philosophy for manual extract/expand deployable devices.
5. Axial versus radial scientific instrument removal/replacement.
6. Major scientific instrument module size and configuration.
7. Spares: quantities, stowage, and volume (if in the payload bay).
8. Crew time during an EVA; the use of an integrated suit versus strap-on/umbilical environmental control and life support systems. Manipulator versus extravehicular crewmembers (or a combination of both capabilities).
9. On-orbit checkout modes and verification plans.

## Basic crew system requirements

Using extensive analysis, verification of mockup designs, simulation tests (from neutral buoyancy, 1-g simulations, and past EVAs) several fundamental crew requirements were established:

1. Operations by one EVA crewman.
2. Translation aids designed as part of the telescope.
3. Component/module changeouts designed for one gloved hand or hand-held tools.
4. Replaceable modules (ORUs).
5. All hardware and spaceframes designed for crew safety.
6. All access doors sized to permit module, scientific instrument, and component withdrawal/insertion in space [or on the ground].
7. All component/modules sizes to be manageable by one or two persons [specifically for ground maintenance].
8. Adequate internal volume for ground or extravehicular crew access for maneuverability with components, modules, and scientific instruments.
9. Least contaminable crew ground/EVA translation routes.
10. Shortest route to spares pallet and work platforms.
11. Minimum need for special crew support equipment in space.
12. Direct uncluttered escape route and rescue access.
13. Options for RMS use.
14. Maximum use of shuttle crew/system capabilities.

From over 4 years of human factors research, a preferred maintenance approach emerged in favor of unaided manned extravehicular operations for servicing the telescope. The primary maintenance mode for scientific instruments was by accessing the equipment section from the outside, radially, one bay at a time. The astronauts would use a combination of the RMS and mobile work platforms in order to access various locations, assisted by a network of handrails across the exterior of the telescope. The inclusion of access doors would permit easier access on-orbit and during ground maintenance.

An essential requirement for on-orbit telescope servicing was the ability of the shuttle to rendezvous, grapple (or dock with) the telescope, and secure it into the payload bay for the period of maintenance prior to either releasing it back into orbit or returning it to Earth. By 1975, extensive industry studies had demonstrated that the simplest and most cost-effective technology to achieve this was to soft-berth the telescope into the payload bay. Evaluations had assessed the positioning of the spacewalkers on various work platforms, the location of spares in the bay and the proximity of various work platforms to the telescope. Other driving factors were the operating envelope of the RMS and the ability of the other crewmembers to observe and support the progress of an EVA from the aft flight deck of the orbiter. As early as mid-1972, a program plan had been developed for a full scale high-fidelity mockup to be used for water-immersion in a neutral buoyancy man-in-the-loop simulation program. This mockup was constructed by Lockheed in cooperation with the Marshall Space Flight Center as part of a design and verification tool that would be used for each of the telescope's major engineering and system disciplines. The results from these studies supported the theory that EVA installation and removal of hardware located in the equipment section of the telescope would be relatively straightforward. This conclusion was due in no small part to the work of Frank Cepollina and his team at Goddard in developing the modular component concept.

**Summary**

Only a decade had elapsed since the first EVAs had taken place, and apart from the 15 lunar surface excursions and three EVAs in deep space conducted during Apollo there had only been 22 EVAs performed in Earth orbit. Together with the studies for generic EVA by shuttle crews, plans were being formulated to undertake multiple EVAs on a large payload involving not only bulky items but also intricate tasks for maintenance and repair. It was a daunting prospect to develop a credible maintenance program for the telescope, working in effect from a blank sheet of paper, whilst shuttle procedures were themselves under active development and hardware was being built that would influence the way in which on-orbit servicing of the telescope would be carried out.

By the mid-1970s, the studies that had been undertaken thus far had indicated that EVA maintenance of the telescope was not only feasible but was well within the capabilities of an astronaut crewmember. Along with 1-g simulations, the neutral buoyancy man-in-the-loop simulations provided clear evidence that this would be the most efficient and workable way forward. The success of Skylab crews in achieving "seemingly impossible" mission-saving tasks during EVA convinced the planners that such maintenance planning could and should be adopted for the telescope. What was even more encouraging was that the Skylab EVAs were accomplished in locations where there were few, if any, built-in EVA mobility aids or work platforms. This optimism reinforced plans and suggestions that payloads assigned for on-orbit servicing would be a time and cost saving investment if crews could maintain them. This would also benefit the end users of those payloads, because an extended life increases the return on the investment.

While the battle for authorization and funding was being fought in the corridors of power, the genesis of what would emerge as the field of satellite servicing was gaining momentum, with the space telescope as an exemplar. These skills were to have an important application for even larger space structures, including the long held desire to create a permanent space station in low Earth orbit. The commitment to that was still several years away, and no one could have guessed that its completion would take nearly three decades, but the execution of that enormous task would be helped in part by experiences gained from the Hubble service missions.

In addition to servicing the space telescope, by the early 1980s there were other satellites being considered for on-orbit maintenance by astronauts, including:

- Solar Max
- Upper Atmosphere Research Satellite
- Landsat 4
- Gamma Ray Observatory
- Advanced X-ray Astrophysical Facility
- Shuttle Infrared Telescope Facility

and a number of classified Air Force satellites.[6]

## ASTRONAUT OFFICE INVOLVEMENT

It was one thing to have the idea to perform servicing in space and develop the hardware to do it, but unless a robotic strategy was pursed, the act of servicing would require astronauts who were familiar with the procedures to complete those tasks. However, between the final Apollo mission in 1975 and the anticipated commencement of shuttle operations by the end of the decade, very few astronauts were still active. Two astronauts assigned to developing EVA procedures for the shuttle and the space telescope were Bruce McCandless and Story Musgrave.

McCandless had been working on the Manned Maneuvering Unit for some time and was in his office at JSC in August 1978 "when I received a direct phone call from George Abbey, which was odd because he seldom called me, and he advised me 'Those MSFC folks are up to something' regarding an EVA serviceable Large Space Telescope, and that I was 'it' from the Astronaut Office perspective, and that if I knew what was good for me I'd be up there, at Marshall, the next day to straighten things out. I never did get to find out what he meant by 'straighten things out'; put a stop to it, or help with the development. I chose the latter, and ultimately we prevailed. I had brief stints of support from Anna Fisher, Shannon Lucid and Kathy Sullivan, Dave Griggs and a few others. I did try, especially as the 1990 launch date approached, to get as many individuals as possible to acquire hands-on experience with the real telescope in Sunnyvale before it became isolated in orbit.

"I was very close to the Program Office at Marshall and two prime contractors Lockheed and Perkin Elmer. We had quarterly program reviews, which I thought were very valuable. Typically they brought together 200 people in the tenth floor penthouse of Building 4200 at Huntsville, back in the days before 'No Smoking', so the smoke level was usually five feet off the floor, but everybody chipped in and it worked well." McCandless recalled discussing some safety issues, such as ensuring the power was off before a connection was made or broken, or ensuring that an EVA astronaut wouldn't get jammed in the Aft Shroud, unable to get out. "[At] Marshall they had a large water tank and started out with crude mockups, and we were looking at issues like, could it take a US telephone booth size instrument, get it out and back in again, or what was the requirement for a retention mechanism with no straight load paths and very small clearances. It was more like looking at the size of the instruments to fit in the telescope and whether the telescope would fit in the orbiter payload bay. We got down to the controversy of one-tenth of an inch clearance on all sides [of an instrument] or a 50-thou clearance." McCandless was one of the ongoing advocates for on-orbit servicing in the Astronaut Office in those days. "When we finally launched it, I was proud of saying that the only 'non-serviceable' items were the main wire harness and the primary and secondary mirrors—and of course THEY would never need servicing!"[7]

Musgrave recalled being assigned to Hubble even earlier. "1975, I worked in parallel with other space flights and other work such as the simulation work on Spacelab, and they told me I would be looking after [the Large Space Telescope], identifying every possible problem and coming up with a spacewalking fix."[8] He was the point of contact in the Astronaut Office for EVA issues, having taken over the role from Rusty Schweickart in 1972. "Of course, in 1975 we didn't have a telescope, just the idea; we didn't even

Astronauts McCandless (*foreground*) and Lucid (*background*) preparing for an NBS simulation in 1980. (Courtesy Marshall Space Flight Center)

have Congressional approval. But it was decided the telescope should be serviceable by a spacewalker, and so the program had a requirement that it would be designed for servicing. So I started straight away, and used my imagination without drawings and without hardware. I started out at home—I had balloons, balloon telescope, balloon spacewalkers and various other things. And as soon as I started to get components, it just evolved from drawings into hardware into components as the structure of the telescope just moved along." Musgrave explained that it was not as simple as removing a component and sliding in the replacement. The items had to be certified safe for use. "They decided to meet serviceability specifications on some components, and not on others. What I was fighting for was for all components to be 'friendly' as opposed to 'certifiable'; there's a difference. There were some which were certified and they went very well, [but] there were others that were not even touched and they weren't friendly for the spacewalker. So that took a huge amount of doing with STS-61, to do things that had not been designed for servicing."

**Combined team work**

Robert Trevino of the EVA Branch at JSC recalled the early efforts from the different teams across NASA and the contractor prior to the launch of the telescope. "Lockheed at Sunnyvale had a few engineers that were very knowledgeable of EVA and were meeting NASA's EVA standards and requirements. Marshall had project management requirements, and they had some good engineers who had worked on Skylab. In the 1980s Lockheed had a preliminary design showing which components would be ORUs. They also had a preliminary idea where EVA handholds, handrails, and portable foot restraint (PFR) sockets should be located. I was in the EVA Operations Branch of the Mission Operations Directorate and was part of the NASA team that was tasked to plan neutral buoyancy testing to the Hubble mockup and verify the location, size, and number of all the EVA handrails, handholds, and PFR sockets needed to accomplish ORU changeouts. Luckily, we had a long lead time to do all the testing and add and reposition EVA translation and restraint aids as needed. The long lead time proved to be a godsend because once we had the main ORUs well under control, Goddard and Lockheed added a whole new set of new ORUs that were not part of the original plan."[9]

## EVA SUPPORT EQUIPMENT FOR SERVICING

Once the shuttle had rendezvoused with Hubble, the RMS arm would grasp it and securely stow it vertically on a platform in the aft end of the payload bay, whereupon the arm would withdraw. This arrangement allowed the EVA astronauts to carry out their maintenance and repair tasks on a stabilized payload, with the RMS yielding assistance. In fact, for service missions the orbiter carried a variety of Space Support Equipment (SSE). The telescope was held in place on a rotating maintenance platform called the Flight Support Structure (FSS). The Orbital Replacement Units (ORU) were on the aptly named Orbital Replacement Unit Carrier (ORUC) with additional storage positions available on Axial Carriers. In addition to the standard EVA tool kit and support equipment carried on every shuttle mission, servicing tools and hardware, some unique to Hubble, had been developed to support servicing tasks.

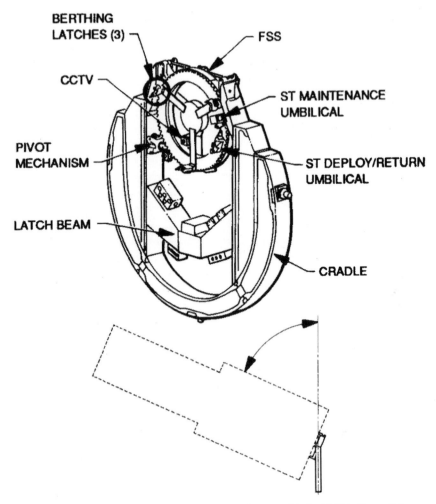

Diagram of the Flight Support Structure configured for use with the Hubble Space Telescope.

## Flight Support Structure

The Flight Support Structure (FSS) hardware, which held Hubble safely and firmly in the orbiter's payload bay, was not unique to the service missions. It had being developed in a number of configurations to accommodate large payloads in the shuttle program. Prior to the deployment of the telescope the FSS had been utilized on the STS-41C Solar Max Mission (SMM) repair in 1984. It consisted of two major components: the direct interface with the payload bay was by a supporting latch beam (which also provided structural and electrical connections with the orbiter) and a horseshoe shaped cradle. The telescope was placed into a circular ring device called the Berthing and Positioning System (BAPS) which stabilized the telescope in the bay. It could pivot (tilt) and rotate (turn) to ease the astronauts' workload by simplifying access to all sides of the telescope and enhancing the visibility of the non-EVA crew members on the aft flight deck monitoring the progress and safety of their colleagues.

For launch and landing the BAPS ring was locked in the pivoted down position, but for orbital operations it was pivoted 45 degrees to allow it to receive the berthing facility on the base of the telescope. A TV camera on the FSS provided a convenient point of view for the RMS operator lowering the telescope onto the ring. Three pins on the base of the telescope, resembling a towel rack, were grabbed by three remote-controlled latches on the ring. While attached to the ring, a remote-controlled electrical umbilical connector on the FSS engaged the telescope to supply electrical power to prevent the telescope from draining its batteries, and with the high gain antenna stowed a communications link provided telemetry data to the Space Telescope Operations and Control Center at Goddard and relayed uplinked commands. The ring could be tilted at the appropriate angle to enable Hubble to endure orbiter re-boost maneuvers and, at the end of servicing, to redeploy the telescope. During the first EVA on each service mission the astronauts would install the BAPS Support Post (BSP) that provided an additional linkage to both support and to isolate the telescope during EVA operations and subsequent re-boosts. By maintaining the BAPS immobile against high torques, the BSP also dampened vibrations. The BSP was retained in position after the telescope was released and was not re-stowed for landing.

All mechanical control was conducted from the aft flight deck, from where the crew was able to command the berthing latches, umbilical connection, pivot, BSP lock, rotator, and ring down-lock. As an extra safety assurance, each mechanism in the system featured full electrical redundancy incorporating manual overrides and backups.

**Orbital Replacement Unit Carrier**

The Orbital Replacement Unit Carrier (ORUC) used a modified Spacelab unpressurised pallet outfitted with both shelving and protective enclosures to accommodate a variety of ORUs, as required for the mission. All scientific instruments and replacement units were located in dedicated protective enclosures to prevent contamination and to maintain their temperatures within flight or design limits. To simplify the design and for ease of use, the instruments were mounted by the same manually driven latch mechanisms that held them inside the telescope. Delicate instrumentation was mounted on spring systems in order to reduce the vibrations of ascent and landing. There were two other sizes of ORU enclosures available, logically known as the Small ORU Protective Enclosure (SOPE) and the Large ORU Protective Enclosure (LOPE). The ORUs in these carriers were mounted in Transport Modules and surrounded by foam or visco-elastic material to help reduce launch and landing vibrations.

As equipment was exchanged during servicing, the extracted scientific instruments were temporarily stored on a bracket on the ORUC. The new instrument was then removed from its protective enclosure and installed on the telescope. The final action was for the astronaut to relocate the replaced item from the bracket into the appropriative protective enclosure for return to Earth.

Power for the thermal control subsystem on the ORUC came from the FSS, which also provided temperature telemetry throughout the unit for readout both aboard the flight deck and on the ground.

An Orbital Replacement Unit Carrier undergoing processing at KSC.

## Second Axial Carrier

Based on the Goddard Pallet Assembly, the Second Axial Carrier (SAC) for use on Hubble missions was capable of being adapted to carry servicing hardware such as solar arrays and (later) the NICMOS instrument. It received electrical power directly from the orbiter via the Enhanced Powered Distribution and Switching Unit (EPDSU), a modular avionics unit for shuttle payloads with unique requirements for power, commands and telemetry. The carrier spanned the cargo bay and was attached to the vehicle structure at one keel trunnion and four longerons. Active isolators (similar to shock absorbers found on automobiles) were provided to minimize vibrations at launch and landing.

## CREW AIDS

As the EVAs were to be conducted from the shuttle, most of the spacewalking apparatus and procedures were developed by that program, with some items being adapted for the Hubble missions.

### Spacewalking from the shuttle

The capability to perform spacewalks from the shuttle was integral to its design. The orbiter was, from the start, promoted as a vehicle from which space-suited astronauts could venture outside to undertake a variety of tasks, including the servicing and repair of satellites. These operations would be made by two of the crew in full view and supported by their colleagues inside. To achieve this, an airlock was developed to allow pressurized operations to continue whilst a spacewalk was underway. The capacity to perform EVA from the shuttle existed on every mission whether planned, unplanned, or in an unforeseen emergency situation. Not all missions called for an EVA, but at least two crewmembers were EVA trained on each crew, and support apparatus was available to complete at least a contingency spacewalk should the need arise.

The categories of EVA from the shuttle were:

- *Scheduled EVAs* were planned prior to launch and included in the nominal mission timeline.
- *Unscheduled EVAs* were not included in the nominal timeline, but these unplanned tasks could be undertaken in response to an event or failure during the mission in order to achieve operational success.
- *Contingency EVAs* were "emergency spacewalks" in response to a serious threat to the safety of the flight crew or the vehicle.

During the shuttle program, a range of EVAs were accomplished under the following categories:

- *Experiments and evaluation*: STS-6 made the first shuttle EVA to evaluate the new model of spacesuit called the Extravehicular Mobility Unit (EMU), STS-41B which evaluated the Manned Maneuvering Unit (MMU), and STS-64 which evaluated the Simplified Aid For EVA Rescue (SAFER) addition to the EMU backpack.
- *Mission success repairs:* STS-114 and the removal of the loose gap-fillers from the thermal protection system on the belly of the orbiter.
- *Spacecraft servicing:* STS-41C (Solar Max repair), STS-51I (Leasat repair), STS-49 (Intelsat repair), and of course the series of Hubble service missions.
- *Space structure construction:* STS-61B and the Experimental Assembly Structure in EVA (EASE) and Assembly Concept for Construction of Erectable Space Structures (ACCESS) experiments, plus STS-49 with the Assembly of Space Station by EVA Methods (ASEM) series of spacewalks for the development of large space structures including the potential of constructing a space station.
- *Contingency and unplanned:* STS-51D's unplanned "flyswatter spacewalk" which attempted to activate a failed Leasat satellite, and the unique STS-49 three-person EVA to retrieve an Intelsat satellite and fit it with a propulsive stage.

## BASELINE EQUIPMENT

For the Hubble EVAs, the astronauts had at their disposal a suite of apparatus that had been developed for the general shuttle program. In fact, the complexity of developing the service missions for Hubble was directly associated with development of the shuttle EVA system.

### Extravehicular Mobility Unit

The Extravehicular Mobility Unit (EMU) was a totally independent system which allowed the wearer to conduct a nominal EVA from the shuttle lasting up to 8 hours. It provided the necessary environmental protection, appropriate mobility for orbital operations, an adequate life support system (15 minutes for egress, 7 hours for useful work, 15 minutes for ingress, and 30 minutes for reserve) and communications to fellow EVA astronauts, the crew aboard the orbiter, and Mission Control. The complete EMU was an integrated system consisting of two subsystems: the Space Suit Assembly (SSA) and the Life Support System (LSS).[10] The suit itself consisted of: the Hard Upper Torso (HUT) with its integrated sleeves; a pair of outer EVA gloves; the Lower Torso Assembly (LTA) with the legs; the Helmet and EVA Visor Assembly (EVVA) with its integrated lamps and cameras; the Liquid Cooling and Ventilation Garment (LCVG); the Operational Bioinstrumentations System (Biomed), the 'Snoopy' Communications Carrier Assembly (CCA Comp Cap); a Disposable In-Suit Drink Bag (DIDB); the Maximum Absorption Garment (MAGS); and the Waste Management System.

The system was developed by Hamilton Standard and ILC Industries of Dover, Delaware during 1974–1977, then the contract was awarded to Hamilton Standard of Windsor Locks, Connecticut. Its first flight was on STS-1 in 1981 but in the emergency mode and it was not used. An on-orbit test of wearing the suit, but without performing a spacewalk was made on STS-4. An EVA attempt was aborted on STS-5 owing to problems prior to venturing out, so the first EVA occurred on STS-6 in 1983. The original configuration of the suit was phased out during 1998–2002 and replaced by an improved version to support ISS operations. This was used exclusively on shuttle flights from 2002 through to the end of the program, and it continues to be used on the station.

The Life Support System included a Primary Life Support Subsystem (PLSS) to supply oxygen, electrical power, communications, and cooling to the user. The PLSS also included the controls for maintaining suit pressure and the internal thermal environment, and it could circulate oxygen as well as eliminating nitrogen, humidity and trace contaminants from the system. The control of contaminants was by means of a Contamination Control Cartridge, a lithium hydroxide (LiOH) cartridge (or METOX cartridge) which removed carbon dioxide and trace contaminants. Though never used during a mission, in an emergency a Secondary Oxygen Pack (SOP) would be automatically activated during an EVA to provide a maximum of 30 minutes of oxygen in an open-loop purge mode. It also had a Space-to-Space EMU Radio (SSER), and a chest-mounted Display and Control Module (DCM) to provide the astronaut with caution and warning messages, EMU parameters, and EMU controls. During all EVA operations, the Real Time Data System (RTDS) would transmit EMU status parameters and biomedical data to the EMU flight controller console in Houston.

Since the spacewalkers carrying out the telescope servicing would remain tethered to the shuttle at all times, they didn't wear the Simplified Aid For EVA Rescue designed to return an untethered astronaut to the safety of the spacecraft in the event of becoming detached and adrift in space. As the Hubble EVAs were carried out in close proximity to the telescope, to have added more equipment to the already bulky EVA suit would have hindered access and mobility. With one astronaut on the RMS and the other tethered to the shuttle the chances of an astronaut coming adrift were considered low, and in any case the shuttle with Hubble on board would in principle (albeit with difficulty) have been able to maneuver to effect a recovery. The SAFER is used by astronauts making spacewalks from the ISS, which has no means of giving chase, and the task of the unit is solely to allow an astronaut to close in and regain a grip of the structure to restore a tether. It is not a substitute for the MMU "flying backpack".

## Airlock

The provision of an integral airlock in the orbiter afforded space-suited astronauts the ability to move between the crew compartment and the payload bay without having to depressurize the cabin. The airlock could be positioned either in the aft middeck or in the forward payload bay dependent upon the requirements of the mission. For the first three service missions it was located in the middeck area, but on STS-109 the airlock was installed outside to provide more room on the middeck. The airlock supplied the EMU with oxygen, power, and cooling during EVA preparations and post-EVA operations. It would also replenish the consumables (oxygen, battery, and feed water) in between EVAs. The facility could easily accommodate two space-suited crewmen and an unsuited crewmember to assist in preparations and post-EVA activities. In one case, the airlock enabled three astronauts (during STS-49) to conduct a simultaneous spacewalk.

## EVA equipment

Shuttle crewmembers were afforded a range of specific equipment for special tasks and the following generic items available for each EVA:

- *Tethers*—There was a 55 foot safety tether, two waist tethers, and wrist tethers. The protocols were "all equipment and crewmembers tethered at all times" and "always make a connection before breaking a connection".
- *Mini Workstation* (MWS)—A small framework was attached to the front of the EMU to carry small tools. An MWS end effector and retractable tether provided restraint to the crewmember at the worksite.
- *Body Restraint Tether* (BRT)—This was attached to the MWS, and also had an end effector to act as a semi-rigid restraint for the crewmember at the chosen worksite. This was useful in that it took less time to set up than a Portable Foot Restraint and was found to be more stable than the end effector of the MWS. It was also used for the translation of small objects during an EVA.
- *Foot restraints*—These were attached to a structure via a socket, and provided an EVA crewmember with a rigid restraint at the chosen worksite. Only the shuttle's Portable Foot Restraint was available for servicing Hubble.
- *Remote Manipulator System* (RMS)—The RMS provided crucial support during each of the Hubble servicing EVAs. The RMS operator on the aft flight deck coordinated their actions with the EVA crew, other crew members, and controllers on the ground, in relocating a crewmember on a foot restraint at the end of the arm or tethered to the arm in order to remove, retrieve, or replace items of hardware. The arm could also be used to give astronauts access to the upper part of the telescope, thereby saving their precious EVA time and physical energy.

## Tools of the trade

The unique nature of the Hubble service missions meant that tools and items of equipment had to be invented, designed, fabricated, tested, and approved for specific missions.

## EVA PROFILE

Each shuttle EVA was divided into three phases: the preparatory tasks, the actual spacewalk, and the subsequent tasks.

*Pre-EVA activities*: The work to prepare for an EVA began with checkout of the EMUs and the airlock, and had to be performed during the first and second flight days prior to the first EVA. On subsequent days the amount of pre-EVA preparation was able to be reduced by the efficiency and experience of the crewmembers as the spacewalking program progressed. All suits and small items of equipment were checked and prepared prior to each EVA. Lowering the cabin atmosphere early in the flight assisted in purging nitrogen to prevent the "bends"; it would be restored to normal pressure after the final EVA of the mission.

*EVA operations*: It took about 30 minutes to fully depressurize the airlock and a further 15 minutes was allotted to allow the astronauts to transfer into the payload bay. Once they were out, the astronauts would follow the prepared EVA timeline supported by colleagues inside the shuttle and controllers in Houston. In addition, support staff at Houston and Goddard as well as the Space Telescope Science Institute and contractors in rooms supporting the main control rooms and back at the contractor main locations, would be on standby during each spacewalk. At the end of the EVA, the entry into the crew compartment was reversed, with 15 minutes allocated for ingress to the airlock and 30 minutes for repressurization. It was a requirement for all shuttle extravehicular crewmembers to be trained to perform a variety of contingency tasks as well as for scheduled and unscheduled tasks. Contingency EVA tasks for all shuttle flights included:

- Failed airlock hatch latches or actuator tasks
- Failed Remote Manipulator System tasks
- Manual stowage of the payload bay door radiators or the Ku-band antenna
- Manual closure of the payload bay doors
- Installation of payload bay door latch tools.

*Post-EVA operations*: Following the completion of entry to the airlock and repressurization, the EMUs were doffed and the astronauts re-entered the crew compartment. A program of EMU maintenance included recharging the oxygen tank and the batteries, refilling the water tank, regenerating the METOX, replacing the LiOH, cleaning the suit, and (if appropriate) resizing of suit components for the next user.

## GROUND SUPPORT

Three directorates at the Johnson Space Center (JSC) in Houston were responsible for EVA support and management. The EVA office (mail code XA) was responsible for long range planning of EVA activities; purchasing flight hardware; and a general overview of shuttle EVA operations and planning the Hubble EVAs in that broader program. The Engineering Office (EC5) was responsible for the development and testing of EVA hardware for flight. The Mission Operations Directorate (DX32 and DX35) was responsible for the development of operations, the training of the flight crews and flight control teams, and real-time mission operations.

## EVA TRAINING

There was a range of training equipment available to support Hubble EVA operations from the shuttle.

Jeff Hoffman training for STS-61 on the air-bearing floor at JSC.

At the Johnson Space Center:

- The full-size shuttle mockup trainer included the crew compartment trainer and the full fuselage trainer
- Precision Air-Bearing Floor (PABF)
- EVA communications trainer
- Virtual Reality Laboratory
- Vacuum chambers, including the 11 foot Environmental Test Article (ETA) Shuttle Airlock Vacuum Chamber in Building 7 that was part of Crew and Thermal Systems Division
- The Weightless Environment Training Facility (WETF) was 24 meters in length, 9.8 meters wide, 7.6 meters deep, and contained 1.82 million liters of water[11]
- The Neutral Buoyancy Laboratory (NBL) of the Sonny Carter Training Facility at Ellington Air Force Base was 61.56 meters long, 31.06 meters wide, 12.18 meters deep (6.09 meters above ground level and 6.09 meters below ground) and contained 28.18 million liters of water[12]
- EMU Caution and Warning System (ECWS) trainer

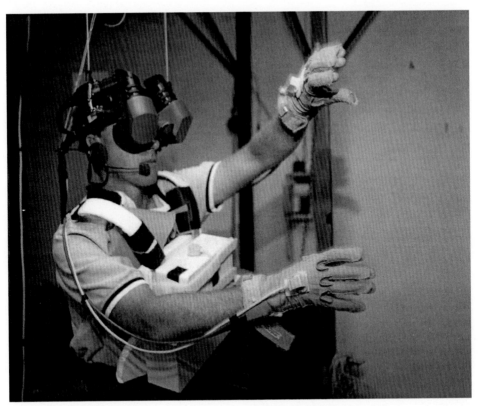

Mike Foale training for STS-103, using early Virtual Reality devices.

- KC-135 parabolic training aircraft
- Hubble 1-g mockups
- Hubble WETF mockups.

At the Marshall Space Flight Center:

- Neutral Buoyancy Simulator (see below)
- HST Deployment Crew Trainer
- HST Maintenance Trainer
- Hubble 1-g mockups.

At the Goddard Space Flight Center:

- The High Fidelity Mechanical Simulator (HFMS) was a mockup of the HST Focal Plane Assembly bays for the changeout of axial and radial scientific instruments
- The Vehicle Electrical Systems Test (VEST) was a mockup of the electrical support system equipment section which also contained a number of smaller ORUs such as computers, electronics boxes, and tape recorders
- The Exterior Simulator Facility (ESF) mockup replicated the forward shell of the telescope for training on exterior components and the application of insulation

210 **Simulating servicing**

The WETF facility, Building 29, JSC, Houston, during STS-82 EVA training.

EVA training 211

- The Aft Shroud Door Trainer (ASDT) mockup allowed the astronaut to train on the door latches, door opening and closing techniques, as well as various contingency operations
- The Power Control Unit Trainer (PCUT) was a high-fidelity trainer built specially for the fourth service mission, designated SM-3B, that replicated the very tight working space available when replacing the 36 electrical connectors during of the PCU.

HST 1-g mockups including the "balloon" mockup (*right*) used in conjunction with RMS training simulator at JSC. (An enlargement of a low resolution scan issued by NASA)

HST 1-g RMS trainer, Building 9 JSC.

Astronaut simulator on the end of RMS trainer, Building 9 JSC.

## EVA potential

During the development of the LST, and with ground maintenance the preferred option, the astronaut involvement in maintenance was expected and planned to be limited. As a result it was decided to mount components in modular units or trays to simplify servicing either on the ground, or in space by RMS or by astronauts. By 1973 experience from Gemini, Apollo and Skylab gave a deeper appreciation of the likely complexity of a Hubble service mission. Further evaluations led to the decision to abandon robotic servicing (which was still in the early stages of development) and by 1985 servicing on the ground had been deemed too expensive. Thus maintenance of the telescope became the sole responsibility of the astronauts flying on the service missions.

Consideration had to be made when mounting equipment on the telescope to ensure that there would be sufficient room between components to permit access to astronauts. Though the pressurized glove would afford some dexterity, it would still be difficult to manipulate a small item readily and efficiently. The downside of this requirement, was that adding room for movement also added to the workspace volume, which was restricted by the dimensions of the telescope, so each servicing task involved making a certain trade off. Working out the sequence of servicing tasks in advance and practicing them would be like choreographing a complicated dance routine on stage, but with the added benefits and issues of doing so in a weightless environment and inside a pressure suit. Ensuring that the fixings, tools, restraints, and support equipment were designed to help ease the astronauts' work, and not to add to it, was important. The timeline, choreography, and procedures were refined by trial-and-error during high-fidelity rehearsals. Factors to be taken into consideration included the choice of tools, lighting conditions, whether it was a one or two person task, was the RMS required in that task, and what were the consequences of an action taken which may hinder another task farther down the EVA schedule. Another thing to be taken into consideration was time. Each shuttle mission had a limited time available in space. This had to accommodate the planned EVAs, with each operation also limited by the design of the hardware and the capabilities of the astronauts. The lessons from Gemini through Skylab showed that a straightforward task on Earth could take far longer to complete in space than it took in training.

Since the mid-1960s the use of huge water tanks had been evaluated and exploited in the training of astronauts for spacewalk activities. Specialist pools were constructed at JSC and MSFC to support this work. They became crucial in developing the EVA techniques for the Hubble service missions. Many of the lessons learned were of benefit in planning the EVA program for the ISS. Foremost in the use of underwater methods to simulate EVA activities was the Neutral Buoyancy Simulator (NBS) Facility of the Marshall Space Flight Center.[*]

## TAKING HUBBLE UNDERWATER

In 1979 the Essex Corporation began its involvement in the development of a mockup of the Space Telescope and a series of underwater simulations to evaluate on-orbit servicing tasks, including the changeout of ORUs. Originally, the Hubble ORUs would feature components and subsystems that were designated as mission critical and designed for early replacement. Later, other items in the Optical Telescope Assembly Equipment Section (OTA-ES) and in the Support System Module Equipment Section (SSM-ES) were also considered for orbital replacement because they employed EVA-compatible electrical connectors and mechanical fasteners in their design and were accessible to EVA astronauts.

During meetings between Essex personnel and MSFC in the autumn of 1984, preliminary initial test objectives and mockup design requirements were established, as were a number of ground rules for conducting the tests:

- The primary objectives were to evaluate crew access to potential ORUs and their electrical connectors and mechanical fasteners.
- ORUs would be mounted to a bolt plate which in turn was mounted to the structure of the appropriate equipment section. This would allow for the removal of a specific ORU for modification, repair, or replacement without requiring to remove the entire equipment section in which it was housed.
- Long lead time components that were unavailable for the tests would be substituted with "like items".
- Electrical connectors would be without their pins.
- Lockheed would supply drawings and photographs of the OTA-ES and SSM-ES to assist in the development of the mockup unit.

In December 1984, the contract was awarded to Essex for the design and fabrication of 14 non-ORU mockups, for the purpose of evaluating changeout tasks in the Neutral

---

[*]The Marshall tank was the Neutral Buoyancy Simulator (NBS) Facility. The tank that was later built at the Sonny Carter Training Facility is known as the Neutral Buoyancy Laboratory (NBL).

Buoyancy Simulator (NBS) at Marshall. The contract was later modified to include three additional non-ORU mockups. The design, fabrication, and testing of these units were only meant for evaluating potential ORUs, but they had realistic interfaces and were suitable for being worked on underwater. Wherever possible the mockup hardware would be made as close as possible to the flight versions, with "flight like" hardware connections with cables that closely simulated those intended to be used from the telescope to the relevant piece of hardware. Other items of hardware that were not previously designated as ORUs were still evaluated for changeout studies in the MSFC NBS. Called "high-fidelity mockups", these were developed by Essex and were identical to those intended to be used on the telescope. The mockups were installed in positions as close as possible to the real case, using similar fitting and cable connections.[13]

The simulated telescope components for use underwater included the following:

- There was one Data Management Unit (DMU) attached to the door of Bay 1 by a total of 22 fasteners. It included 38 connectors with accompanying cabling.
- Two Multiple Access Transponders (MAT) were on the door to Bay 5. Each used a total of 18 fasteners and included 15 cables and interface connectors.
- Two Solar Array Drive Electronics (SADE) were located on the door of Bay 7 and used a total of 12 fasteners, 10 connectors and cables (per unit).
- There were three Tape Recorders. Two were mounted in the back of Bay 5, each of which was held in place by four fasteners. Each unit featured three connectors and associated cabling. A third tape recorder was fabricated for changeout purposes on the mockup to reproduce accuracy with the flight vehicle.
- Two Data Interface Units (DIU) were installed in the back of Bay 7, each of which had 6 fasteners, 20 connectors and cables.
- Power Distribution Units (PDU). Two units were built to a high-fidelity standard and another pair at a low fidelity. Each was held in place by 10 fasteners, and included 14 connectors and cables.
- One Deployment Control Electronics (DCE) unit was located on the door of Bay 7. It was attached by 6 fasteners, and included 14 connectors and cables.
- One Electrical Power/Thermal Conditioning Electronics (EP/TCE) was located at the rear of Bay H in the OTA Equipment Section. It included 4 fastening screws and 28 connectors/cables.
- One Optical Control Electronics (OCE) was on the door of Bay C and was attached by 4 bolts with 13 connectors and cables.
- One Single Access Transmitter (SAT) was located on the door of Bay 5 and used a total of 5 connectors and 8 bolts.
- One Mechanism Control Unit (MCU) was located on the inside of Bay 7 above the DIU and had 4 bolts and 6 connectors/cables.

From request to evaluation, a four step process was created for each component:

- In Task 1, members of the Essex staff analyzed Space Telescope drawings supplied by NASA of the non-ORU items and their mountings. Wherever there was a crew interface—connectors, spacecraft mountings, module attach points—a high-fidelity mockup was made. Wherever possible, electrical connectors were ordered (without pins), and if the desired connector couldn't be sourced it was substituted by a close approximation.

- Task 2 involved making mockup drawings from the information gained by Task 1. Engineering drawings were prepared and, in several cases flight drawings redlined for particular attention to crew interfaces. Each of these drawings was reviewed for approval by a NASA representative.
- Task 3 involved the fabrication of hardware, but only after the completed item had been checked and approval by a project manager in order to verify the dimensions and configuration. The non-ORUs were manufactured with 6061-T6 aluminum and stainless steel fasteners. Their crew-operated mechanical fasteners were identical to the flight configuration and fabricated from 300-series corrosion resistant steel. The electrical connections were made with the same characteristics as the flight units, in that the cables were similar in terms of the number of wires and the size of the wire bundles.
- Task 4 included final delivery and installation of the hardware, and a test program. For delivery, Essex installed the non-ORUs in a full-size mockup of the equipment section. Company engineers escorted the mockup to Marshall, and company divers assisted NBS personnel in placing it in the water tank and thereafter supported the test activities.

**Attention to detail**

On the whole the test program went well, but in its final report Essex highlighted a number of problems that had not been foreseen at the beginning.

In November 1984 Essex had requested drawings and photos from Lockheed for each non-ORU because such drawing were not available at MSFC. But when these were received they often lacked key dimensions, and some drawings were different from the photographs. There were two further attempts made to request more detailed drawings. When the final drawings were compared with photos, it was found that the Lockheed and supplier drawings were not always compatible in all areas.

For example, in the case of the MAT, the dimensional difference between Lockheed and Motorola was no more than 0.2 inch (5.08 mm) but this was still sufficient to cause fastener problems and tool/fastener misalignment. The recommendation was that should a component be fabricated from "as built" drawing files and installed on the flight vehicle as a candidate for maintenance or repair then "care should be taken to acquire not only Lockheed drawings but those of subcontractor and vendor as well". Interestingly, there were no discrepancies in the report relating to the Perkin-Elmer drawings.

After the contract was underway, an evaluation of connector tools was added to the work schedule. This involved obtaining high-fidelity connectors and the associated cable bundles. Fortunately, Essex reported that these late additions did not impact overall cost, lead time or information. Nevertheless, supplier delivery had delayed the arrival of some connectors for evaluation.

Owing to questions of accuracy of the information provided, and potential modifications applied to hardware after the drawing were completed, Essex did not express confidence in the mockups of the MAT, SAT and DCE. It also expressed concern about the fidelity of co-axial connectors on the DMU. In all cases the mockups were produced to meet the NBS test schedules and relied on the drawings that were available at the time.

This issue demonstrates, in the era before the introduction of Computer Aided Design (CAD), the frustrations arising from having different suppliers prepare drawings for the same item, with small variations in dimensions. These incidents underlined the importance of ensuring that drawings made by different sources were compatible, and that a "master" set was defined as the source for the eventual production of hardware.

## PROGRESSIVE SYSTEMS ANALYSIS

The underwater simulations conducted at MSFC continued to investigate ST mechanisms, and how compatible and functional they were for crew interfaces with the goal of defining the type and location of crew aids to be incorporated into the telescope to support servicing such mechanisms.

The NBS testing commenced with low fidelity mockups that were progressively evolved to higher fidelity over a 24 month period from August 1979 to August 1981. The later NBS testing and training benefited significantly from these early demonstrations and findings in identifying the need for crew tools through a series of acceptance or rejection tests. What is more, entire subsystems originally believed to be replaceable on-orbit during scheduled or contingency maintenance were tested for whether they were indeed potential ORUs.[14]

### Significant results

The results of this testing program proved enlightening to areas not specifically associated with the Space Telescope. For example, one of the foot restraint devices evaluated that met requirements for flexibility, low weight, and ease of operations, resulted in a device which could be easily adapted for a wide variety of EVA operations.

The varied and complex maintenance program for Hubble revealed the need for several versatile tools, and the tests conducted at the NBS yielded a unique manual ratchet wrench, but also identified the need for power tools to relieve crew fatigue and assist in one-handed tasks. From successes and failures revealed by the tests and simulations in the water tank, a valuable starting point was identified for the development of new tools and technologies for EVA activities far beyond those envisaged for servicing the telescope.

Over the 24 month period, the most significant results from the water tank evaluations included:

- Extended use of hand tools to accomplish tasks over a period of time resulted in hand and finger fatigue; the development of a power tool would eliminate these problems.
- Any foot restraint must be adjustable in elevation, pitch, roll, and yaw.
- Special tools were going to be developed for specific functions at the telescope, but it was suggested that these should incorporate features that would make them useful for other future EVA tasks.
- The transfer devices used to relocate ORUs from the carrier in the payload bay to the telescope should be provided with adequate restraints during transfer operations.

- Devices with threaded fasteners that would be operated during an EVA should have sufficient friction built into the fastener to permit wrenches to perform a ratcheting action without suffering back-turning.
- When a scientific instrument alignment mechanism was fully seated and engaged, a feedback device was needed in order to indicate this alignment because the capture mechanisms were often hidden from the direct view of the operator.
- Guide rails must be provided to slide massive/bulky ORUs in or out of their flight environment. These rails must provide restraint without either binding or imposing high dynamic loads on the rails.
- The RMS was a useful device for jettisoning unwanted items.
- A Portable Foot Restraint (PFR) would provide an extremely versatile and portable workstation from which a crewmember could carry out assigned tasks. A simple 12-point socket would enable it to be mounted in any location on the telescope (if that was suitably equipped).

**The simulation program**

By using the NBS in "real-time conditions", as EVA planning developed it was possible to incorporate full-size mockups to verify hardware concepts and gain experience in planning actual tasks to be performed on-orbit.

The opportunity to review crew aids also helped in evaluating planning for contingency operations. If all went well on a mission, the probability of undertaking a planned task was 100 percent. A major effort of the simulations was therefore to determine which crew aids could be used, and to select the best locations to place them for any specific task. This was particularly important for the ORUs of telephone booth size. The precise type, number and location of crew aids could be estimated in advance of the simulation and then evaluated in the tank, thereby refining the sequencing and timing of tasks. On the other hand, where an unplanned event occurred, the remedial action would only be undertaken after a real-time estimate of the best way to employ the available crew aids was demonstrated in the water tank. However, a wide variety of these events could be identified and a range of solutions devised in advance. In a contingency situation, the failure could occur in a location which was not readily accessible to the crew and thus was not laden with fittings for crew aids. In these cases, planning was more of an analytical evaluation in which the simplest and most unhampered approach was best. In these cases, a low probability of such an event actually occurring didn't warrant the expense of creating a mockup and then supporting a full scale NBS simulation in real time.

The dives in the tank were also used to determine whether a relatively simple low fidelity mockup used in NBS simulations could be progressively upgraded and refined to a level of accuracy that reflected flight hardware. If this was the case, then perhaps the mockup could be used as a crew trainer, then revised as plans changed and improvements were introduced to the systems and hardware. Hence the same hardware used to evaluate early systems and procedures in the tank became the basis for a higher fidelity mockup for crew training. This made the most of the training hardware dollar in a limited budget. It was also beneficial to crew training, because making the mockups as close to flight hardware as possible ensured that the hardware and tasks performed in the tank closely matched operations on-orbit. The more familiar a crew became with the most accurate hardware and procedures, then the less likely they were to be surprised in space. This was the rationale for Essex seeking drawings that were sufficiently accurate to reproduce "perfect" replicas for tests and training.

## THE NBS TEST FACILITY

The Neutral Buoyancy Simulator (NBS) at the southwest corner of NASA's Marshall Space Flight Center in Huntsville, Alabama, was a complex of metal structures which collectively became known as Building 4705. It featured a large steel water tank and ancillary apparatus as well as an adjacent control room and a range of support facilities. The tank was on the site of a shelter designed to accommodate mockups of the Saturn IB launch vehicle. Large doors and removable roof panels provided easy access for large items of hardware. Its construction was completed in 1968 and the resulting tank, 75 feet (22.86 meters) in diameter and 40 feet (12.19 meters) deep, was large enough to submerge a full-size mockup of the space telescope in its 1.4 million gallons (5.2 million liters) of crystal clear water. Support systems allowed the use of both air and nitrox gas in the simulations. Adjacent to the tank were support and control rooms to supervise the tests. Each simulation was recorded on video tape with audio voice communication for later analysis, and also for evaluating the timeline, the events, and any corrective actions.

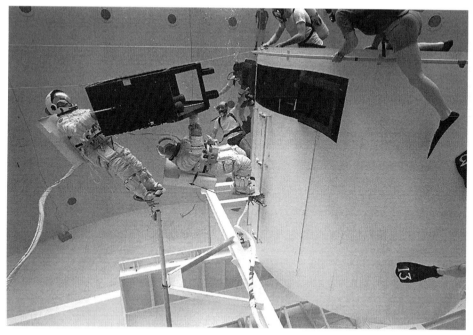

Neutral Buoyancy Simulator Complex, Building 4705, Marshall Space Flight Center. (Courtesy Marshall Space Flight Center)

# The NBS test facility

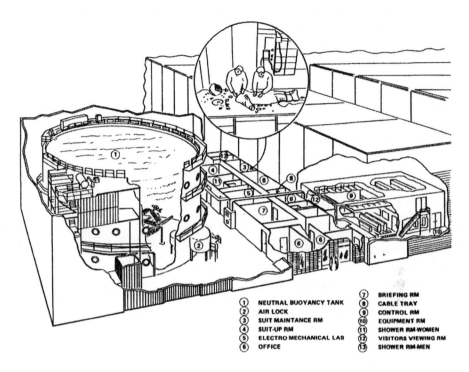

| | | | |
|---|---|---|---|
| ① NEUTRAL BUOYANCY TANK | | ⑦ | BRIEFING RM |
| ② AIR LOCK | | ⑧ | CABLE TRAY |
| ③ SUIT MAINTANCE RM | | ⑨ | CONTROL RM |
| ④ SUIT-UP RM | | ⑩ | EQUIPMENT RM |
| ⑤ ELECTRO MECHANICAL LAB | | ⑪ | SHOWER RM-WOMEN |
| ⑥ OFFICE | | ⑫ | VISITORS VIEWING RM |
| | | ⑬ | SHOWER RM-MEN |

A cutaway diagram of the NBS building at MSFC, Huntsville, Alabama. (Courtesy Marshall Space Flight Center)

The test subjects, including MSFC engineers and astronauts, were supported by a team of scuba divers and support engineers. Subjects involved in early space telescope development wore an Apollo era A7LB spacesuit to perform the simulated weightless environment tasks. This suit, previously used on the Apollo 'J' series, Skylab, and ASTP missions, was slightly bulkier and less flexible than the EMU intended for the shuttle. It was reasoned that if a task could be performed underwater wearing a restrictive suit, it would be possible for a person wearing a suit with greater flexibility (the shuttle EMU) to accomplish the same task on-orbit.

When Ed White made the first American EVA in June 1965, just weeks after cosmonaut Alexei Leonov conducted the world's first spacewalk, it became evident that a human could perform an EVA and survive. However, several questions remained on the effectiveness of working outside the spacecraft, such as handling large objects possessing considerable mass and inertia but no apparent weight. Even from these simple excursions outside, it was found that imparting just a small force on an object would cause it to move in a steady motion, but stopping it proved difficult. Would this prevent astronauts working safely with large objects or prevent them from carrying out meaningful tasks with small objects?

Clearly something had to be devised to enable simulations to explore the practicality of working in space, but without the added dangers of actually testing new ideas out in space without adequate preparation. As it was almost impossible to reproduce weightlessness on Earth for any length of time, it was found that by adopting diving principles and carefully weighting a spacesuit, it was possible for test subjects to evaluate techniques for achieving mobility and stability intended of spacewalking. Using neutral buoyancy, test subjects and astronauts accompanied by safety divers have since 1965 demonstrated the ability to master the techniques of spacewalking here on Earth by using water tanks such as Marshall's.

In the late 1960s, astronaut training was synonymous with the Manned Spacecraft Center (later Johnson Space Center) in Houston; an association that was guarded passionately. They expected the neutral buoyancy training facility would be built there but Congress had started to reduce NASA's budget and the management at Marshall, realizing that the next program, using Apollo systems for a space station, would benefit from a tank to assist in preparing for long duration missions, set out to get that facility for Huntsville. Although they were unable to officially build a new 'facility', there was funding available for 'tools'; therefore the NBS became a new 'tool' for the preparation of what became Skylab. By the time the accountants recognized this subterfuge it was too late and the NBS was ready for use. After Marshall was awarded responsibility for the Skylab program, astronauts were soon using this tank because it was the best facility that the space agency had.[15]

From the late 1970s, the NBS tank was being used to conduct the earliest simulations of space telescope servicing. In September 1980, as the first shuttle flight loomed, a simulated RMS was added to the NBS facilities to join a mockup of the shuttle payload bay. Salvaged from pieces of earlier manipulators that were used during tests in the late 1960s, this 'RMS' was operated from a control panel located outside of the tank. As rudimentary as this was, it was a fairly accurate simulation of an RMS operator on the shuttle flight deck. It could not simulate the full movement of the real arm, since it was difficult to keep neutrally buoyant, but it proved the concept. In 1993, with preparation for the first Hubble service mission in full swing, RMS-II was inaugurated at the NBS and operated from a duplicate of the shuttle aft flight deck at a control station on Level 3 of the tank. Later, a video camera was added to the arm. A system to record an entire simulation precisely as it was intended to be done on-orbit allowed engineers, trainers, and astronauts to analyze the procedures after the test. As astronauts were already too busy in their various simulations to handle the filming, and safety divers were there for that reason (safety), the camera added to RMS-II relieved the astronaut of this task and demonstrated that using cameras on the arm would be a useful task in space to evaluate the exterior conditions of the spacecraft which, following the Columbia accident, proved to be the case.

The use of the tank proved invaluable for Hubble service missions, and also supported the Solar Max repair and the evaluation of early EVA techniques for a space station. In 1980 JSC was finally able to acquire funding to convert the former Flight Acceleration Building (#29) which had hosted the centrifuge, into the Weightless Environment Training Facility (WETF) for training shuttle astronauts. However, the Marshall tank could handle larger mockups and so simulations for the space telescope continued there until the late 1990s. In January 1997 a new, larger Neutral Buoyancy Laboratory at the Sonny Carter Training Facility near JSC was completed to take on the increasing role of training ISS EVA astronauts and the final Hubble servicing crews. As a result, the tank at Marshall that had been awarded the status of a National Historic Landmark on October 3, 1985, was closed and drained.

Neutral Buoyancy Laboratory, Sonny Carter Training Facility, Houston.

## Mockups and support equipment

The initial telescope test (designated the 23rd neutral buoyancy test, NB-23) in the NBS in August 1979 utilized a full-size mockup of the Support System Module (SSM) Aft Shroud positioned in such a way as to resemble its relationship in the payload bay. In these tests the equipment included:

- Focal Plane Assembly structure
- Fixed Head Star Trackers (FHST) with separate light shields
- Rate Senor Units (RSU)
- Equipment Shelf
- SSM Aft Shield
- Orbiter payload bay
- Support Structure (mockup peculiar)
- Adjustable Foot Restraints

## 222 Simulating servicing

Space Telescope Mockup in the NBS at MSFC, circa 1979.

- Crew Aids (handrails and tethers)
- Tools (ratchets with various extensions and sockets)
- Wide Field/Planetary Camera (WFPC)
- Fine Guidance Senor (FGS)
- Clothesline (ORU Transfer Device).

*NB-23, August 13–17, 1979: This test included simulations on a mockup of a modular section of the HST in removal/replacement of scientific instruments.*

Three months later, in November 1979 the tests continued on a modified RSU/FHST and the axial science instruments were added to the site. The pneumatic (power assisted) ratchet wrench was also introduced. The axial science instruments featured four unique mechanisms associated with their removal and replacement: electrical connections, instrument guide rails with Focal Plane Assembly and Aft Shroud extensions, a pre-load subsystem, and a latching system.

During that same month, the final mockups were delivered for solar array tests and these were attached to the mockup; it was a tight fit because when they were fully erected the top of the mockup was just 5 feet below the water level. These new additions included:

- SSM Equipment Section
- Forward Shell
- Light Shield (not the full length, as that would have protruded from the water)
- Solar Array
- Jettison clamp
- Aft and Forward latches

*NB-34, November 1, 1979: In this test, astronauts Bruce McCandless and George "Pinky" Nelson trained on a mockup of a modular section of the HST in removal/replacement of scientific instruments.*
*NB-38, May 6–8, 1980: Six months later Bruce McCandless and Shannon Lucid trained on the axial science instrument changeout procedure. Some sources say that both Anna Fisher and Joe Kerwin participated in these tests.*[16] *In 2013 Kerwin confirmed he did not personally complete any Space Telescope simulations in the tank. The images used were probably publicity shots and incorrectly captioned as several telescope dives were being conducted at the time and Anna Fisher did conduct dives in the NBS.*

George "Pinky" Nelson recalled his experiences in working in the tank at Marshall during these tests. "I did quite a bit of work on the development of the HST mission hardware along with McCandless. Henry Waters was the prime person at Marshall that we worked with. We did simulations to help develop the mechanisms for changing out the axial instruments, star trackers, WFPC, fine guidance sensors, gyros and other components. This included choosing locations for handholds and foot restraint receptacles. Marshall had a pretty good high-fidelity mockup that was later transferred to JSC, I believe."[17]

Nelson began working on the Space Telescope around 1979–1980, on the positioning of foot restraints and handholds and how easily their positions could be changed. The tie down bolts for instruments had already been designed and they were very specialized, held within millimeters, so the work focused on how well aligned they were. As Nelson recalls, at that time the open issues included how long each instrument was and how far it slid in or out of the compartment on extending rails.

During sessions in the water tank at Marshall, the astronauts practiced planned activities and then evaluated handholds and facilities to see what worked or what didn't work and if they could be changed. They investigated each hardware sequence as a process to determine "how you get this giant box out of that big hole, or into to it, while wearing the spacesuit, or on to the arm, or where you wanted it". The sequence began with a series of 1-g bench tests before progressing to the tank. "We would look at hardware on deck and Bruce McCandless would take the lead: talk and walk us through what we thought we'd do. I'd say, 'Let's get this tool and try that', receiving any help from scientists on what instruments might require work." The system was worked out from the very beginning to put specific instruments in certain places, and they would evaluate these ideas. They would then move to the 'what if' contingencies or what to do when things went wrong. They serviced batteries, star trackers, and tested tools on replacement items. The driving force behind all this activity was Frank Cepollina, who Nelson says did a marvelous job of thinking the issues through.

The time spent in the water tank depended on the task or dummy EVA setup required, or what could be supported at Marshall. Usually a session lasted about 3 hours, with a debrief afterwards. Nelson recalled attempting to work out the right tools to use at a time when there were too few tools available. "It was pretty amazing at this point, as I remember once trying to figure out which tools to use," because the tools were not defined as they are today. "We flagged down a company that made hand tools to work on automobiles. They sold from out of a truck, so one day we flagged down their truck and went through their inventory to order the extra tools we needed. For Solar Max, I just went to a local hardware store and bought a little electric screwdriver, brought it in and said I want to use this!"

The work carried out in the tank on ST helped in developing what they were trying to do on Solar Max, developing techniques to handle combination work with big and heavy items which had very small tolerances. The work undertaken by Nelson and others in this manner helped to define activities not only for Hubble but other EVA missions. Nelson praised the sterling work on early shuttle-based EVA procedures by fellow astronaut Jim Buchli, whom he said was "a forgotten hero of the space program". Buchli headed procedures development for shuttle contingencies such as stowing radiators, closing payload bay doors, and stowing the Ku-band antenna (which Nelson says was required on one mission) that carried over to Hubble. It was Buchli's suggestion to add the T-bar device to the front of the suit, and this proved to be a useful tool caddy.

Noting that it was harder to work underwater than in space, Nelson recalled an incident whilst conducting one of the underwater tests. "The support divers usually keep out of way, mostly for safety reasons, but I had to use them once. I was inside the axial instrument bay when the umbilical cord came off and all of sudden everything went silent, the air stopped, the radio was quiet. It doesn't take long for carbon dioxide to build up in the suit so I swam out, and [the divers] grabbed me and brought me to the surface. Now it wasn't a dangerous situation, but it was certainly uncomfortable."

## FULL SCALE SIMULATIONS

All the mockups were used in the initial series of simulations and were the first opportunity for full scale simulations to evaluate the locations of proposed mechanisms and equipment configurations. This first look provided engineers with clear objectives for designing future test programs and simulations.

### Subsystem development

During these tests the latest developments in the telescope's subsystems were evaluated, reporting on any changes that would be required as a result of testing the hardware in the tank.

### Axial Science Instruments

The four axial science instruments were the largest ORUs on the telescope and, at that time, the most massive objects planned to be handled by astronauts in the history of the program. Each box was 86 inches (2.10 meters) in length with a volume of 62 cubic feet (18.90 cubic meters) and a mass of up to 700 pounds (317.8 kg). In handling these units,

Axial instrument exchange simulations in NBS, MSFC, circa 1980. (Courtesy Marshall Space Flight Center)

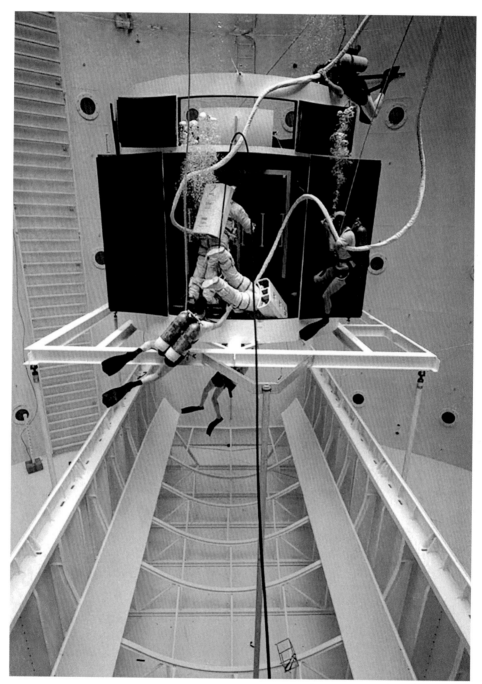

Early Space Telescope simulation with Bruce McCandless and George "Pinky" Nelson in the NBS, MSFC, circa 1980. (Courtesy Marshall Space Flight Center)

the delicacy of their components had to be taken into account. The EVA test subjects couldn't simply grab any part of the instrument. The only locations where an instrument could be handled or slid into its guide rails were the four vertical handrails for use by astronauts and the hard points for ground support equipment. Changing out a unit involved a choreography whereby both test subjects made use of the RMS and foot restraints to undertake a lengthy sequencing of actions.

**Radial Scientific Instruments**

This task was made easier by lessons learned in developing the technique to install the axial instruments and by the fact that the only instrument of this type, the WFPC, was smaller and provided both better visibility and unrestricted access.

**Fine Guidance Sensor**

Access to these elements was much harder since they were inside the Aft Shroud and doors had to be opened to gain access. Underwater simulations found that the task was simplified when a foot restraint attachment point was provided, because this kept the astronaut's body clear as the door was opened. It also allowed the astronaut to be close enough to the surface of the Aft Shroud to reach and work on the FGS located about 12 inches (30.48 cm) inside.

**Fixed Head Star Tracker/Rate Sensor Units**

This was an example of how the underwater simulations helped to develop procedures for items which were not initially designated as ORUs. The FHSTs were on the inside of large doors on the −V3 axis. The RSUs on these units were treated as ORUs and thus candidates for replacement if required. However, to do this the star tracker light shields had first to be removed. The problem identified in the underwater simulations in August 1979 was that the covers on the light shield were difficult to access due to their location, and hence were only suitable for a single handed operation. As this area was not large enough for a crew-member to work from a foot restraint, it became necessary to grasp the light shield with one hand for stability and manipulate the fasteners using the other hand. Bruce McCandless was the test subject for this exercise. He designed a 'J-hook' which proved to be useful in removing the shield, but with difficulty. The issue which he encountered was in trying to hold the RSU in place while simultaneously torquing the hex bolts. Trying to reach the position to work the tools proved difficult. The preferred solution to the problem (moving the RSUs to the most convenient location) was deemed too expensive, so a less attractive solution was adopted in which the RSUs were moved to another location which was workable but not optimum, yet had no cost impact.

**Support System Module Equipment Section**

Underwater simulations were also conducted on the SSM-ES, for which initial testing had indicated additional handrails were necessary for opening the doors on Bays 2, 3 and 10. In addition, concept latches were built and tested for one-handed operation; this was worked out quite early in the simulations as it was thought an astronaut may not be able to use both hands to accomplish a task. As a result, items were developed for single handed operation if the astronaut would need to grasp something for stability, or if the restricted mobility of the pressurized suit would prevent using both arms in the available workspace.

**Optical Telescope Assembly**

The simulation of maintenance tasks involved removing and replacing the electronics ORUs for a Fine Guidance Sensor on each of the three doors to determine the best place to position crew aids and the accessibility of the fasteners on the ORU.

**Solar Arrays**

The first series of SA tests in the water tank were in December 1979 and were designed to evaluate contingency mechanisms for both the SA and Diode Box assembly, and also the potential for jettisoning an array. The tests focused upon the action of the forward and aft latches to verify the jettison action. However, by the time that the tests on the low fidelity mockups were conducted the equipment was out of date because more up to date designs were being introduced sooner than the simulation mockup could be upgraded. The fact that this was not achieved until February 1981 highlighted a general problem of obtaining flight standard equipment as soon as it was produced, so as to test it in the tank. As a report said, "This condition was the result of not having flight type connectors and reflected a constant problem which was evident throughout the testing program: namely, that of being forced, through scheduling constraints, to conduct testing before equipment design is finalized by Critical Design Review (CDR)."[18]

Testing of the high-fidelity SA and associated apparatus held in February 1981 led to a considerably more complex review of the SA/Diode Box EVA tasks being planned for the service missions. As these were all contingency operations, it was imperative to have the latest version of the hardware available for testing so that the crew could receive at least some training on apparatus that was as close as possible to the flight model. That way, if a contingency occurred during an EVA they could avoid unnecessary (or indeed dangerous) actions. Fortunately, the superior quality of the hardware used in the tests led to very few surprises, termed "brush fires". The report acknowledged that the success in this stage of testing was due, in part, to having a contractor representative from ESA/BAe to provide a "valuable technical insight to the workings of the SA mechanism". A little over a decade later, this experience led members of the first servicing crew and an EVA team from JSC to visit BAe in England to examine the flight hardware for the second generation solar arrays that were to be attached to the telescope several months later.

**Tools**

"During the series of tests, it became apparent that there was a need for the definition of a baseline tool complement for the ST," so wrote Fred Sanders of the System Analysis and Integration Laboratory at Marshall in a 1982 report.[19] The tool kit which was being created for the shuttle would be available, but this was designed to cover planned and contingency

EVAs like shutting the payload bay doors, re-stowing the antenna dish, strapping down or jettisoning a failed RMS, and lowering the payload support apparatus. The design of the ST and the tasks involved in servicing it, required a more defined approached to providing an adequate tool kit which was essentially a set of sockets, extensions, and a manual wrench. The simulations defined the specific tasks that various types of tools could be used for and the most suitable length of extension required to reach the worksite (e.g. in the case of the Fixed Head Star Tracker/ORU highlighted above). The simulations showed that often there was insufficient room to operate a wrench unless it was done by one hand, which made the manual ratchet wrench a valuable asset. This was another lesson that was learned the hard way. During the early stages of the testing program when there were inadequate restraints, test subjects experienced discomfort. During a 2 hour simulation, the strain and associated fatigue was seriously tiring.

The solution suggested in the report was to tether the person at the waist to the relevant worksite in order to restrict movement. But this would need significantly more attachment points on the telescope, which was not considered practical. Attempting the operation with one hand did not work either, since firmly grasping a handhold over a prolonged period of time resulted in hand fatigue from resisting the glove's natural pressure level to maintain a grip. Securing the person's feet in one position using a foot restraint was deemed to be the most desirable solution, but the balance between placing limited attachment points on the telescope whilst still being able to provide a useful range of reach options became the new challenge. The report stated that with inadequate foot restraints, crewmembers "rotated their feet upwards towards the head, thus contracting the muscles at the front of their legs between the knee and ankle. By holding this position for 15 seconds the degree of fatigue that can be induced during a 2 hour simulation can be appreciated."

Clearly more work needed to be done in both the support of adequate restraints and more suitable tools—a challenge first identified in 1966 during the Gemini missions.[20] Following the underwater tests, the report identified the need for a power ratchet wrench. The challenge to develop suitable tools to address not only the needs of the telescope but also the comfort of the astronauts required (as will be explained in the next chapter) a major investment of time, resources, and engineering ingenuity.

**High Gain Antenna**

No mockup of the HGA was available at the time of the tests but there was sufficient data to determine what could be done and where to place the crew aids even "without the benefit of the crudest mockup". Using data from the simulations on the solar arrays, it was possible to evaluate the location of foot restraints to support work on the HGA, and later confirm these attachment points by astronauts using a "basic" mockup.

**Aperture Door**

As with the HGA, no mockup of the Aperture Door was available. This wasn't an oversight. The 7.5 foot (2.28 meter) Light Shield was omitted from the mockup because it would have projected over 2 feet (0.60 meters) above the surface of the water, making neutral buoyancy simulations impossible. In this case an analytical approach to the

230 **Simulating servicing**

The STS-125 EVA crew discuss the dive program for NBL tests at the Sonny Carter Training Facility.

placement of crew aids was devised. As the door was at the very end of the shield, that area was free of "exotic or hidden fasteners to access" and it was just a matter of requesting handrails and foot restraint sockets to be incorporated on the flight telescope by the contractor in order to gain access to the door area in flight.

**Portable Foot Restraints**

The search for a suitable portable foot restraint (mentioned above) became a personal quest for the engineers. Two years into the simulation program, many designs had been fabricated and tested, becoming progressively more ingenious in design and subsequently infamous in complexity, but the development was still far from complete.[21]

The problem was not the restraint itself; it was pretty simple to supply an item that would interface with the telescope and accommodate a spacewalker's boots. The problem lay in the design between these two ends—the foot restraint plate and the connection receptacle—and still make it acceptable for a myriad of chores by ensuring that the crewmember was not so near the telescope that their faceplate almost touched the structure, yet not so far away as to prevent a comfortable reach; and all the time retaining the structural integrity of the design.

Given the mass restrictions imposed on the telescope by its constrained budget, it was not feasible to insert foot restraint sockets on the telescope at will. The task facing the designers was to emplace the restraint attachment fittings at strategic points where they would be most effective. This may not seem such a difficult task, but as the effort to eliminate unnecessary mass increased, attention turned to items such as the additional restraint locations. The foot restraint structure was not the issue, that was a separate unit which would be moved around; it was the attachments into which the foot restraint was inserted that were of concern. Each socket weighed 3 pounds (1.36 kg), including the support structure to hold it in place. They were easy targets for the weight-saving effort. When 8 sockets were deleted, this yielded an immediate saving of 24 pounds (10.89 kg).

The work conducted in the NBS by engineers and astronauts was critical in determining the optimum locations for the remaining sockets, without giving up the flexibly to reach as many workstations as possible. The solution was to incorporate a 2 foot (0.60 meter) long shaft between the foot restraint plate and the retention stud that fitted into the socket on the telescope. This added a corresponding reach in any direction by use of a cantilever structure adjustment on both ends of the strut in the pitch (up or down) axis, adjustments to the shaft for roll (left or right) and to the foot plate restraint in yaw (forward or backwards). This resulted in a suitable foot restraint with four degrees of freedom that could access any area within about 6 or 7 feet (1.8 or 2.1 meters) of the socket location. Consequently, a single foot restraint socket could be used to access four different locations for maintenance without requiring the foot restraint to be moved. It was a major advance.

**Handrails**

It would be reasonable to assume that with more handrails available, the easier would be the task of moving across the face of the telescope and stabilizing a position at a worksite. This was certainly the case in the early trials in which the telescope was covered in

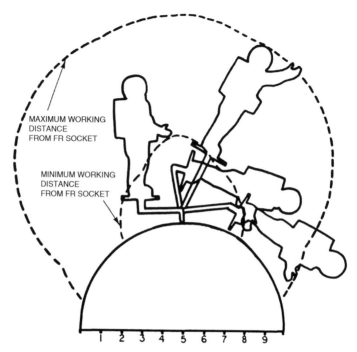

A two dimensional representation of the astronaut's reach using a portable foot restraint.

handrails. But simulations in the tank established that some translation paths were never used. Furthermore, it was often the case that when a handrail was required for a specific task there was not one available. So the unwanted handrails were removed and others added. The weight reduction effort that reduced the number of portable foot restraints also affected the number and total length of handrails installed on the telescope. As the handrails were omitted, so too were the associated structural supports and mounting brackets. The philosophy used to decide which handrails were essential was very simple. Handrails that were not used to transfer from one workstation to another were deleted. So too were those that were never grasped for stability whilst working. Then the handrails that remained were shortened in length and networks of "paths" were created for moving across the structure.

Most of the work focused on positioning the handrails for the planned service missions. This accounted for the majority of the handrails. In other areas, where a spacewalker would be required to work during a contingency EVA, there was only a partial circumference of handrails around the top of the Light Shield, a single path between the +V3/−V2 axes with which to reach the Aperture Door, and the longitudinal handrails terminated at the forward latch of the HGA. Another efficiency was made in identifying the need for handrails on the bay doors for the axial science instrument and FHSTs. Here a single length of handrail was installed to provide access to the upper part of the telescope but it doubled its function as a handle for opening and closing the large doors.

## Latch design

Every ORU on the telescope except the series of WFPCs were located behind a door which had to be opened and closed by an astronaut for access. As a result, the door latch operation became a normal part of the ORU simulations in the water tank. The latches had to be of a simple design that was capable of one-handed operation using baseline tools, and yet strong enough to withstand launch vibrations and stress from applied torques over several cycles. Three latch designs were evaluated during simulations. An adjustable grip latch in which a rotating table applied compression force to close and lock the door from the 90 degree open position was used to secure both the OTA-ES and SSM-ES doors. It could be operated one-handed using the powered or manual ratchet wrench. A similar design was used for the Aft Shroud doors, but using a simple T-bolt force. The Hatch Latch Assembly was a variation on this design, with an over-center locking fixture that enabled the doors to remain closed prior to applying the final locking torque.

## Fasteners

For the NBS tests, a ratchet wrench and all extensions used a 12-point 7/16th inch (11 mm) socket that could fit the standard fasteners for all items that required a crew interface.

## RMS related activities

The RMS used in the tests was the one fabricated from spare parts from earlier projects, and was intended only for local use owing to its limitations. It was used in support of solar array jettisoning and in developing innovate crew aid concepts. The full-size unit was mounted on the sill of the payload bay mockup. Its end effector mockup mated with the portable grapple fixture which was used in simulations of jettisoning a solar array. A portable foot restraint of limited movement was also installed on the end of the arm. After the first flight of an RMS on STS-2 in 1981, the operational fidelity of the RMS mockup against film of the Columbia RMS in operation showed it to be comparable, including the oscillations which occurred in space. This provided a greater sense of realism to the arm installed in the tank.

At the time of the report (June 1982) the initial RMS had not been formally man-rated in the tank owing to safety aspects of a space-suited test subject being underwater on the end of a 40 foot robotic arm. But steps towards this qualification were made, and "without fanfare, astronaut Bruce McCandless ingressed the RMS mounted foot restraint and translated a 700 pound mass with relative ease".

## Conclusion from the initial NBS tests

Without doubt the opportunity to perform actual simulations of planned telescope activities provided designers, engineers, managers, contractors, and astronauts a chance to witness the activities for themselves, together with areas of difficulty that were not so readily identified by those whose attention was focused on only one particular process or piece of equipment.

## 234 Simulating servicing

The RMS simulator in the NBS tank at Marshall Space Flight Center. (Courtesy NASA MSFC History Office/US Library of Congress Historic American Building Collection; HAER ALA,45-HUVI.V,7B-11; 1995, by Jet Lowe)

Though these early EVA simulations used baseline data, significant information was gained and input directly into the design of the flight hardware and the development of procedures, guidelines, and contingencies which would be valuable when the time for flying the service missions approached. Certainly having the ability to see test subjects, whether engineers or astronauts, going through a process offered a real-time view of what could be expected on-orbit. This became far more valuable than the traditional post-simulation debrief, or reading about the activities in a report, or from viewgraphs in a presentation weeks or months later.

After only 2 years of NBS operations with Hubble hardware, the results and experiences were encouraging. The underwater program continued for the next few years, supported by the development of new tools, techniques and better hardware. By using the availability of engineers and astronauts in a series of dive programs, a valuable source of data was created that could be applied not only to servicing the telescope but also in support of other future programs and concepts.

The NBS RMS simulator control room at MSFC reproducing identical RMS controls and panels to the shuttle orbiter aft flight deck. Wearing glasses, an operator could view 3-D video projection in the aft flight deck windows as the operator could not view operations in the water tank while at the control panel. (Courtesy NASA MSFC History Office/US Library of Congress Historic American Building Collection; HAER ALA,45-HUVI.V,7B-14; 1995, by Jet Lowe)

## NBS TESTS 1983–1990

In March 1983 the Essex Corporation was awarded a contract to design and fabricate Hubble Space Telescope EVA Training Hardware in the NBS at Marshall and at the new Weightless Environment Training Facility (WETF) at JSC in Houston. Conducted between March 1983 and December 1990, the objective of this program was to design and build training devices, develop new hardware, and employ the services of various utility divers, test conductors and test subjects to make a database of materials to assist the HST management with both design and program decisions.

Essex's association with the NBS at Marshall began in 1979. Not only did the company design, fabricate, and install every piece of Space Telescope hardware in the tank during the development years, it also supplied utility divers, test conductors, and suited test subjects to supplement the work conducted by NASA astronauts in the tank.[22]

## Upgrading the NBS trainer

To round out the contract, Essex conducted a final refurbishment and upgrade to the HST mockup training devices used in the tank. There were two configurations: the Deployment Crew Trainer which supported the contingency EVAs planned for the deployment mission and the Maintenance Crew Trainer configured to support the first service mission.

By 1990 the *HST Deployment Crew Trainer* consisted of: Diode Boxes, the Aft Shroud, the Support System Module (SSM) Equipment Section, the Optical Telescope Assembly (OTA) Equipment Section, the OTA Forward Shell, the Light Shield, the Aperture Door, the High Gain Antenna, the Solar Arrays, and an Astronaut Control Panel. It also had a mockup consisting of Emergency Umbilical Disconnects, Umbilical Retract Mechanisms, and an Umbilical Tower.

The *HST Maintenance Crew Trainer* consisted of the Aft Shroud, the Aft Shroud Orbital Replacement Units, the SSM/Equipment Section along with its ORU envelopes, electrical connectors, electrical wiring, decals and fasteners, the OTA Equipment Section with its ORU envelopes, electrical connectors, electrical wiring, decals and fasteners, the Diode Boxes, the Focal Plane Assembly, three Fine Guidance Sensors, and the Wide Field Planetary Camera. It also had Manipulator Foot Restraints, crew tools, two Portable Foot Restraints, and several ORU Carrier configurations. Typically the ORU Carrier consisted of two Spacelab Pallets. These pallets featured ORU Storage Shelves with supports and shelf adapter plates, the Keel Latch Support Structure, the Wide Body Keel Latch Assembly, the Load Isolation System, the FGS Scientific Instrument Protective Enclosures, the Flight Support Systems Cradle A, Solar Array Carriers in various configurations, spare ORU envelopes for the ORU Carrier, and exterior ORUs such as the Low Gain Antenna and three Coarse Sun Sensors.

## The underwater test program 1984–1988

Between March 1984 and December 1988 at least 12 separate test programs were conducted using the space telescope training hardware in the water tank at Marshall.

Two programs were completed in 1984. In the NB-45B simulation in March, test subjects evaluated astronaut access to the deployment/return electrical umbilicals, including access to and visibility of indicator markings on the first generation disconnect mechanisms. They also tested access to the cryogenic vents of the Axial Science Instruments and Equipment Bays 1, 4, 6 and 9 of the SSM. Then in simulation NB-45E in July, the Preliminary Design Review Configuration of the ORUC was verified and its ORU mechanisms evaluated, together with the RMS Manipulator Foot Restraint serving as a work platform and as a transfer system for ORUs and crew transfer. All crew aids were also evaluated, as were the temporary parking locations for instruments extracted from the telescope for return to Earth. Refinements were then made to the intended timelines for instrument changeout.

Four test programs were conducted in 1985. In May, access to the ORUs in Bays 1 and 7 were assessed. In September the PFR was evaluated, as was the Essex-supplied wrench that had extensions and a drop-proof feature to prevent it sinking to the bottom of the water tank if accidentally released. Another item that was verified was the "clothesline" manifested for the deployment mission. This test provided an opportunity to establish a baseline of elapsed time to assist deployment mission crew training. The following month's test evaluated how accessible the Block II ORUs were. Finally, in NB-45J in November Bruce McCandless and Kathy Sullivan undertook an end-to-end timeline drawn up for the Mission Timeline Profile Analysis. This simulation included using the RMS to transfer the FGS replacement from its carrier in the payload bay to the Aft Shroud. Other work performed in this session included an investigation of crew aids for future-concept ORUs, the management of tools and crew aids, and the preparation of worksites.

Two NBS sessions were held in 1986. In February, the month following the Challenger disaster, McCandless and Sullivan were back in the tank to determine whether the ORUC Foot Restraint Receptacle was acceptable for accessing ORUs. They also defined the EVA umbilical operations of the "station 100 mockup" and a high-fidelity mockup of the FSS. In addition, the placement and verification of crew aids for all FSS contingency activities were carried out that month. Both astronauts were also able to confirm that access to the LGA of the telescope was possible from existing Foot Restraint Receptacles on the FSS.

Two further simulation programs were completed in 1987. In February, the Solar Array Carrier concept was validated, and the placement of crew aids on the SA carrier identified. That month also saw the development of the first generation SA removal and replacement timelines. In September, the second session of the year evaluated the proposed SA carrier design.

This program of water tank simulations in the NBS at Marshall was completed with two sessions in 1988. In April the designs of the flat base and T-frame SA carriers were verified. Suited test subjects evaluated the potential of an electronic latch device, and whether it was necessary to have two mirror-image latches. The tests evaluated if additional EVA-installed handholds would be required for SA tasks, and evaluated the multi-layer insulation and the FGS retention device. The final session was in December, and featured end-to-end timeline runs for a simulated HST M&R mission. This consisted of two 6-hour EVA days in the tank. Tests also changed out the NiCd to NiH2 batteries, conducted reach and access tests for 12 new Block II ORUs, and assessed manual translation of an SA as a contingency operation in the event of RMS failure. This session also included the removal, parking, and reinstallation of a WFPC fitted with the extended handrails, tests of the EVA interfaces of the Wide Body Latch Keel, and an evaluation of an insulation blanket for use with the small ORU Carrier.

**Achievements from the tests**

The neutral buoyancy EVA test program was driven by the need to develop techniques and equipment to service the telescope at a time when it was still in development; conversely it was found that the simulations themselves made significant contributions to the refinement and improvement of flight hardware on the telescope, revealing a full circle of investment.

The feedback inputs from the tests included:

- *Axial SI Guide Rail Spring*: Often some of the suggestions and changes which were made to equipment were quite small, designed to make a larger task much easier to accomplish. During the development of the Axial SIs, it was necessary to have the instrument roll out of the telescope on guide rails to allow the astronauts smoother motion in and out of the tight confines of that instrument's locating bay. One of the evaluations in the underwater tests was to establish the best spring rate to retain the given instrument in the upper guide rail. As a result, Essex fabricated five different springs, each with a different rate. When the tests were completed, the results were sent to Perkin-Elmer to determine the best spring rate to incorporate into the flight hardware.

- *Flight Ratchet Wrench:* As the underwater simulations progressed, various working scenarios for servicing began to emerge. It soon became evident from the NBS tests that to overcome the limitations of working in a pressure suit, underwater or in space, a ratchet wrench would be helpful. Essex fabricated many different designs of ratchet wrench for evaluation. The best design (which received the logical title of the Essex Ratchet Wrench) was then prepared for flight.

- *EVA timelines:* Taking into account the viscosity of the water and the limitations of the NBS in reproducing a real EVA, the use of the tank and its mockups allowed for developing servicing scenarios and determining the actual time that each task should take. As the servicing objectives for each mission were finalized, the simulation crew were able to develop fairly accurate beginning-to-end timelines with which to refine their training skills.

- *Assorted Crew Aids Development:* The availability of the water tanks, primarily at Marshall but also at JSC, afforded the opportunity for simulations of all or part of a planned EVA. This was especially useful in the years leading up to the deployment flight, and later the first service mission. It allowed both engineers and astronauts to directly participate in simulated Hubble EVA tasks and operations in an environment on Earth that was as close to the weightless conditions of space as it was possible to reproduce. The simulations also ran for a prolonged period of time, offering realistic EVA-duration simulations. The opportunity for contractors and managers to directly view these operations also improved understanding of the tasks and challenges, and why some equipment or procedures worked better than others, and how experiences from the tank led to new ideas for tools and other equipment such as the door stays, connector removal tools, and ratchet wrench socket extensions.

- *Neutral Buoyancy Trainer Paint Specifications:* Often in the development of space hardware an off-shoot idea was developed from an original concept, process, or item that then found application in other areas of technology, especially outside the space program. These Space Spin-Offs are visible evidence of "technology for the general public" attributable to the enormous investment placed in NASA over the years. One example from the work conducted in the NBS was in the preparation and painting of the NBS mockups by Essex workers. Because the training mockups were gradually deteriorated through sustained immersion, in order to prolong the useful lives of the mockups the company researched paints that were both resistant to water and suitable for the NBS staff to prepare and paint on the mockup hardware. These paints proved to have commercial potential.

Some of the contributions were small and not reported outside of the contractor or NASA, but each was critical to the success of the missions flown. Many of the ideas, developments, and techniques were dead ends, or were not adopted owing to cost, time, and complexity. Of course, other ideas did make it through and the experience and procedures developed in part from work carried out in a water tank in Alabama enabled astronauts to confidently maintain a telescope in space.

**Additional NBS tests**

After the Hubble Space Telescope was deployed in 1990, the neutral buoyancy simulations performed a range of checkout procedures on the HST mockup in readiness for supporting the development of scheduled service mission EVAs. Several astronauts participated as part of their routine technical and support assignments, including Mark Lee and Rich Clifford on June 27, 1991 and Jim Voss and Jay Apt on August 21, 1991.

In 2013 Story Musgrave recalled the challenges of working in water to simulate EVA.[23] "It worked even though water is not a good simulation, water is a pretty bad simulation. It's good for reach, for visibility, and for moving large objects and what the track will be. It's not good for suit work, you go upside down in a suit, you soon realize you are not in zero-g. If you have 170 pounds (79.38 kg) [the mass of the pressure suit] resting on your collar bone you really can't work at all, and so in the water you chose less than optimum body positions to get the job done—you have to in the water. The thing about your suit in the simulations is that it has 1000 hours on it, it's lubricated by water, and the suit you use in the tank is not the suit you fly with. The water does not restrict you, you go ahead and do the whole job so you choreograph the whole thing, but you know how rough it is going to be when you get to do the real thing in terms of suit geometry and the choreography of moving tools around where things are going to be. The choreography stays the same, but the water isn't so good for suit dynamics."

Musgrave explained that the viscosity of the water could be a problem, "It is easy to tell yourself how easy it will be in flight. The viscosity holds you more stable, so your body is going to be more unstable in flight. So you have to tell yourself all the time in terms of these issues. But mostly it's going to be easier, much easier to move the big objects, so you soon get in a hurry doing that out there [in space]".

**GAINING EVA EXPERIENCE**

In the 1980s the long lead time in getting Hubble into orbit helped to achieve a number of EVAs in order to gain experience. "On STS-41B we were able to test many EVA tools and equipment that would be used later on HST," noted Robert Trevino at JSC. "This included the EVA Power Tool, the Multiple Foot Restraint, the EVA ratchet, etc. That allowed us to work out any bugs in the hardware."[24]

Prior to the first service mission in December 1993, there had been only 20 EVAs from shuttles during the period April 1983 to September 1993, just 10 of which directly involved satellite retrieval, maintenance, or repair. Work was continuing in the NBS at Marshall, but between 1986 and the end of 1992 only two shuttle missions had included EVAs, and both had undertaken contingency or unplanned EVAs.

On STS-37 in April 1991 a contingency EVA was inserted ahead of a planned EVA when an antenna on the Compton Gamma Ray Observatory failed to deploy; the EVA pair of Jerry Ross and Jay Apt went out and freed it.

In May 1992, the STS-49 EVA astronauts encountered difficulty in securing the stranded Intelsat VI satellite into the payload bay of Endeavour. In the end, it required three EVAs to complete the operation, the final one marking the first time that three people—Tom Akers, Richard Hieb and Pierre Thuot—went outside simultaneously. The original plan was for the crew to conduct three EVAs with Thuot and Hieb performing the first and third, and Akers with Kathy Thornton conducting the second EVA to allow their colleagues a day off. What actually happened was that Thuot and Hieb made three outings, two of which were back to back on consecutive days. Though this naturally tired the men, it supplied valuable data on human engineering and endurance in planning the more intensive EVAs for Hubble and the ISS. With the first service mission planned for 1993, clearly more experience was needed in order to investigate spending long periods in the payload bay on successive days. As events transpired, Thornton and Akers would later team up for EVAs on the first service mission.

## DTO-1210

During the 1986–1991 lull in shuttle EVA operations, partly as a result of the grounding of the fleet following the loss of Challenger but also in the absence of EVA plans in the Return To Flight program, a team was created to develop techniques and procedures and test new hardware in support of the Hubble service missions and the space station's EVA program. Hence on November 25, 1992 NASA announced the decision to assign a series of EVAs to shuttle missions over the next 5 years in order to train astronauts and flight controllers, gain experience of specific EVA operations, and evaluate space construction techniques.[25] This was to be undertaken within the Detailed Test Objective (DTO) program. The objectives of DTO 1210 (EVA Operations Procedures/Training Program) were to broaden knowledge of EVA though both planning and practice.[26]

On STS-54 (January 17, 1993) Greg Harbaugh and Mario Runco made the first EVA for DTO 1210. One task was to evaluate carrying a large object, but because no hardware was added to the mission the astronauts took turns carrying each other. The pair also manually positioned the TDRS tilt table, and evaluated their ability to align a bulky object (i.e. each other) in a bracket in the airlock which otherwise supported the EMU. Following the EVA the pair responded to very detailed EVA questions, and once back on Earth they repeated their EVA tasks in the WETF at JSC to help improve future EVA training. In the case of Harbaugh, this was valuable experience because he was also in training as a backup EVA astronaut for the first service mission later that year, and subsequently flew in 1997 as an EVA crewmember for the second servicing mission.

STS-57 (June 25, 1993) saw the first EVA from an airlock built into an extension tunnel linking the middeck to the Spacehab module in the payload bay. On this mission DTO

1210 was supplemented by DTO 671 (EVA Hardware for Further Schedule EVA Missions) that tested EVA apparatus meant for both the first Hubble service mission and the space station. David Low and Peter Wisoff rehearsed the replacement of the faulty optics on Hubble. This involved Low manipulating his partner as the "large bulky object". With Nancy Sherlock operating the RMS at its 50 foot (15 meter) fully extended length, and Low riding the foot restraint on the end effector, he picked up Wisoff and, through instructions to Sherlock, was maneuvered on the arm to a specific location in the rear of the payload bay. This duplicated some movements planned for the first service mission. Similar tests had been conducted on other shuttle missions to evaluate the capability of the RMS to move bulky objects, but this exercise was to provide data on using the arm in the small fine movements which would be required on the service missions. A new torque wrench proved easier to use in space than in the water tank. This gave confidence that they were on the right track. While working on the RMS, both astronauts assessed moving each other about, working with tools, and the use of safety tethers. They also found that while facing away from the payload bay, into space, they became cold enough in their suits to shiver and their hands became numb or painful. In their post-flight debriefing both men called the exercise "time well spent". It also proved valuable experience for Nancy Sherlock, who 9 years later (as Nancy Currie) was the RMS operator on STS-109, the fourth service mission, designated SM-3B.

The next EVA in the series was on STS-51 (September 16, 1993). Jim Newman and Carl Walz evaluated tools on the Provisional Stowage Assembly (PSA) bolted to the payload bay floor. They both also evaluated a glove-warming technique that involved holding the gloved palms of their hands against one of the payload bay lamps. Tasks directly linked to the first service mission were evaluating work involving high torques and low torques, as well as a Portable Foot Restraint (PFR) that was to be flown on SM-1. In their post-flight debriefing the astronauts stressed the importance of thermal vacuum tests as part of the EVA training and testing program. The heat-lamp warming exercise warmed Newman's hand, revealing that the EVA gloves weren't such a good thermal barrier as first thought. As a result a new glove design was requested, but it would not be ready in time for the spacewalks on STS-61 due in less than 3 months, so for that mission the astronauts would wear an outer glove. In comparing their training with their experience on-orbit, Newman and Walz said working in the WETF was more difficult than in space. This raised confidence that the preparations for the first service mission were sound and that the demanding tasks could be accomplished.

In December 1993, the STS-61 crew spectacularly proved this expectation correct in the successful execution of the first Hubble service mission over five long EVAs conducted by Story Musgrave and Jeff Hoffman and by Kathryn Thornton and Tom Akers.

The work on STS-51 would also prove useful to Jim Newman, who later participated in the first ISS assembly EVAs with Jerry Ross on STS-88 in December 1998, after which he participated in the EVAs of STS-109 in March 2002 for the fourth Hubble service mission.

The STS-64 EVA (September 16, 1994) was more associated with the ISS than Hubble, with the crew performing the first untethered excursions since the MMU flights of a decade previously. Mark Lee and Carl Meade tested the Simplified Aid For EVA Rescue (SAFER) device designed to enable an untethered astronaut to return to the vehicle under emergency conditions. It was a successful test, but as Hubble required both astronauts to be tethered or fixed in foot restraints at all times the SAFER was not flown on Hubble service missions because the shuttle could, in theory, pursue a stranded astronaut. In contrast, the Electronic Cuff Checklist (ECC) planned to replace the paper cuff checklist that had been in use since the early Apollo lunar landings, didn't perform so well and required more evaluation. And during the EVA Meade reported his feet were cold, again indicating the severe variance of temperature encountered when conducting a long EVA. Lee later carried out EVAs during STS-82, the second Hubble service mission, designated SM-2.

The flight of STS-63 featured the first spacewalk (February 9, 1995) under the EDFT program in preparation for the ISS assembly, but it also provided valuable information for planning Hubble servicing. To follow up on the reports of astronauts feeling cold during EVAs, Mike Foale and Bernard Harris wore EMUs that had thicker underwear, improved glove insulation, and a bypass switch to allow them to reduce the water flowing through the Liquid Coolant and Ventilation Garment worn next to the skin. With both men on the RMS, the arm was raised 30 feet (9.1 meters) above the payload bay for a "cold soak" intended to last 15 minutes. The plan was to make both men as cold as possible by aligning the orbiter away from the Sun during the day pass and turning it "out" to space in the night pass. Data recorders on the MFR and in the gloves recorded conditions for later comparison with the astronauts' own recollections and comments. As the EVA progressed they worked with the 2500 pound (1100 kg) SPARTAN 204 free flyer, with Foale lifting it to Harris in order to evaluate both an astronaut's ability to handle large loads and an EMU's ability to cope with the added exertions. It was during this task that both men reported becoming "unacceptably cold". The temperature in Foale's gloves fell below 20 degrees Fahrenheit (minus 6 degrees Centigrade) and Harris's feet felt cold through the contact with Discovery's structure as that dipped to minus 130 degrees Fahrenheit (minus 90 degrees Centigrade). The EVA was curtailed and the payload bay warmed by the Sun before the two men concluded their challenging spacewalk. Foale would later carry out an EVA during STS-103, the third Hubble service mission, designated SM-3A.

On STS-69 (September 16, 1995) Jim Voss and Mike Gernhardt evaluated EVA tools for the ISS and also tested power tools, removed thermal blankets from a task board on the side of the payload bay, and manipulated fasteners, ORU boxes, electrical connectors and tethers in preparation for Hubble service missions. In addition, each man completed a cold soak 30 feet (9.1 meters) over the payload bay for 45 minutes. They were kept comfortably warm by EVA gloves that had fingertip heaters powered by 3.7 volt lithium batteries. The pair also tested new EMU helmet lamps and various restraints.

Hubble servicing mission EVA simulation in NBL Sonny Carter Training Facility. Note the presence of at least seven support divers in scuba gear.

Two of the remaining three EDFTs—STS-72 (January 15, 1996) and STS-87 (December 3, 1997) carried out tasks primarily associated with the forthcoming ISS but this experiences of assessing tools, procedures, and thermal conditions within the payload bay was applicable to Hubble servicing. When the outer hatch could not be opened on STS-80 (December 1996) and the ISS development EVAs were canceled, this was particularly concerning since this was the only exit into the payload bay from the crew compartment in the event of a contingency EVA. As a result, NASA added a set of hatch door tools to the shuttle inventory for freeing the hatch from inside the airlock, should the situation occur again (which it never did).

The EVA demonstrations conducted during this period provided additional experience of working conditions in the payload bay. By December 1997 the first two service missions—STS-61 and STS-82—added to this knowledge. All the hard work in developing, simulating, and practicing telescope servicing techniques over the previous two decades in a 1-g state, in the various water tanks, and in space had been worthwhile.

## STS-125 postscript

Right through to STS-125, the fifth and final Hubble service mission, designated SM-4, the training underwater fully supported the preparations for each service mission. In December 2006 the 58 NBL sessions at the Sonny Carter training facility that would be required for the final mission commenced. These had to be spread out, because it took some time to alternate the facility between Hubble and ISS training that was being supported at the same time. The NBL training runs for the development and verification of tasks assigned to SM-4 provided an 11.6:1 ratio of NBL time to actual EVA time.[27] It was this thorough preparation that made the operations in space appear almost effortless.

**REFERENCES**

1. Frank Cepollina, NASA Headquarters Oral History Project, June 11, 2013
2. *Skylab, America's Space Station*, David J. Shayler, Springer-Praxis, 2001
3. Mr. Fix-It, Frank Cepollina takes repair calls to new heights, Robert Zimmerman, Air & Space Magazine May 2010 on-line version www.airspacemag.com/space/mr-fix-it-10234989/?no-ist last accessed October, 20, 2014
4. A quote from David Martin, who worked for Cepollina from 1988 to 1994 in Mr. Fix-It Frank Cepollina takes repair calls to new heights, Robert Zimmerman, Air and Space Magazine, May 2010
5. Orbital Crew Extravehicular Maintenance Operations, H.T. Fisher, Lockheed Missile & Space Co. Inc. In The Space Telescope, NASA 1976, NASA SP *392 Author's summary of paper presented 21st annual meeting of the American Astronautical Society, Denver, Colorado, and August 26–28, 1975*
6. Full details of these and the latest developments in human and robotic satellite servicing will feature in a forthcoming title: *Satellite Servicing: Applying the Skills of EVA*, by David J. Shayler, Springer-Praxis in preparation. A follow on proposal to the 2004 title *Walking in Space*.
7. Email from Bruce McCandless to D. Shayler, March 27, 2004; AIS interview with Bruce McCandless, August 17, 2006
8. AIS interview with Story Musgrave, August 22, 2013
9. AIS E-Mail from Robert Trevino, December 2014
10. *U.S. Spacesuits* (2nd Edition) Kenneth S. Thomas and Harold J. McMann, Springer-Praxis, 2006 pp271–340, 438–439, 444–445
11. *Walking in Space*, D. Shayler, 2004, Springer-Praxis, pp133–136
12. *Walking in Space*, D. Shayler, 2004, Springer-Praxis, pp136–138
13. *Development of Space Telescope Non-ORU Hardware*, Final Report, December 31, 1985, Essex Report Number H-85-10, Contract NAS8-36384 (for NASA MSFC), NASA CR 178713 (N86-20492), Essex Corporation, Space Systems Group, Alabama, Prepared by Kem B. Robertson, David E. Henderson
14. *Space Telescope Neutral Buoyancy Simulations – The First Two Years*, By Fred G. Sanders, System Analysis and Integration Laboratory NASA, MSFC, Alabama NASA TM-82485 (N82-30942) June 1982
15. MSFC Neutral Buoyancy Facility, Historic American Engineering Record, National Park Service, Department of the Interior, Washington, HAER No. AL-129-B, 2006
16. mix.msfc.nasa.gov/abstracts.php?p=1753 last accessed September 14, 2014, MSFC negative number 8004566 dated May 6, 1980
17. Emails and AIS interview with Pinky Nelson, July 2013

18. *Space Telescope Neutral Buoyancy Simulations – The First Two Years*, By Fred G. Sanders, System Analysis and Integration Laboratory NASA, MSFC, Alabama NASA TM-82485 (N82-30942) June 1982
19. *Space Telescope Neutral Buoyancy Simulations – The First Two Years*, By Fred G. Sanders, System Analysis and Integration Laboratory NASA, MSFC, Alabama NASA TM-82485 (N82-30942) June 1982
20. *EVA Operations in Gemini: Steps to the Moon*, David J. Shayler, Springer-Praxis 2001 pp269–302
21. *The design, fabrication and delivery of a Manipulator Foot Restraint Mockup for Space Telescope Development Testing,* Essex Corporation Final Report Contract No. NAS8-36366, April 11, 1986, Essex Report No. H-86-05 Prepared by Michael A. Hollingworth
22. Final Report for Contract NAS8-35318, December 31, 1990, '*The Design and Development of the Hubble Space Telescope Neutral Buoyancy Trainer*', covering the period Match 1993- December 1990, Essex Corporation, Huntsville Alabama
23. AIS interview with Story Musgrave August 22, 2013
24. Email from R. Trevino December 2014
25. NASA News, 92-066 dated November 25, 1992
26. *Walking to Olympus: An EVA Chronology,* David S. Portree and Robert C. Treviño, Monographs in Aerospace History Series #7, October 1997, NASA
27. Email from Chuck Shaw, December 2014

# 6

# Tools of the trade

> *In general the tools are functionally equivalent to tools used on the ground. Some are standard items in the shuttle contingency EVA tool set while others are unique to the telescope.*
>
> NASA (1986)

Developing the Hubble Space Telescope in a form which would produce the desired science was just the start, because not only did it have to be suitable for launch and possible retrieval by the shuttle, it also had to be capable of being serviced—the initial expectation being that this would occur either on Earth or on-orbit, but eventually solely on-orbit. This requirement added complexity to an already advanced project, creating the need for crew aids to support the servicing work. Furthermore, the development of the large hardware items was only part of the story. The provision of the Flight Support Structure (FSS), Orbital Replacement Unit Carrier (ORUC) pallets, the actual ORUs, the science instruments, the Remote Manipulator System (RMS) and other large items of equipment on the shuttle would not in themselves be sufficient to enable astronauts to changeout components on the telescope or solve some of the more intricate repairs. A suite of tools would be needed in order to fully support the service missions.

Considerable time, labor and ingenuity was devoted to developing the smaller hand tools to enable each service mission to conduct its EVAs with confidence and efficiency. There were naturally difficulties along the way, but the creation of alternative tools, work-around processes and backup contingencies enabled the majority of issues to be quickly overcome. The lessons learned would steer the development of tools for the EVAs planned for the ISS assembly missions and, in the longer term, will influence developments in both human and robotic operations far beyond Earth orbit.

## THE BEGINNING

Like the development of the telescope, the story behind the creation of the servicing tools is a fascinating one, influenced by strict requirements and limitations. Many individuals, some of whose names have been lost to history, devoted almost their entire careers to the invention of implements or procedures they would never use themselves operationally. Others were at the forefront of the drive to ensure that the service missions were provided with the best tool kit, and some of their recollections are recounted here, as are those of one individual who made the transition from mechanical engineer to astronaut and had the rare opportunity to witness one of his own tools being used first in servicing Hubble and later to personally manipulate an improved model in the assembly of the ISS.

### Robert Trevino

Engineer Robert Trevino joined NASA in 1977, the year in which the Space Telescope was authorized. He was assigned to the Engineering and Development Directorate at the Johnson Space Center which was then headed by the legendary Max Faget, the leading designer of the Mercury capsule.

Robert Trevino. (Courtesy Robert Trevino)

At that time, development work was underway for conducting EVAs from the shuttle, so Trevino was assigned to the Design Section of the Directorate, tasked to develop new EVA tools and equipment. One of his early tasks was to design the tools that would be required to manually close the payload bay doors. This task was a last minute request from the Shuttle Program Office when it was realized that if the doors failed to close automatically then the orbiter would not be able to re-enter the atmosphere until they were closed manually by an EVA. The Office was also working with Ed Whitsett on the MMU concept that was being developed for the shuttle and under consideration for use on the Solar Max mission. A short time later Trevino was reassigned to the Mission Operations Directorate at JSC, in which he worked as an EVA trainer and EVA flight controller for all the early shuttle EVA missions, including the Solar Max repair by STS-41C and the retrieval of both the Palapa and Westar satellites by STS-51A. Trevino also served as a member of the EVA team which supported the deployment of Hubble by STS-31.

Around 1980, Frank Cepollina at Goddard contacted the JSC Design Division to request assistance in supporting the early planning for the proposed repair of Solar Max. As a result, Trevino worked with Goddard in developing suitable tools for this task. No spacewalk had been made by an American since the Skylab program in 1974 and the shuttle had yet to fly, but NASA was planning to have spacewalkers retrieve and repair satellites! In parallel with the planning and preparation for the Orbital Flight Test phase of the shuttle, the agency was preparing for EVAs that would test the MMU and demonstrate various tasks that would be required during satellite service missions.

Trevino's work included dives in the WETF in Houston and the NBS at Huntsville to test equipment in conjunction with the EVA operations staff in the Crew Training Directorate. A full-scale orbiter airlock and payload bay (without an operational RMS) was intended for the WETF but this would not be ready in time for the planned series of Detailed Test Objectives (DTO) leading up to the Solar Max mission. To resolve this, Trevino designed a test fixture that acted like the RMS in that it could correctly position crewmembers during their training sessions to assist with preparations for STS-41C.

In addition to all this, the Hubble telescope was in its final design stages and Trevino was asked to work with Lockheed engineers in Sunnyvale, California. He participated in working out the locations for EVA interfaces (i.e. handrails and portable foot restraint sockets) to ease the motion of astronauts across the main surfaces of the structure during their servicing tasks, avoiding touching its delicate surfaces. Trevino also worked on the design of the ORUs and the important elements of developing a standard design for fixings and their associated tools. According to Trevino, astronaut Bruce McCandless was extremely instrumental in ensuring that the 7/16th inch (11 mm) hex became the main bolt standard across the program.

## Ron Sheffield

Another key figure in the early development of tools for Hubble servicing was Ron Sheffield. On retiring from Army aviation as a lieutenant colonel with three tours of Vietnam, he joined Lockheed as an engineer. After participating in the development of the

Air Elevator Support Trailer and the MX Peacekeeper missile, in 1985 he was assigned to the Hubble project with the task of making as many as possible of its components replaceable ahead of launch. The decision to conduct all servicing in space instead of on the ground led to a steady growth in the number of orbital replacement units. However, cutbacks during 1980–1983 reduced the ORUs to 14 types. By 1985 there were only a total of 24 individual ORUs on the telescope. The delays in re-qualifying the shuttle after the loss of Challenger in 1986 gave Sheffield's team another 4 years to accomplish their task.

One difficulty faced by Sheffield was in authorizing conversion to an ORU on the actual telescope. "A major hindrance in accomplishing my task was that I could not disconnect or remove any component from the telescope unless it was in need of testing or repair." But he prevailed, and when Hubble was launched in 1990 "there were a total of 94 types of ORUs that had been upgraded to have EVA enhancements. That was a total of over 120 ORUs and spares by launch time."[1]

Engineers Ron Sheffield (right) with Steve Leete make new entries in the flight log while on-console in the Flight Control Room at JSC during SM-4 (STS-125).

Before launch, Sheffield and his team had to fit-check, verify, and document all the spares and ORUs on the telescope. They also took over 25,000 photos of individual components and interfaces, and these proved invaluable in planning the service missions. "We also had over 3 hours of various video of HST activities and I later made a 1 hour video of all the interfaces on the HST components," he recalled. The video was used for training astronauts assigned to the service missions. Sheffield worked closely with Bruce

McCandless and Kathy Sullivan during many of the early HST component fit-checks, and his team wrote all the initial EVA procedures, all the way from the deployment mission to the final service mission. The team also built all the HST NBS mockup components and did the setup and removal for the daily NBL dives, with many of these simulations running for between 2 to 4 days. Over the years, Sheffield also planned and wrote the procedures, conducted, and documented the numerous crew familiarization activities using flight hardware and both high-fidelity mockups and the actual flight tools.

For each mission Sheffield personally conducted the hardware walk-down with the flight crew at KSC prior to the payload being installed in the shuttle. On each service mission he also planned and conducted the final walk-down of the HST hardware after its installation in the payload bay. His team wrote the initial EVA procedures for all of the service missions, built and maintained the HST component mockups, and conducted the water tank activities. During a mission Sheffield worked in the support rooms at JSC, having developed, updated, and published training aids to help not only the flight crew but also HST program personnel at Lockheed, Johnson, Marshall and Goddard. This work familiarized the support teams and management on EVA events and hardware for the upcoming flight and provided a reference for each mission. The work included making over 500 HST scale models, the HST outboard profile and ORU inboard component layouts, as well as payload bay configuration diagrams.

Many astronauts have applauded the work of Ron Sheffield and his team for aiding their preparations for Hubble servicing.

**Design standards**

"More than 80 tools, including common wrenches and screwdrivers were available to service and maintain the observatory. In general the tools are functionally equivalent to tools used on the ground. Some are standard items in the shuttle contingency EVA tool set while others are unique to the telescope," wrote the Space Telescope Projects Office at Marshall in November 1986.[2]

As the launch of Hubble approached, the tools that had been developed for the astronauts were widely compared to those of general industry, or to those sold by a home improvement store or those found in any domestic DIY tool box. Of course, there was some truth in these comparisons but the selection of tools for the astronauts ranged from some which were very familiar, to others which were specially designed for the task in hand. And naturally they all had to be usable by an astronaut in a pressure suit working in space.

One of the most important considerations was the compatibility of the tools to the ORUs that were to be installed, repaired, maintained, or replaced on the telescope. Originally each ORU was designed to use standard fittings, so that only a limited number of tools would be needed to install them. But as the list of ORUs grew, so did the methods of fixing them and more tools were conceived to work the fittings. With the decision to service the telescope in space, the range of ORUs had to be re-evaluated in terms of the number of tools that would have to be carried in order to accomplish on-orbit servicing. For ground servicing, the tools had not been required to be used by a weightless astronaut wearing a pressure glove. It was too late in the design to completely reconfigure these orbital elements, but there was time to amend some of the associated tools.

The most common type used in mounting the ORUs were 7/16th inch (11 mm) bolts with double height hex heads. For this option there were so-called J-hooks, captive fasteners, and keyhole fastenings. To work on these ORUs, the astronaut would use an extension socket on a ratchet wrench. There were also extensions up to 2 feet (0.60 meter) in length to facilitate operating in otherwise inaccessible areas, plus torque limiters to prevent the over-tightening of a fastening. Shrouded-tip sockets and screwdrivers were developed to capture bolts and screws as they were unscrewed. In addition, wobble sockets and flex-tip screwdrivers were added in order to aid access and grip in hard to reach positions.

**Special tools**

Despite all of the reviews aimed at matching standard tools with the increasing number of ORUs, there remained some fastenings, mostly electrical connections, which simply could not be manipulated other than by special tools. In order to loosen standard D-connectors an astronaut had first to release two slotted screws and then move them back and forth. A tool with variable jaws which resembled a pair of levers (pliers) that was usable wearing gloves was created to assist in loosening the connector without touching the wire harness. A pair of grippers ensured a straight entry back into the housing when reinstalling the connector. Any circular connectors were simply rotated for mating.

Some of the electrical ORU boxes featured connecters without wing tabs due to the close proximity of other connectors, which made it extremely difficult to manipulate an individual connector with bare fingers, let alone when wearing a pressure glove. A modification would require a redesign of the tabs and connections and the provision of additional volume for the fit, which would result in an increase in the cost. To overcome this, a special pair of grippers (connector pliers) were adapted from another EVA tool set and these essentially functioned as an extension to an astronaut's fingers in order to achieve a firm grip and allow rotation of the connector sleeve. This cleverly eliminated the need to redesign the connector.

## THE HST SERVICING TOOL KIT

The Hubble tool kit also featured a variety of carrying and handling aids, including portable handles to allow extraction of ORUs, supplementing those provided for replacing the Solar Arrays and the Aperture Door. There were also tool caddies, tethers, transfer bags, and a range of covers that would be used to protect replacement instruments and star trackers from contamination. No mission carried a complete inventory of tools. A given mission required only between 150 and 200 tools, and they were selected to match the ORUs carried and the tasks assigned. The separate shuttle EVA tool kit was available to supplement the specialist tool kit, and in the event of unforeseen situations the ingenuity of the astronauts, engineers, and flight controllers usually achieved a workable option.

The standard EVA tool selection for a Hubble mission included:

*Screwdrivers*:

- Shrouded flex screwdriver, 4.8 and 8.6 inch shafts (12.1 and 21.8 cm)
- Shrouded rigid screwdriver, 3.8 and 8.3 inch shafts (9.65 and 21.0 cm)
- Torque-set tip tool #10, 10.3 inch shaft (26.16 cm)

## SHUTTLE EVA TOOLS COMPATIBLE TO HST SERVICING MISSIONS (1993)

The document from which this selection was taken lists approximately 350 tools available in the Shuttle EVA tool inventory. Not all tools were carried on every mission and those presented in this table represent those which supported the first HST Servicing Mission. The Shuttle tool manifest constantly changes over the course of the program but these represent a sample of what could be adopted from the standard Shuttle EVA tool kit to any of the Hubble Servicing missions in support of the mission specific tools listed in the text.

| | |
|---|---|
| Bag, HST Tool | PFR Socket (HST) |
| Bolts, EVA Captive | PFR Socket Converter, 90° |
| Connector Demate Tool, D | Portable Foot Restraint (HST) |
| Connector Mate Tool, D | Power Ratchet Tool |
| Connector Pin Straightener, Multisize | Power Tool, EVA (HST) |
| Connector Tool, Circular (HST) | Power Tool, Rotary Impact |
| Connector Tool, Coax | Ratchet, 3/8-inch Drive McTether |
| Connector Tool, Locking Electrical | Screwdriver, Shrouded Flex |
| Connector Tool, Round Coax | Screwdriver, Shrouded Rigid |
| Door Stay, Adjustable, SSM | Socket, 5/16-inch (HST) |
| Door Stay, OTA | Socket, 7/16-inch Adjustable |
| Drive Unit Preload Tool | Socket, 7/16-inch (HST) |
| Fuse Transfer Container | Socket, Extension and 7/16-inch (HST) |
| Grapple Fixture, Portable Flight Releasable | Stowage Box, HST Tool |
| Handle, Jettison/Transfer | Tether, Adjustable Equipment (HST) |
| Handle, Large Portable ORU | Tether, Semirigid |
| Handle, Primary Deployment Mechanism | Tool Board, Cushioned |
| Handle, Small Portable ORU | Tool Caddy, McCaddy and Ratchet |
| Hook, J | Torq-Set Tip Extension (HST) |
| Light, Portable Flood | Torque Limiter, Multisetting |
| Mechanical Finger | Wrench, 7/16-inch and ½-inch Box End |
| PFR Extender | Zip Nut |
| PFR Ingress Aid | |

Taken from: *EVA Tools and Equipment Reference Book (Formerly EVA Catalog Tools and Equipment) R.K. Fullerton, NASA-TM-109350, JSC-20466 Rev B. November 1993*

Shuttle EVA tool list.

*Wrenches and sockets*:

- Right angle drive tool
- 5/16th inch (8 mm) rigid hex capture tool, 10.3 inch (26.16 cm) shaft
- 5/16th inch (8 mm) wobble hex capture tool, 10.3 inch (26.16 cm) shaft
- 5/16th inch (8 mm) non-capture wobble socket, 7.3 inch (18.54 cm) and 10.3 inch (26.16 cm) shaft
- 7/16th inch (11 mm) and 0.5 inch (12.7 mm) box end wrench
- 3/8th inch (9.5 mm) drop-proof tether ratchet and caddy
- 7/16th inch (11 mm) open-ended ratcheting wrench
- 7/16th inch (11 mm) rigid hex capture tool, 10.3 inch (26.16 cm)
- 7/16th inch (11 mm) wobble hex capture tool, 10.3 inch (26.16 cm)
- 7/16th inch (11 mm) socket extensions, 6, 12, 18, 24 inch (15.2, 30.4, 45.7, 60.9 cm)
- 0.5 inch (12.7 mm) box ratchet wrench
- Torque limiters, 6.5, 9.0, 35 foot-pounds (1.98, 2.74, 10.6 meter-kg)

## The HST servicing tool kit 253

*Electrical connectors*:

- Circular connector tool, 0 and 90 degree jaws
- Coax connector tool, hex (with and without shoulders) and round
- Installation and internal and external removal tools
- Multi-size pin straightener

*Lighting*:

- HST portable work lights
- EVA flashlight

*Handling and positioning aids*:

- Preload tool (High Gain Antenna/Aperture Door)
- Mechanical finger
- Shepherd's hook
- Adjustable door stays
- Portable Foot Restraint, plus extender and socket
- Manipulator Foot Restraint
- Portable EVA grapple fixture
- Portable handhold plate
- Portable ORU handles
- Tool boards
- Assorted tethers
- Caddy, with French hooks
- Standard caddy
- Jettison handle

*Transfer gear*:

- "Clothesline" transfer aid
- Transfer bag
- Trash bags
- Multiple transfer system
- Fuse transfer system

*Power tools*:

- Power tool, high torque/low rpm

*Cutting tools*:

- EVA scissors
- Cable cutter

*Protective covers*:

- Wide Field/Planetary Camera mirror cover
- ORU electrical connector covers
- Fixed Head Star Tracker delta plate cover
- Fixed Head Star Tracker light shade cover
- Fine Guidance Sensor mirror cover
- 7/16th inch (11 mm) non-capture wobble socket, 3 inch (7.6 cm).

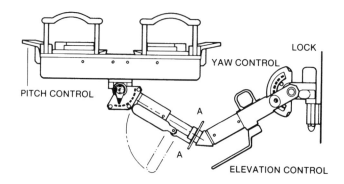

Mobility detail for the Portable Foot Restraint (PFR).

**Power tools**

Three power tools and a torque limiter were flown on the first Hubble service mission.

The *HST Power Tool* supplied by JSC was a modified battery operated tool with torque and rpm control. Its design included a 3/8 inch (9.5 mm) drive which had both forward and reverse drive options. Its torque could be varied from 50 to 300 inch-pounds-force (6.913 to 41.48 cm-kg-force). It came with a bracket to enable it to be installed on the spacesuit. JSC also provided the battery operated *Mini-Power Tool* to act as a screwdriver when the larger ratchet was not needed and where the available space was limited. It could be operated with or without power and there was a lock on the output shaft for manual operation.

Goddard developed the *Power Ratchet Tool* (PRT) for SM-1. It was of a far more robust design than the JSC power tools. This titanium and aluminum tool measured 17 inches (43 cm) and had a mass of 8 pounds (3.62 kg). It applied between 0.5 and 25 foot-pounds (0.0069 to 3.456 meter-kg) of pressure when powered by the 28 volt silver-zinc battery, or 75 foot-pounds (10.37 meter-kg) in the manual mode. The drive speed was variable between 10 and 30 rpm. In contrast, about 2 to 5 foot-pounds (0.2765 to 0.69 meter-kg) would be exerted by hand for a regular screwdriver. This tool was to be used on tasks which required controlling speeds, torque, or turns. It featured an option that enabled it to be applied at right-angles, for increased flexibility to the user. The unit consisted of the main power wrench, the controller, umbilical, and battery module. A spare ratchet was also carried on the orbiter.

The success of the PRT on SM-1 led to the development of a smaller and more efficient *Pistol Grip Tool* (PGT). This was evaluated successfully on several shuttle missions before being used on the later Hubble service missions and carried over to ISS assembly missions. The PGT was a self-contained computer controlled and battery powered 3/8th inch (9.5 mm) drive tool featuring a pistol grip handle. Being computer controlled it offered several torque, speed, and turn limits. These could be pre-programmed into the tool for a range of mission-specific applications. A light emitting diode (LED) on the tool indicated the torque that was being applied, at what speed, and how may turns the motor had made.

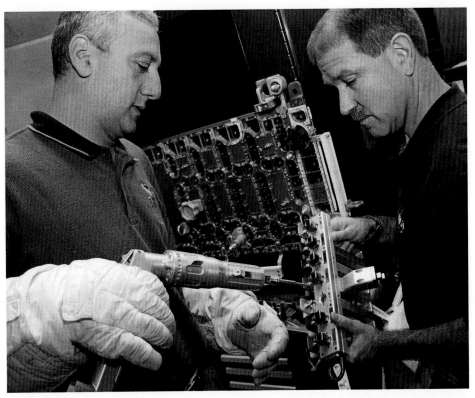

Mike Massimino and John Grunsfeld evaluate the power ratchet capability during 1-g EVA bench test simulations for the final servicing mission.

Michael Good (STS-125) practices installing a battery module into the Hubble High Fidelity Mechanical Simulator, in the cleanroom at NASA's Goddard Space Flight Center.

256  **Tools of the trade**

The LED could also display a range of error messages. The motorized torques ranged from 2 to 25 foot-pounds (0.2765 to 3.456 meter-kg), the speeds from 5 to 60 rpm, and the number of turns from 0 to 99. In the manual mode it could apply a torque of 38 foot-pounds (5.26 meter-kg).

**Tool storage**

Along with other crew aids, the tools were stored on (or inside, as appropriate) the FSS, the ORU carriers, the HST Tool Box, adapter plates on the payload bay side-wall, the Adaptive Payload Carrier (APC) and the Provisional Stowage Assembly (PSA). There were also a few stowage provisions for spares in the crew compartment, mainly in mid-deck lockers.

The following is an example of the mission-specific tools for a service mission, in this case SM-1:

| The first service mission tools and crew aids | |
|---|---|
| Tool/crew aid | Stowage location |
| Axial Safety Bar | ORUC (Orbital Replacement Unit Carrier) |
| BAPS Support Post | FSS Latch Beam |
| Co-Processor/DF-224 | – |
| Stabilization Post | ORUC |
| COSTAR Contamination Cover | ORUC (launch): starboard PSA (return) |
| Delta Plate Cover, FHST | Starboard PSA |
| FSS Handrails | FSS |
| FSS PFR Sockets | FSS |
| Light Shade Cover, FHST | Starboard PSA |
| Low Gain Antenna Cover | FSS |
| ORUC BISIPE Handrails | ORUC |
| ORUC BISIPE PFR Sockets | ORUC |
| ORUC Pallet Handrails | ORUC |
| ORUC Shelf Handrails | ORUC |
| ORUC LOPE Handrails | ORUC |
| ORUC SOPE Handrails | ORUC |
| ORUC PFR Sockets | ORUC |
| PFR Socket Converter (90 degrees) | Starboard PSA |
| PFR Ingress Aid | ORUC |
| Portable Grapple Fixture | Solar Array Carrier |
| Solar Array PDM Handle | Solar Array Carrier |
| Solar Array Spines | Solar Array Carrier |
| Solar Array Transfer Handle | Solar Array Carrier |
| Solar Array AFR | Solar Array Carrier |
| Solar Array Carrier Handrails | Solar Array Carrier |
| Solar Array Carrier PFR Socket | Solar Array Carrier |
| COSTAR Handling Aid | ORUC |
| High Speed Photometer Handling Air | FSS (launch): HSP (return) |
| Umbilical P105, P106, Connector Cover | Middeck Locker |
| WF/PC Portable Handhold with zip nuts | ORUC |
| WF/PC Guide Stud Adaptor | Middeck Locker |
| WF/PC Mirror Cover | ORUC |

(continued)

| The first service mission tools and crew aids | |
|---|---|
| Tool/crew aid | Stowage location |
| HST Radial Bay Cover | Solar Array Carrier |
| 7/16-inch Socket Extension, Long | Middeck Locker |
| 7/16-inch Socket Extension, Short | Middeck Locker |
| Connector Locking Tool | Middeck Locker |
| Multi-setting Torque Limiter | HST Tool Box and Middeck Locker |
| Power Ratchet Tool Wrench | Middeck Locker |
| Power Ratchet Tool Controller | Middeck Locker |
| Power Ratchet Tool Battery Module | Middeck Locker |
| Power Ratchet Tool Transfer Bag | Middeck Locker |

Decisions were made to delete, add, or refine the tools available for a service mission based on experience in space and in the NBL simulations. After Hubble was deployed in 1990, and as preparations continued towards the first service mission, several adjustable extensions shafts were made available to reduce the time that an astronaut would require to complete a tool changeout. These replaced several earlier fixed-length extensions, and they featured adjustments from 12 to 16.5 inches (30.4 to 41.9 cm), and 15 to 24 inches (38.0 to 60.9 cm). When carried fully retracted, these shafts were far more compact than the earlier units and hence were less likely to cause damage to hardware and instruments, or injure an astronaut.

Servicing the larger items of hardware on Hubble required a variety of specific tools. For example, changing out the Wide Field/Planetary Camera with its successor required WFPC handholds, WFPC guide studs, quick-release zip nuts, WFPC pick-off mirror cover, forward fixture, aft fixture, and HST radial bay cover. The changeout of the High Speed Photometer with COSTAR required the COSTAR contamination cover, COSTAR handling aid, an HSP handling aid, forward fixture, aft fixture and Axial Science Instrument Protective Enclosure (SIPE) safety bar. The replacement of the solar arrays required an articulating foot restraint, solar array primary drive mechanism handles, solar array temporary stowage brackets, solar array transfer handles, solar array jettison handles, solar array spines, Portable Flight Release Grapple Fixture, and a Marmon clamp. Servicing the Gyro Rate Sensors required a Portable Foot Restraint socket convertor (90 degrees), Fixed-Head Star Tracker (FHST) light shade covers, and a FHST delta plate cover.

In addition, there were multi-setting torque limited adjustable extensions with 7/16 inch (11 mm) sockets, ingress aids, portable light receptacle, locking connecting tool, Low Gain Antenna covers, umbilical connector covers, the Berthing and Positioning System (BAPS) support post for the Flight Support Structure (FSS), and a multi-layer insulation repair kit.

These "tools of the trade" for astronauts assigned to Hubble missions have a fascinating story of development and acceptance behind them.

## RUSSELL WERNETH, MANAGING THE TOOLS

When Russell L. Werneth announced his retirement from Goddard Space Flight Center in early March 2007, many of his co-workers were surprised, having presumed that he would wait for the SM-4 mission of STS-125 in 2009. But no, he left at the end of the month after almost 43 years in government service, the last 15 of which were spent working on Hubble,

Russell Werneth HST EVA Manager with a model of Hubble. (Courtesy Goddard Space Flight Center)

primarily on EVA tools for the telescope. For Russ this was not just work, it was a passion, and the decision to leave was not an easy one because he knew he would miss "the Hubble people". But it was time to move on and he wished to do other things, maybe even return to teaching. His boss, Frank Cepollina, said that Russ's expertise would be sorely missed, and was surprised to see him go prior to the final service mission.

After retiring, Russ couldn't keep away from Hubble for long, and today he is kept busy with public and education outreach activities as the Education and Public Outreach Engineer for the Hubble Space Telescope Project.[3]

Werneth was a mechanical engineer by training, but NASA preferred to consider him an aerospace engineer. In 1990 he was assigned as Hubble's Crew Aids and Tools Manager at Goddard, and later moved up to become EVA Manager under Cepollina. Prior to this, in the 1980s, he had worked on space station robotics, then transferred to astronaut training before heading up the work on developing EVA tools for Hubble servicing.[4]

**Test, test again, and then re-test**

Werneth explained that when the need for tools for Hubble servicing was first established, the logical place to begin was the existing JSC tool kit for shuttle EVAs. By design, these were mostly generic to cope with specific contingency tasks that could occur on any mission, such as closing the payload bay doors. The difference on Hubble was that although the structure of the telescope remained the same, each payload task could be different and therefore require a set of custom-built tools. The team did not want to invent new tools unnecessarily, and often a standard tool could be adapted for use on Hubble, but the majority of tools did end up being uniquely designed or adapted for the tasks to be accomplish. It was a magnificent team effort by all concerned.

As Werneth acknowledges, there was some pressure to get the job done, and done well. Initially his team set up standard criteria, knowing the launch loads and the environment in which they were to operate. Safety factors were observed for each and every tool, together with rigid specifications for testing and certification. Werneth clearly recited one of Frank Cepollina's primary objectives: "Test, test again, and then re-test." It was ingrained into the approach of the team. This was an important lesson to learn, because the time spent on the ground in developing the tools would save valuable EVA time on-orbit, and cut costs. The tools had to work correctly, safety, and efficiently. Each shuttle mission had a given mass allocated to the overall payload, and the assigned tools were a small percentage within that. According to Werneth, there was never any concern in his team that the chosen tools might cause a substantial increase in the overall payload mass.

The way that the tool selection worked out for a given mission commenced with a team coming in from NASA Headquarters to Goddard to discuss the mission priorities for what had to be done. That discussion identified the structure of what was planned, specifying the order of importance. This prioritizing influenced what the crew would attempt to do by EVA during the mission, which tasks would be assigned to which EVA, and the sequence in which they would be attempted. This in turn led to the "mission success" targets for each day of the mission. Given an outline schedule, the set of tools for each EVA's tasks could be identified and assembled in the agreed sequence they were to be used. The tool catalogue was used as a baseline reference but this was constantly updated, with redundant tools being deleted as new ones became available.

The number of staff on Werneth's team varied depending upon the work to be carried out, but for those at Goddard it was between 15 and 20 members. To gain further information or insight into the specific hardware that the crew would work on during a service mission, the team would "lean" on scientists, engineers, and the designers who built the carriers and other apparatus, or ran the simulations in the NBL. There would be some 50 people from Goddard in the support rooms at Mission Control in Houston to support each EVA, but it could be quite fluid, depending upon how things played out in real time.

**Team work gets the job done**

For Werneth, part of the satisfaction of working on Hubble was the tremendous teamwork in getting the job done. Success was due to everyone pitching in, and all working hard not only on their own assignment but also supporting their colleagues. The part of Werneth's team in Mission Control "working consoles" was split into three groups in support of the three-shift operation of the main flight control teams, referred to as Orbit 1, Orbit 2 and Planning Shift. Supporting the Goddard group at the same time as they supported the flight controllers was a matrix of people known as "systems engineers". Being focused on the task in hand, "no one actually looked at a name badge—they could be specialists in EVA, systems, in thermal, or mechanical, it didn't matter—as they came together on shift of 10 to 12 hours, overlapping the outgoing or incoming shift for an hour each side of their own tour of duty," he explained. The members of his team occupied support rooms adjacent to the main MCC room, tracking the flight controllers in the MOCR and working closely with the EVA group there. The team effort extended not only to the actual missions but also the many training sessions. Although the crew training program was known, it was not closely followed by the people at Goddard, though they did integrate into the wider team, especially for the EVA training sessions. They also participated in the post-flight debriefings and evaluations to identify the lessons learned and follow up on items that needed further attention.

One area of training for Hubble EVAs that Goddard staff did not really contribute to was the development of "virtual reality" (VR). Introduced in the early 1990s, this was only really used to some extent on the final two or three service missions, primarily focusing upon the larger scientific instruments, and was under the control of JSC because, as Werneth put it, "size, mass, and tool interface issues were mainly a JSC training function".

**Retaining the "team"**

With almost 20 years between the Hubble deployment mission and the final servicing flight, the Astronaut Office and the support teams suffered attrition. The challenge was to preserve the experience and knowledge. According to Werneth, the departure of astronauts who were highly experienced in EVA did not really impact the work at Goddard. "This was a credit to JSC and their system of crew selection, which firmly remained a JSC function." NASA was good at ensuring that new astronauts learned from their predecessors. Goddard looked after the servicing payload, payload/EVA training, and the NBL.

As one service mission was wrapped up, the early stages of the next one were already in progress. It was the management at JSC that determined which astronauts would fly the next service mission. A number of factors went into the decision, and several candidates might be under consideration for a single slot. In fact, even during the 2 weeks usually reserved for a series of NBL tests, the Astronaut Office would typically assign a number of astronauts to participate in the tests for experience and possibly to evaluate their potential for a particular mission. Neither Werneth, nor, he suspects, the astronauts themselves, had any idea whether these support roles would transition to actual Hubble flights. This is borne out by the fact that several astronauts who had participated in NBL tests related to Hubble development over the years were never selected for a service mission; e.g. Shannon Lucid, George 'Pinky' Nelson, Jim Voss, Jay Apt, and Rich Clifford. Other astronauts, including Anna Fisher, may also have been involved.

The Astronaut Office has always selected astronauts by a variety of known and unknown criteria. Obviously some of this was past experience and compatibility for the role and crew, but very few people knew the full criteria used. It was strictly a management function at JSC. Werneth observed no impact of astronaut rotations into or out of Hubble support roles; most moved to the ISS. "Hubble missions remained very popular with the astronauts," he recalled. A number of astronauts who flew on Hubble missions were invaluable for their participation in, and feedback concerning tool development. According to Werneth, John Grunsfeld, who was an astrophysicist, "had an enormous mechanical ability and provided a lot of feedback used in the development of tools and procedures, as did Mike Massimino and Scott Altman".

Throughout the program, the work in the water tanks at Marshall and Johnson continued. Werneth's team reported on the insights gained from these exercises and, where appropriate, the results were incorporated into design or procedures changes. It was inevitable that things would change over the years, as experience was gained in managing and operating missions. As the focus at JSC moved towards work relating to the assembly of the ISS and less on the independent scientific or payload deployment flights of the shuttle, Goddard maintained its focus on Hubble.

One clear example of how the experience of Goddard affected its role in developing tools for Hubble was with the power tools. At first the power tool developed by JSC was used, but then Goddard developed its pistol grip tool that matched specific requirements. It was not so cumbersome and had a computer chip which recorded the torque, speed, direction, turns, etc. This data was analyzed post-flight, indicating which tool-bit was selected, when it was used during the EVA, for how long, and at what rate. It was then possible not only to measure the effectiveness of a specific tool in a given task but also to determine the degree of wear that it suffered in performing a task so that the tools could be made more efficient.

Werneth mentioned that the early scenes of the 2013 feature film *GRAVITY*, in which the character played by actress Sandra Bullock is depicted working on a tray attached to Hubble was indeed "great entertainment and good animation but that concept could not be done on the real telescope". But one point he applauded—which is often explained to others desiring to learn about "seeing space 'tools' in action"—was that the tool featured on the movie was numbered, and that replicated the real tools which were also numbered so that the engineers on ground could see by video coverage of real EVAs which tool was being used and how it performed.

Returning to the pistol grip tool, Werneth explained that the plan for which tool should be used for which task was included on the internal chip, stored alongside the real-time data of in-flight usage. The pre-flight plan and the in-flight data was used in post-flight analysis by the tool team while debriefing the flight crew about which tools worked, which didn't, and whether things could be improved. The data from the tool, video of it in use in space, and a recording of the astronauts' personal and subjective interpretations were analyzed to obtain useful feedback on the ergonomics of the tool. Werneth and his team then worked out any improvements. If necessary, prototypes were built for testing underwater in the NBL and in 1-g simulations. If the modifications could be introduced before the next service mission, then the astronauts would benefit from better tools to make their EVAs even more efficient and safer.

When things didn't go well in space, there were procedures and systems in place to deal with failures and contingencies; all created by Werneth's team, usually months prior to that flight. The team went through various scenarios for each mission, initially step by step for a nominal mission in which there were no problems, then they identified likely contingencies, bearing in mind that, as Werneth explained, "no one could predict every contingency, but a favorite approach was 'what if a tool broke?' or 'what if they had a problem with torque?'"

The answer was to analyze as many potential contingencies as they could. Some sessions involved simply sitting around a table and discussing what might be done in any number of situations that were deemed to be lower tier contingencies. Backup procedures were always documented for future reference. Other contingencies featured the use of these backup tools, alternative operations, and even spare tools. This measure of flexibility in the EVA planning allowed for changes to the written procedures in real time. Hence if the power grip tool did not work, its torque limit could be changed. If a bolt was stubborn the response would be to drop a torque factor and continue the task, whilst remaining within the safety rules and not risking damage to the tool. Another option was to select an alternative tool or process to achieve the desired result. Wherever there was a problem, the communication between the crew on EVA, the flight controllers in Houston, and Werneth's team in the support rooms at either JSC or Goddard, along with contractors, offered a valuable and important network to address and solve the problem, ideally promptly but always safely.

Part of this support was the extensive library of photos and drawings of the telescope that had been built up over the years, thanks mainly to Ron Sheffield and his team at Lockheed. Because Hubble was fabricated in the early 1980s and launched in 1990, most of Werneth's team had never seen the real hardware for which they were developing tools, so there was a tremendous requirement for drawings, diagrams, photographs, and models. They ended up with a vast amount of highly detailed documentation on Hubble. For example, for the latch configurations the tool designers contacted the engineers who had built that portion of the telescope. The scientific community was an invaluable source of reference material for the instruments.

As regards surprises encountered while using his tools, Werneth recalled one thing from the first service mission. At Goddard they had built a 1-g Aft Shroud Door Trainer in which the astronauts practiced the ordering for the latching sequence and the forces needed. As it turned out, there were some differences between the training devices and the flight devices. Fortunately, from their training and awareness of the available contingencies the crew knew enough to overcome the problem. Werneth remembers that the best feedback to his team by the crew during their post-flight debriefing was, "Everything looked exactly like

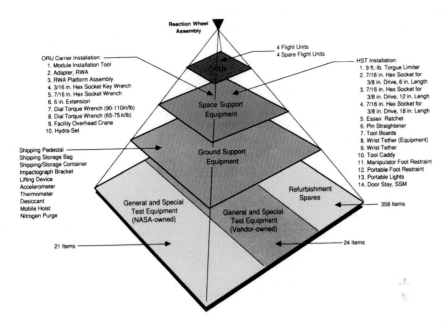

Typical ORU Support Pyramid for the Rate Sensing Unit.

you trained us in 1-g, or underwater, but you guys weren't there in scuba and we didn't see any bubbles coming up." In other words, the simulations were so similar to the real thing that there was little to distinguish between them, apart from their surroundings.

Some of Russ Werneth's lasting highlights of the program were the final walk downs of the shuttle prior to each service mission and the pride he felt as each EVA was successfully completed. When asked, after so many years of working at Goddard and on Hubble if there was anything he would have done differently in his career, he replied, "I would've started at Goddard sooner." Clearly, he was a man who loved his job.

## MECHANISMS ENGINEER PAUL RICHARDS

In the late 1980s Paul Richards was a satellite mechanisms engineer at Goddard working on the flight robotics servicer for Space Station Freedom. After the cancellation of the servicer, Richards decided to work on something which had a higher possibility of actually making it into space. In 1990 he joined the Hubble team under Russ Werneth, just after the telescope had been found to suffer from spherical aberration.

Richards was given a part-time assignment on the multi-setting torque limiter tool which was intended to ensure that bolts were not over torqued. He soon found design flaws in the mechanisms, and began to redesign it. He also investigated why the original power ratchet tool wasn't working to its perceived requirements. This work eventually led to a full-time position as a team leader, overseeing the redesign of many of the tools. He even invented a few tools himself, including the first adjustable extension tool that was used by the crucial first service mission. At around this time, Richards started to participate in the underwater tests at the Huntsville NBS, which he reflects on as "one of the really fun parts of job".[5]

His team developed mockups and flight versions of the tools, then evaluated them in the NBS. As a result of this work, he became associated with the astronauts training in the tank and, having aspirations to venture into space himself, he reckoned his scuba experience in the tank would assist in his selection as an astronaut, which it did several years later.

After each simulation, Richards would write a report in which he detailed the events and activities, explaining the use of the tools in great depth; a detail which others apparently did not do. As he has observed, this work became much more important the closer the servicing and ISS missions came. The Hubble service missions needed several EVAs back-to-back on successive flight days owing to the short duration of a shuttle mission. And things would be even more intense on the ISS. As Richards explained, "If NASA and the astronauts couldn't mount four or five EVAs at Hubble and do them back-to-back by successive teams following a strict and complete timeline with intricate tasks which had never been done before, then the premise of the ISS was flawed." At that time, there was significant pressure upon the service missions not only to resolve the mirror problem and to maintain the telescope flying but also to provide EVA feedback that could be applied to preparations for the space station. Within the ISS EVA program, the unprecedented challenge of making a series of multiple EVAs on successive assembly flights in order to create an independent facility, and then expand it into a huge orbital complex, became known as the "Wall of EVA".

Richards acknowledges the depth of experience held at Goddard. Coming to the Hubble team as a freshman to EVA equipment and spacewalking procedures, "I was learning a lot from what were known as 'grey beards', engineers who had over 20 or 30 years' experience in designing satellite mechanisms—having lunch with them every day, traveling with them, and consulting with them. I was sort of an apprentice, and I learned quite a lot about how to design satellite mechanisms and how the tools of Hubble were designed." Hubble was a new challenge for everyone, involving ways to devise new techniques and operations for an item of space hardware who's mission was planned to last 5, 10 or 15 years on-orbit. "Previously in mechanisms, the case was you did certain things a certain way. Typically the mechanisms present the highest probability of failure on a satellite because you have to get something to move in that harsh environment." Richards had to learn new "tricks of the trade" to prevent these failures, then he applied that experience to the design of the tools.

**A frosty welcome**

Goddard was responsible not only for maintaining the payload on the telescope, but also for any new hardware that would be installed. Logically, therefore, they developed the tools for servicing the telescope with that new hardware. The problem was that JSC had tools of their own, developed for the shuttle program, that it believed could be adapted to servicing tasks. This duplication naturally created a rivalry between the two centers. Richards, who was new to this, received a very frosty welcome to the tool world, because it was thought that all the tools ought to be developed at JSC. The expectation at JSC was that the power ratchet tool and a torque limiter that he had been developing would be carried by STS-61, tested, prove unsatisfactory, and be deleted from future manifests, but events turned out very differently.

As Richards recalls the difficulty in getting JSC to accept the Goddard tools for flight—shortly prior to the first service mission he and a colleague attended a conference at JSC at which they were told that their tools were no longer needed, even though they hadn't flown yet. The shuttle would carry them, but only in the airlock and they were not scheduled to be used. Richards was confused, since not to test the tools would make a post-flight debriefing pointless. Nevertheless, even though his hard work and dedication were acknowledged, the recommendation was that his tools "were too complicated, too expensive and were just not needed at JSC, which had plenty of tools that could be used on the Hubble missions". As a mechanisms engineer, Richards was dumbfounded. Having personally built the power tools at Goddard he pointed out that a tool that JSC was proposing to fly was only a "plastic tool wrapped with aluminum tape, and as soon as you pull the trigger that tool is going to knock out your communications because of the brushed motors". The motors would interfere with electromagnetic frequencies on the UHF band and, furthermore, the particular design of the brushes in the motor would "deteriorate in a vacuum and they would jam the bearings, and that was if the batteries didn't cease first". But the JSC contingent was insistent: the mission would carry the Goddard tool as a backup that would be left in the airlock during the EVA. Richards left the meeting a little aggravated, to say the least.

However during STS-61, as soon as the trigger of the JSC tool was pulled the astronauts couldn't determine whether it was working correctly. Then the first tool failed. A short time later the second tool failed. An hour into the first EVA, the astronauts returned to the airlock to fetch the power ratchet designed by Richards. It functioned perfectly, enabling the crew to finish their tasks. After the EVA, Jeff Hoffman, speaking from orbit, thanked Paul Richards for the ratchet, and said that it was simple to use on-orbit, easily "stops on a dime" and was "probably the best power tool I have ever used". This was praise indeed for a tool that JSC had said was "never going to fly".

With that type of endorsement from the astronauts, things changed rapidly. Richards was amazed at the post-flight debriefing, in comparison to his experience at the pre-flight shake down. All of a sudden Houston wanted him to redesign all their power tools and asked if he could make them smaller. So he set to work and invented the pistol grip tool for Hubble. He also pointed out that if he was going to continue designing tools then he really ought to have some experience of wearing an EVA suit underwater in order to better understand the issues. This gave him even more exposure to the astronauts and EVA experts from JSC, explaining to them each tank run prior to their simulation. He also offered his thoughts on which tools worked best or how a modification might enable them to work better. He became proficient in wearing the EMU and conducted his own tests.

The question of how a given tool was selected for Hubble and fitted into the system was explained by Richards as a combination of the need to adapt and use existing tools but also devise a specific tool for a particular role which could also be adapted for other uses where possible.

Richards saw one change as the program developed. Both the power ratchet tool and the torque limiter had been fit-checked with the telescope before that was processed for launch, but after STS-31 the development of any new tools which would directly interface with the telescope could not be fit-checked, because the telescope was in space. The development of these tools was a little risky, as it was possible they would not fit and function when applied during an EVA. As Richards explained, "It took a lot of detective work to ensure we made the tools with known tolerances, being careful not to design something that was too large a diameter or too small, too short, or too long." As the tolerances could be in thousands of an inch, it was a fine line between a useful tool and a wasted element of hardware. Fortunately, most of what was provided for Hubble worked as designed, reflecting the skills, dedication and hard work of people like Richards, Sheffield, and the others involved in preparing tools for flight.

Interestingly, although the upgraded power ratchet tool looked like the earlier versions, it was of a completely new design inside; one that fitted into already defined dimensions but of course worked much better. The same approach was taken for the torque limiter. Other tools were developed from experience, such as a 24 inch (60.96 cm) extension that was planned to be used only once and in one location, in order to changeout an ORU inside the Aft Shroud. This tool was carried on the chest area of the EMU. In evaluating this tool in the Huntsville water tank, Richards found that owing to its length it tended to become tangled with tethers and generally got in the way when he moved about. That made him think: "What if the tool extension was smaller, but was able to be extended out to the required length?" There was a machine shop at Huntsville that the team often used to amend tools quickly without the need to go through proposals, costing, mockup evaluations and endless reviews. There was also a small local engineering company close by that had a group of hands-on engineers who only needed to see a simple sketch and have a basic explanation in order to understand what was required, and then create it on a short timescale. There were other examples, Richards said, where an astronaut in the NBS might suggest that it would be great if there was "a hole here in the handle to loop a tether". As a utility diver supporting the astronauts, either Richards or one of his colleagues would get out of the tank and, still in his wetsuit but without the scuba gear, literally run down to the machine shop, drill a hole as required and, on returning to the tank, give the improved item to the astronaut who was still working on the simulation. The astronauts became used to the "lack of bureaucracy that the Goddard folks seemed to have. They were kind of do'ers, responding quickly to requests on the fly, which was not usually possible at NASA."

As Richards has wryly pointed out, once he was himself an astronaut at JSC, some of his suggestions for tools for the space station involved a lengthy series of committees, changes, drawings, revised drawings, more meetings and reviews, and so on. "It took a long time to change things, which is good in one respect," he acknowledged, "but not so good in others." During Hubble simulations, "a quick turnaround was required so that the flight crew would get used to the responsiveness of the designers, where they [the engineers] would go to the machine shop and change exactly what they [the astronauts] wanted and bring it back to the pool for immediate use".

Richards said that he not only learned from designing mechanisms at Goddard, he also "went on the road" visiting trade shows across the United States. He also read all the trade magazines and bought power tools that he took to Goddard in order to take them apart and assemble them again. "So I was doing something opposite to NASA's spin-off program, I was creating a 'spin-in' to see what private industry was doing, and then improve on it if I could." He even supplied feedback to the power tool companies to see whether they would incorporate his suggestions into the tools, which they did with a couple of sets of gears and different motors. The result was "pretty successful designs" for the power ratchet tool and pistol grip tool on the second service mission, which was the first flight of the lithium ion batteries. So successful was the pistol grip tool that Richards eventually 'sold' it to the ISS program because the non-recurring engineering on the unit was complete. Richards and his colleagues made a further eight units which are used on the station to this day. Indeed,

Paul Richards on the flight deck of Discovery during STS-102.

as a member of the STS-102 crew that visited the station in 2001, he was able to gain personal experience in using this tool in space.

Richards bought DIY tools from Home Depot, as well as from companies that provided higher end, commercial, or industrial tools. He attended the large Construction and Assembly Show at the Chicago McCormick Center, where the latest tools were introduced to the trade and home improvement markets. It was from these experiences that Richards' pistol grip tool evolved, mimicking tools designed for the automotive industry. The original tool had a rack on one side to accommodate the electronics. Richards changed it by reducing the electronics down to two boards, one analogue and one digital. "I took what they were doing, and did it better," he said. To protect the re-design, a utility patent was taken out on the newer version of the software and a design patent was taken out for the tool itself. This ensured that if ever the unit were to be commercialized, the patent would allow all of the rights to revert to the space agency, instead of the licensing and commercialization going directly to the General Treasurer of the United States.

**The mechanism engineer becomes a 'Sardine'**

Richards feels it was an incredible experience and privilege to work directly on the Hubble missions. This became a personal stepping stone to his selection as an astronaut in not only gaining valuable experiences but also allowing others to see what he could achieve. "People say 'it's not what you know but who you know' but I believe it's more of 'who knows what you know' that is more important."

Richards was one of 35 candidates selected by NASA on May 1, 1996, in the 16th astronaut selection. Along with nine international candidates this became the largest group selected by the space agency. They reported for training in August that year, and following qualification he was assigned as a mission specialist. Due to the limited training facilities it was a challenge to squeeze them through the training program in the nominal 2 years. This tightly packed situation naturally prompted the group to call themselves The Sardines.

In 1997 Richards supported STS-82, the second service mission, as an auxiliary member in the back room, and then advanced to train for other missions. His manager, Russ Werneth, recalled the day he learned Richards had been selected for astronaut training and thus would leave the Hubble team—it was the day that he seemed to "float" into Werneth's office.

## NO MANNED MANEUVERING UNIT

One 'tool' not available to the Hubble service missions was the MMU developed by Martin Marietta during the 1970s and early 1980s. Astronaut Bruce McCandless had worked on the project from the beginning and was the first to fly it in space, on STS-41B in February 1984. Pinky Nelson used it during the Solar Max repair by STS-41C 2 months later. And it was used for two satellite retrievals by STS-51A in November of that year. The MMU was often depicted in early artistic impressions of what the servicing EVAs during the shuttle program might look like, but this was not to be and the device never flew again.

Paul Richards putting his design for a pistol grip tool through its paces during an EVA on the ISS assembly mission on STS-102.

In the buildup to EVA from the shuttle there were many studies related to the evaluation of using the MMU in various servicing roles, including at the Space Telescope. One such study commenced in June 1974 and lasted 7 months.[6] It reviewed a number of scenarios in which the MMU could enhance EVA safety and efficiency, and identified developments in MMU controls and ancillary equipment. The MMU Mission Definition Study was conducted under contract to the URS/Matrix Corporation's Life and Environmental Science Division in Houston, Texas. It was sponsored by the Bio-Engineering Division, Life Sciences Office, NASA Headquarters. The management of the study was performed by the Center Training and Procedures Division of the Flight Operations Directorate at JSC.

The study suggested that the MMU could be utilized as a backup for jettisoning a payload should the RMS suffer a systems failure. It was also suggested that the use of the MMU may also be of assistance in support of "LST mechanisms (e.g. solar cells, antenna, aperture doors, Sun shields) prior to or following orbital release". The study went further, suggesting that an astronaut wearing an MMU could be added to an existing mission to accomplish servicing of the telescope as a "piggy-back" objective to its main goal, "thus avoiding a dedicated Shuttle launch". This obviously required EVA access to the telescope's support system module. In addition to inspection roles, the study suggested that "servicing tasks such as replacement of batteries, recorders, digital processors, and gyros may be accomplished". The proposal was to install a temporary MMU stowage/donning station on the exterior of the telescope, with an MMU located inside an unpressurized area for use during EVA operations. It was hoped that such EVA operations, both inside and outside of payloads, "could be accomplished without interrupting the experiment". If an MMU was used during a separate flight, then appropriate performance and control requirements would have to be devised, including (if required) de-spinning the telescope (an issue encountered a decade later with the use of the MMU on STS-41C Solar Max and STS-51A communication satellite recoveries), the total travel distance from the orbiter to the telescope and back again, and the availability of fuel on the MMU to traverse this distance and still have enough for station keeping. As the report indicated, much more work was needed in these areas, notwithstanding the fact that the MMU was still nearly a decade away from its first test flight.

Despite various studies, artistic impressions and proposals the prospect of using the MMU for orbital construction work, rescuing wayward astronauts, and even maintenance of the Space Telescope simply didn't materialize.

As McCandless explained, "The principal consideration for the use of the MMU was if the telescope was tumbling uncontrollably, you could go over with the MMU and stabilize it, you could get up close and grab hold of one of the two grapple fixtures [in a similar manner to the attempt during the Solar Max mission]. But the concept of servicing work using the MMU was never pursued very far because what you really needed to do was to tether yourself at the worksite, so that you didn't use the MMU propellant all the time just to stay in position. The theory never progressed to fitting a Trunnion Pin Attachment Device as there was no docking facility on Hubble. We never got that far, but most likely it would've been a matter of flying over and grabbing a piece of it, probably a handrail, putting a tether onto one of the handrail support points and then another, and incrementally tying yourself down. If the decision was made in advance of that, then some device could

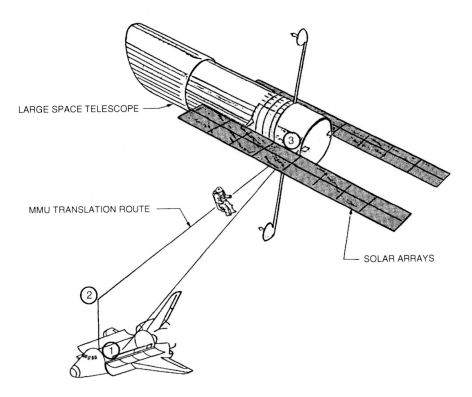

An early (1975) artist's impression showing the use of the MMU at the LST.

be rigged up for HST [such as the 'stinger' device of STS-51A] as it was a lot 'friendlier' than Solar Max because it had handrail and hard points."[7]

Pinky Nelson noted that the RMS was an easier way to capture a stabilized satellite and STS-49 had shown that it was possible, if not ideal, to grab a satellite by hand. "The MMU was a great idea that just turned out not to be that useful. The MMU arms got in the way for satellite servicing, and though the arms could fold down, you couldn't then fly the unit. You would fly to the satellite and secure yourself in foot restraints, fold down the MMU arms and conduct the work, then detach, raise the arms and fly away. During the actual servicing, the MMU would be mothballed, and then you had the great mass on your back while attempting to work, which was just not practical in the long term."[8]

## TOOLS OF THE TRADE

The tools and aids developed over five decades for EVA operations include the lunar surface tool kit for Apollo astronauts, contingency tools for the shuttle, assembly and repair tools for space stations, as well as the servicing tools for Hubble. The developer needed to address the specific requirements of the EVAs that they supported. Ironically, although the materials and technology have changed, most of the more common tools have remained the same: the pliers, ratchets, wrenches, extension handles, and carriers. Gradually, power tools

272  Tools of the trade

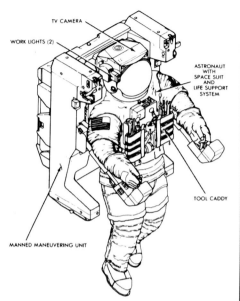

An illustration of the MMU's role as an aid to servicing is demonstrated by Bruce McCandless modelling the unit with a mockup of the Trunnion Pin Attachment Device (TPAD) developed for capturing a satellite. He tested the system on STS-41B in 1984 using a TPAD in the payload bay of Challenger as a precursor to STS-41C, which was to capture Solar Max. A similar option was considered for Hubble, but not pursued.

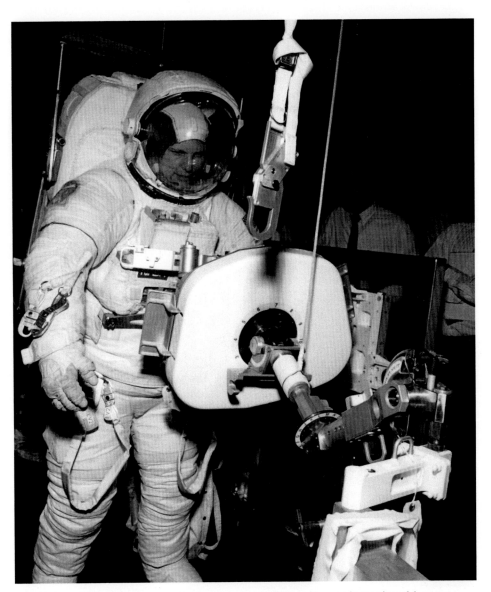

Kathryn Sullivan evaluates early (1990) tools and EVA equipment in monitored 1-g tests.

supplemented the 'brute force' type of tool in order to offer the modern space maintenance man (or woman) a suite of implements to rival any construction or repair worker on the ground.

For the Hubble service missions, the development of these tools and the skills to employ them is one of the great success stories of the program. Though there were some challenging moments during the EVAs, generally the tools performed the job that they were designed to undertake. This was thanks to the insights and efforts of individuals such as Robert Trevino, Ron Sheffield, Russ Werneth, and Paul Richards (and their many colleagues) who devoted a tremendous amount of time, effort, and dedication to ensuring the astronauts were provided with the very best tools and the knowledge, skills, and training to use them effectively.

Armed with a full array of tools, support equipment, and training skills, and supported by a team of specialists on the ground, the hardware for launching the telescope and embarking on the series of service missions was in place. As the focus switched from activities on the ground to operations in space, the missions came to the forefront, but such flights wouldn't have been possible without a huge infrastructure on the ground.

## REFERENCES

1. Email from Ron Sheffield, December 9, 2013
2. *Designing an Observatory for Orbital Maintenance in Orbit: The Hubble Space Telescope Experience*, Space Telescope Projects Office, Marshall Space Flight Center, NASA, November 1986
3. Goddard View, Volume 3, No. 5, March 2007, Employee Spotlight: *Russ Werneth, From EVA Engineering to Retirement—A Hubble Engineer's Journey*, by Susan Hendrix
4. AIS Interview with Russ Werneth, November 11, 2013
5. AIS interview with Paul Richards, November 24, 2013
6. Manned Maneuvering Unit Mission Definition Study, Final Report, Contract NAS-9-13790, Modification No. 15, January 1975, 3 Volumes, NASA JSC
7. AIS interview with Bruce McCandless, August 17, 2006
8. AIS interview with Pinky Nelson, July 23, 2013

# 7

# Behind the scenes

> *I do not consider any one position on the very large team of people that plan and execute HST service missions as more 'important' than any other. Period! The system works because ALL of the parts are 'important', and the system can potentially fail because of the loss of any of the parts to perform their function.*
>
> Chuck Shaw, Shuttle Flight Director, Mission Director SM-4

The visible element of any mission into space is the hardware sent on the journey and, for a human mission, the astronauts assigned to undertake the objectives. With the public interest in the Hubble service missions, the six shuttle crews became some of the most recognized out of the 135 missions. Certainly their public outreach program, both before and after their flights, was an opportunity to spread the word on their experiences and achievements at the telescope, but as with all space missions there was also a large team on the ground, many of whom were not so well known to the public, without whom the mission wouldn't have been able to be flown. If a space mission is compared to an iceberg floating in the ocean, then the flight crew represents the tip which protrudes above water, while the numerous teams on the ground are the bulk that remains underwater and hidden from view.

During the life of the Hubble program, from the 1960s through to its launch and servicing during the 1990s and 2000s, it relied upon scores of skilled and enthusiastic individuals who devoted much more than their professional careers to it. In order to ensure the success of the program they often volunteered considerable passion and a lot of time they would otherwise have spent with their families.

For many, it was never a "just a job" assignment, it became their whole career, their lives, even part of *them*. Talk with many who worked on the Hubble program, especially to those involved in the service missions, and their enthusiasm and passion shines through. Actually, their experiences on Hubble helped to create the people they are today. They became part of the space agency's first Great Observatory. Many were sad to say goodbye at the end of the shuttle phase of the program but they left with the satisfaction

## 276 Behind the scenes

of knowing that it was due to their contributions that Hubble is still operating today, 5 years after the final service mission, 25 years after it was launched on its planned 15 year mission, and almost 50 years since the idea of a large telescope in space began to be considered a real possibility.

While Hubble flew and service missions were dispatched to keep it flying, there was a network of teams on the ground, working around the clock to ensure that each mission was flown successfully and safely, and between the missions these teams were planning the next step.

The primary roles and responsibilities included training the flight crews to complete their missions, the organization of the program office at Goddard, the wider management team at NASA Headquarters, the Johnson Space Center (JSC) in Houston, the Marshall Space Flight Center (MSFC) in Huntsville, the Kennedy Space Center (KSC) in Florida, the astronauts on the support team for each mission, the flight controllers at Mission Control at JSC, and the Space Telescope Operations Control Center (STOCC) at Goddard. In addition, there were several individuals with singular responsibilities, most notably the Flight Director and the Mission Director, the latter post specially created for Hubble.

## TRAINING A SHUTTLE CREW

The training of a shuttle crew began firmly on the ground, with extensive classroom studies setting the scene for the simulators and the 'real' flight hardware. Once selected as a crew, the astronauts divided amongst themselves responsibilities for tracking various elements of the mission, even as they progressed through the formal training program. These elements included the division of specific roles into a primary and backup role, with an emphasis on cross-training and a team philosophy.

In anticipation of an on-time launch, successful mission and safe landing, a fully trained shuttle crew (STS-125) are all-smiles, ready to go, as they board the transfer van to take them to the launch pad at KSC.

## Crew roles

Each shuttle mission featured a mix of pilots and specialists. The Commander and Pilot were responsible for flying the vehicle. They handled the primary roles during ascent, rendezvous, and entry. Generally, a Mission Specialist (MS) was designated the Flight Engineer (FE) to assist during critical phases by providing a third set of eyes on the controls or procedures and reading data from manuals. This trio was normally termed the orbiter or flight deck crew. The other MS normally dealt with the 'science' or 'operational' aspects of their mission, such as operating the arm, deploying payloads, doing spacewalks, and operating experiments and instruments. Whilst some MS were qualified pilots competent to fly the shuttle, they focused more on the operational objectives of their mission. They were designated MS1, MS2 and so on for each mission. Each reflected a seat location on the orbiter and specific roles for ascent and entry. The MS2 role encompassed that of the Flight Engineer for ascent and entry, seated between the Commander and Pilot on the flight deck. The FE also assumed a mission objective role once the vehicle was on-orbit, with many FE doubling as the primary operator of the Remote Manipulator System (RMS). The role of MS1 and MS3 had specific duties during launch or descent, and veteran astronauts usually swapped seats with first time fliers for the ride up or down in order to give a 'rookie' that experience. Other MS had the primary role of making EVAs. Generally, for the Hubble servicing flights, four would be assigned per mission, and they were paired into teams that would undertake EVAs on alternating days and act in support of their counterparts on other days.

The STS-125 crew participate in one of many meeting and briefings, the reality of astronaut training is hours of classroom and conference room attendance. Behind the crew is Steve Hawley (*green shirt*) and, to his left, Bruce McCandless (*dark open jacket* and *white shirt*). (Courtesy Ed Cheung)

Many shuttle missions carried Payload Specialists, who were non-career astronauts who supervised specific payloads; but no PS participated in Hubble missions, all crewmembers were career astronauts, mostly NASA, but with some from the European Space Agency as part of the agreement signed by the two agencies in the 1970s to cooperate on the Hubble program.

Experience helped on Hubble missions, owing to the intensity and importance of each mission. An all veteran crew was picked for the Hubble deployment because of the front-loaded planning of the flight and the need for the crew to rapidly adapt to weightlessness rather than spend the first few days suffering from Space Adaptation Syndrome ("space sickness"). Usually the EVA crewmembers were assigned first, because of the amount of training they would require. As the mission began to take shape, the 'orbiter' astronauts would be named to complete the flight crew. Formal mission training began at that point. Several astronauts flew more than one servicing mission, but often Hubble rookies were assigned. When a novice was assigned an EVA role, it would be in partnership with a spacewalking veteran.

**From classroom to graduation**

The training process for shuttle missions had been operating for well over a decade when Hubble was deployed in 1990. The main center responsible for astronaut training was the Johnson Space Center, just south of downtown Houston, Texas. The training commenced with a series of classroom lectures featuring desktop models, photographs, graphics, and videos. The syllabus included a comprehensive program of baseline ("nominal") situations and emergency ("contingency") scenarios. An astronaut was issued a variety of tasks, and learned the relevant systems, subsystems, and elements in a primary and backup role. The Commander, Pilot and MS2/FE focused on the orbiter systems, including the RMS, while the remaining MS dealt more with the EVA and payload elements.

JSC had various shuttle simulators including those replicating the orbiter, the RMS, and, for Hubble service missions, full-scale mockups of the telescope, the support apparatus and the ORUs. There were also a number of trainers and simulators available for ascent, entry, emergency egress, and the habitability aspects of the mission, but the primary training tool for orbiter operations in association with Hubble was the Shuttle Mission Simulator (SMS). This was computer linked and was used to practice rendezvous and proximity operations at the telescope, as well as its retrieval and re-deployment. Further training for RMS activities was handled by the Manipulator Development Facility (MDF) that refined the training for capturing, berthing, and re-deploying the telescope, plus rotating and maneuvering it on the end of the RMS. There was also a helium-filled balloon mockup with the dimensions of the telescope in the main Shuttle Training Building (#9) and an air-bearing floor to assist with EVA training in 1-g conditions.

The Weightless Environment Training Facility (WETF) at JSC was used for simulated EVA training by the MS (mainly for contingencies) using mockups of elements of the telescope and ORUs. There was also emergency egress crew training in the event of an aborted launch or landing on water. Unfortunately the water tank at JSC was not large enough for full-scale EVA simulations, so the NBS at Marshall was the principal training tool for Hubble EVA familiarization for 20 years until the Sonny Carter Training Facility opened at Ellington, just north of JSC; this was used from 1997 briefly by the

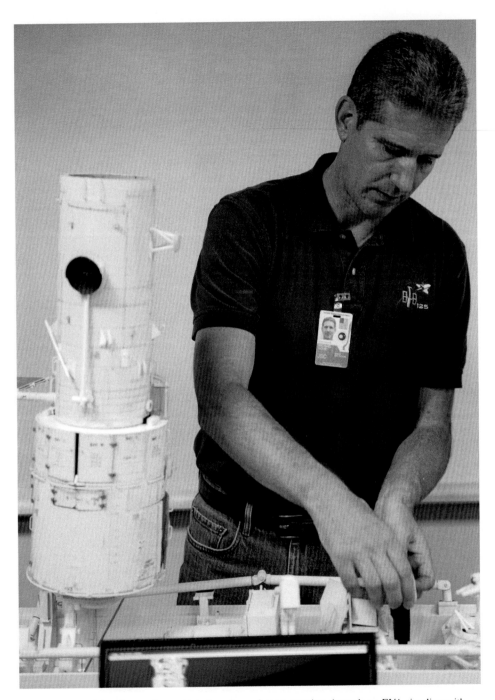
Astronaut training is not always hi-tech and can feature running through an EVA timeline with scale models before moving on to the larger simulators.

Familiarization and practice makes perfect. The flight crew participate in 1-g bench tests and familiarization sessions using flight tools and equipment.

EVA crew of STS-82 and then regularly for STS-103 (SM-3A), STS-109 (SM-3B), and STS-125 (SM-4).

Training on flight hardware was essential, and wherever possible astronauts conducted a variety of fit/function and familiarity tests on the real hardware that they would work on in space. This supplemented the work using various models, mockups, trainers and simulators. A series of standalone training sessions were completed at JSC, and sometimes there were "integrated simulations" in conjunction with Marshall, Kennedy and Goddard. The flight crew made frequent visits to contractors and subcontractors across the continental United States, as well Europe, sometimes working on the flight hardware and being brought up to date on the status of apparatus and meeting the workers who were fabricating it.

All this training was timed to reach its peak just as the hardware was being prepared for launch, enabling the crew to fly the mission soon after their formal training was completed. In reality, there were often delays and cancellations to the mission owing to conflicts in the manifest, changes to the mission, or following problems on an earlier flight that required to be resolved. For the service missions, changes were sometimes the result of the telescope experiencing problems that required a rapid response and a rethink of the planned service mission. If the launch was postponed for a significant time then training would continue in order to maintain the proficiency of the crew, with refresher training to maintain a peak of performance in readiness for a new launch date.

In summary, approximately 1 year before the scheduled date of launch the EVA crew was selected, followed by the orbiter crewmembers. At that time new training hardware was delivered and any old hardware updated. Between 6 and 8 months prior to flight, a series of familiarization briefings were held with both the flight crew and ground support personnel. During this time the EVA training program intensified and the Commander, Pilot and RMS operator began rendezvous and proximity activities, berthing, and deployment. The crew also received training in the support equipment for the mission. When launch was 2 to 3 months away, the Commander and Pilot received payload training and their duties as either backup RMS operator or an IVA crewmember to assist with preparations for an EVA. The Mission Management Team (MMT) provided familiarization briefings, then integrated simulations and training in voice communications protocols started. Joint Integrated Simulations (JIS) were completed 2 months prior to launch. The flight crew would make a number of visits to the Cape as the hardware was being prepared to check apparatus and conduct compatibility and emergency procedures training on the vehicle at the pad. Throughout their training, the Commander and Pilot would maintain their flying proficiency in specially modified aircraft that duplicated the unique handling characteristics of an orbiter gliding home at the end of a mission.

As noted earlier, over the years a cadre of astronauts had conducted a wide range of HST development tests and EVA simulations in water tanks. Not only was this invaluable for the development of the telescope and the skills and apparatus to service it, but it also offered the opportunity to compare training to actual flight experience. After each mission, the training equipment and procedures were modified to reflect the state of Hubble in space.

**Blending training with support**

The Hubble missions were heavily orientated towards EVA operations. As a result there was a point at which the training flow blended with the EVA support team. This support came not only from JSC but also from Goddard, which managed the service missions and the hardware that the astronauts would work on in space, and from Marshall, which was the lead center for the development of the telescope and had the largest water tank for EVA training. In addition, training visits were made to the primary contractors and to the Cape in order to see the flight hardware being processed.

As Robert Trevino explained, "The Astronaut Office (Code CB) had an EVA Branch where crew members would be more involved in EVA testing and evaluation of hardware and operations. The EVA crew was selected before the entire crew was assigned, so they could start the training and get involved as early as possible in servicing tasks. The training team from the Mission Operations side [at JSC] was small, with some involved in the EVA tasks and others involved in spacesuit training. There was a much larger support team from the Engineering Directorate that supported much of the tool development. This had people from Goddard and Cepollina's office, people from the Marshall Project Office, and people from Lockheed, Sunnyvale that had designed the telescope. And there were personnel from Hamilton Sundstrand and ILC Dover responsible for the suits. The team also included over 100 technicians at JSC and Marshall that operated the neutral buoyancy tanks and various other trainers and simulators we used."

Asked about the evolution of the EVA training flow, and how that fitted in with other teams and crews, Trevino replied, "We had an EVA team assigned to each of the shuttle missions, even if there were no scheduled EVAs on that mission we still had generic EVA training, orbiter, and payload contingency training that required EVA support. We had at least two people assigned to plan the training schedule, especially when competing for time in [water] tanks. The Mission Management Team set the priorities for the EVA tasks, and the EVA team would say there would be 6 hours per EVA and then the management team would prioritize those tasks for each EVA and incorporate that into the training plan."

One lesson learned early in the program involved the support by the RMS operator during EVAs. "From previous RMS payload deployment missions we learned that things happened much slower than planned, and that everybody wants to grab a window to see outside. One lesson learned back on STS-41B and STS-41C was that RMS operations required at least two crewmembers: one to operate the RMS, and another to operate the cameras mounted on the arm and in the payload bay. In training the RMS operator controlled both the RMS and the cameras."

Trevino explained there had been a variety of studies completed on the training of HST crews, but it would prove a difficult task to combine this data together into a single source indicating the depth and scope of the preparatory work undertaken before any astronaut was assigned to a service mission. The EVA training began in the classroom and then moved to neutral buoyancy training for generic tasks, orbiter contingency tasks (which made the HST tasks easier to learn), EMU simulator and altitude chamber training, EMU training, EVA air-bearing floor training, the RMS simulator and Building 9 shuttle RMS trainer, the Shuttle Avionics and Integration Laboratory for RMS training, the Virtual Reality trainers, and a few 0-g flights. And *that* was just for JSC. There was the neutral buoyancy training at Marshall and even more training at Goddard and at Lockheed. Even before Hubble, the average was 10 to 20 hours of training in the water tanks for every hour of EVA time on-orbit. "You have to remember that sometimes these training hours also included tool evaluation and procedures development. These things happen simultaneously with the training, especially on a schedule constrained mission. And of course this was just for EVA training."

**GROUND TEAMS**

During the six shuttle flights to the Hubble telescope, dedicated teams of support personnel worked around the clock to monitor not only the systems of the orbiter and the telescope but also the actions, performance and well-being of the flight crew. The two main facilities used to support the service missions were the well-known Mission Control Center (MCC) at JSC, which was responsible for the shuttle and its crew, and the Goddard Space Flight Center and its Space Telescope Operations Control Center (STOCC) which managed the day-to-day activities of the telescope and monitored its onboard systems during servicing periods.

## HUBBLE PROGRAM OFFICE

Goddard Space Flight Center in Greenbelt, Maryland, was responsible for the day-to-day activities of the telescope, managed the development of the payload of each service mission and its suite of tools, and defined a team split between Goddard and JSC during each shuttle mission, focusing mainly on the period when the telescope was in the shuttle's payload bay.

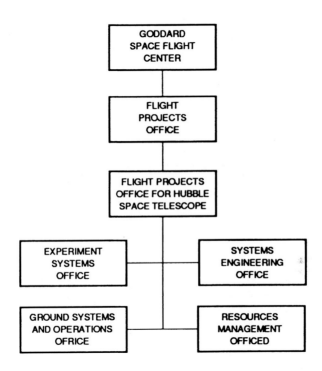

**GSFC RESPONSIBILITIES**

- SCIENCE INSTRUMENTS DEVELOPMENT
- GROUND OPERATIONS SYSTEMS DEVELOPMENT
- SCIENCE VERIFICATION OPERATIONS
- HST OPERATIONS
- M&R PLANNING SUPPORT

Goddard Management Structure (1990).

## Program Manager Preston Burch

Prior to joining the Hubble project in 1991 as Deputy Project Manager for the operational ground systems, Preston Burch had worked to a small extent on Hubble in private industry. Following graduation from high school in 1966, Burch had worked for several years on the Apollo Lunar Module at Grumman and then on some of the OAO satellites which were the precursors of Hubble. While working on OAO-C he learned about ground systems, mission operations, and the skills of a test director. He worked in private industry for over 25 years, and after running his own engineering business for 8 years he decided to move into public service by joining NASA at Goddard.

At that time NASA was encountering difficulty in finding a deputy manager for the HST project, where the qualifications required a broad range of experiences in spacecraft systems, mission and flight operations, science operations, ground systems, and flight software. These were fields in which Preston Burch was well qualified. By then his experience had included work on the Gamma Ray Observatory, ATS-6 and several other satellites. On the retirement of his predecessor in 1999 he was made Project Manager for the HST Operations Project. In 2001 he was asked to take over as overall Program Manager. "That came as a great shock to me," he says. "I thought in time they'd see right through me and I'd be looking for a job as a burger flipper or something. It was [over] 20 years ago. So it all worked out quite well. I've had a lot of fun. I'm very pleased with the outcome, though I never thought I would end up doing that."[1]

Goddard HST Management in Blue Flight Control Room during a Joint Integrated Simulation for STS-125. From the *left*: Randy Kimble, David Lekrone, Preston Burch (all GSFC), Matt Mountain (STScI), Mike Kienlen and Keith Kalinowski (Goddard).

Preston Burch, HST Program Manager. (Courtesy Goddard Space Flight Center)

Precisely how Goddard handled the management of the HST project, Burch explained in some detail. "GSFC is divided up into several major organizational levels called directorates and they are given the title of Code. In my case, Code 400 is the Flight Projects Directorate (FPD). They deleted the name Program [but] it still has programmatic responsibility for all the missions within there. We have all the managers and all the money, so we're everyone's best friend forever. Code 500 is the Applications, Engineering and Technology Directorate (AETD). That constitutes about one-third of the civil servants at Goddard and is responsible for all the engineering support, so the engineers get matrices out to the various projects." He went on to say that Code 300 is Safety and Mission Assurance Directorate (SMAD, formerly Quality Control); Code 200 is the Business and Facilities Directorate (BFD); Code 100 is the Office of the Center Director with all the administrative roles; and Code 600 is the Science Directorate (SD).

The HST Program Office situated at Goddard was originally spread over various offices and buildings but in 1998, just prior to SM-3A (STS-103) it was consolidated in 710-1529, which is actually four separate buildings which have been interconnected via the integrating and test complex. Burch praised this move because it enabled timely and efficient decisions to be made, ensuring that the management team was always aware of what was going on.

As Associate Director (Code 400), the Flight Project Manager for HST, Burch reported directly to the Director of Flight Projects. A role now typically delegated to one of the field centers, although originally they were all located at NASA's HQ in Washington DC, which enlarged the bureaucratic organization to what it is today. The problem of being located in

Washington, Burch said, was that the "managers were not close to the action and what was going on, which created all kinds of difficulties". So it was decided to move some Program Offices out to Goddard.

In the case of the Hubble Program Office, this was classed as a sort of a "super project" that comprised the HST Operations and Ground Systems Project, which would later become the HST Operations Project, and the HST Flight Systems and Servicing Project later known as the HST Development Project. As Program Manager, Preston Burch was responsible for all elements of the Office in terms of technical issues, schedule, and cost performance.

The Operations Project was responsible for the health and safety of the spacecraft, the mission, flight operations, and the science program. It was also responsible for the Space Telescope Science Institute (STScI) that reported to the Program Office contractually. The Operations Project was also responsible for ground systems operation and took the lead on the sustainment, re-engineering, and maintenance of the software for the two main computers of the telescope. In addition to the processors and science instruments and their test beds, it was responsible for service mission operations, planning preparation and all the operational procedures, creating the Service Mission Integrated Timeline (SMIT, pronounced 'Sermit'), contingency planning products, the servicing planning and re-planning tool (a computerized system for generating the SMIT), on-the-fly re-planning during a mission, simulations and training. On top of all that, Burch was also responsible for several HST simulators located at Goddard.

The Development Project was responsible for developing new science instruments. There were the five science instruments on the telescope at launch, but these were superseded over time. The High Speed Photometer was apparently not a popular instrument, with a very few percent of the observation time, so it was the obvious target for removal by the first service mission to accommodate the COSTAR package that compensated for the flaw in the mirror. The instruments developed for installation in space were WFPC-2, STIS, NICMOS, ACS, WFPC-3, COS, and the NICMOS Cooling System. When a "thermal short" caused NICMOS to consume cryogen at thrice the planned rate, exhausting the supply in only 18 months, the team built a reversed back cycle mechanical cooler to restore the instrument to service. They also undertook a significant amount of instrument development and ORU development and created the "black boxes" for all the orbital subsystems. The Development Project was also responsible for the Vehicle Electrical System Test (VEST) facility at Goddard. This became indispensable, ensuring that all items worked correctly before they were taken into orbit. The development group also handled the high-fidelity mechanical simulator to ensure everything fitted together correctly, which in turn revealed whether any item was too large or too small for where it was intended to fit, and most importantly whether it would function or not. The Development Project also handled the enclosures that protected the instruments and ORUs in the payload bay at launch. They made a number of protective boxes to accommodate all the sizes and shapes of hardware, and their carriers, and also had responsibility for the shuttle pallets and the protective enclosures mounted on them, all crew aids and tools, and training the astronauts about the payload. Furthermore, they carried out a comprehensive systematic integration and test program and processed the payload at the Cape.

"The HST Program Office," Burch says, "was responsible for [management] Level 1 and Level 2 requirements for the technical integration of the entire program to ensure, technically speaking, the Operations Project and the Development Project were working together in an integrated fashion. It also had a systems management function which included responding to on-orbit anomalies, forming investigating teams to get to the bottom of an anomaly and then try to figure out what new hardware we might want to put on a future service mission to fix the problem, or what type of workaround we might do on the ground with flight software or ground system software to deal with a failure. [The Program Office] scheduled integration, major technical decision-making by running all the major technical reviews, program and project budgeting and execution, interfacing with center management here at Goddard, as well as HQ, PAO and, during service missions, with JSC."

During each service mission, Burch went with a team to JSC so that they could be on hand in support rooms and make prompt decisions in terms of the work of the astronauts on EVA. If the crew ran into a problem and a decision had to be made to adopt one of the contingency plans, Burch had the final say over those events, although issues relating to crew safety were the responsibility of the commander of the mission.

Although there was a lot of team retention over this period, 1991–2009, there was some staff turnover that meant new team members joining. But as Burch says, "Most workers felt honored and privileged to work on a program of this magnitude, of this degree of visibility, and of national and international importance. Hubble has been an incredible program in terms of scientific discovery, revolutionizing our idea of the universe and our place in it. And all of us just felt incredibly excited coming into work each day to work on something like this. It would take a lot to lure somebody away. Inevitably people get old and retire, [but] some of those did come back and soldier on, some moved area, but overall the turnover rate in terms of the civil service and contractors was extraordinary low."

His role as Deputy Associate Director of Science Operations and as Project Manager for HST required a delicate balance of managing the mission, science, and servicing operations, the development of the ground system and flight software, and the maintenance of both the telescope and the supporting ground infrastructure. This was a varied and complicated role that had to balance the demanding science program with the periods required for servicing. Asked how difficult he found being involved with the science mission and also responsible for suspending that to service the telescope, and then restart it again after the telescope was placed back on-orbit, Burch said, "It was certainly a challenge but one thing to bear in mind, [is that] Hubble servicing development was like a train going down a parallel track to the Hubble science operations, which would be on the other track. There was a lot of sharing of resources and interaction between the two, but the science time never really stopped until it was time to do the servicing."

As Burch further explained, the controllers at STOCC would command Hubble to come out of its science mode shortly after the shuttle entered orbit. By the time the shuttle arrived, several days later, the telescope would be passive, ready for grappling. Then after the EVAs there was some redundancy checking on the science instruments and ORUs in order to verify the electrical connections and basic functions; this was known as the "aliveness test". It was performed while the telescope was still aboard the shuttle, so that the astronauts could attend to any problems. For about 30 days after redeployment, the telescope was re-commissioned by recalibrating systems such as the FGS and checking out the science instruments. A long checkout was required for the science instruments to account for outgassing and ultraviolet polarization contaminants that were Sun and Earth-light sensitive. Then a functional test of the science instruments was completed to align various optics, to focus, and to ensure there were no general degrading and calibration problems. Both to test what the effects the new modifications had been and to promote the telescope to the general public, the Hubble team issued a selection of images dubbed the Early Release Observations (ERO). With this successfully completed the science program was gradually resumed.

On each shuttle mission to Hubble, the scientists involved in the daily research program remained key partners during the servicing phase, since there were new instruments to be installed and old ones to retrieve. "This posed big changes to ground operations at STScI," Burch said. "There were new capabilities to explore, each instrument required calibration, there were further software changes, and even more data to store." Most of the instruments were built by Ball Aerospace, and the data from all the integrated test facilities at Ball was sent to Goddard and STScI for calibration by comparing the archived test data with that coming from the recently installed instrument to verify that they matched.

While all this was going on, various field centers across the country were preparing the *next* service mission. According to Bursch, each service mission took about 12 months to prepare, involving the coordination with each field center, contractors, and investigators. The aim was to have all the hardware and software ready at the same time for launch and to carry out the mission as designed. This also ensured a minimal down time of science while Hubble was locked in the shuttle payload bay or undergoing post-servicing checkouts. The timing had to be just right, with the science continuing right up until the shuttle was on-orbit and confirmed to be in good shape. Burch was appreciative of the heavy training program that faced each flight crew, especially the EVA team who were "selected about 15 months prior to flight, [and had] a lot to learn, and lot to do". Despite their excellent skills there was still much to grasp, especially in mastering the range of tools and procedures to fit or remove hardware on the telescope. He recalled that it drove him crazy to think what might go wrong on-orbit, or what they might not have foreseen in their meticulous preparations. There were periods of crew training conducted at Goddard on mechanical simulators, supplementing the extensive training in the water tank at Marshall, at JSC and at the Cape or out at contractors. Several astronauts—in addition to serving on flight crews—were assigned over the years to support the development of hardware and procedures, and many took to the tasks well, but according to Burch, "some astronauts could not change a light bulb" and they needed more time to come to terms with the challenges assigned to them.

By far the most valuable commodity on a service mission was the 6 hours allocated to an EVA per flight day. Bursch recalled that "developing each spacewalk was a choreographed ballet, trying to eliminate the unwanted trips from the payload bay to the telescope and back again, to save time." He also noted that a lot of Goddard staff had become proficient utility divers in the pools at JSC or NBL, working even the simplest task to help improve the EVA timeline and sequencing, such as retrieving a dropped tool to save an astronaut wasting time in retrieving it. Although they would not encounter this problem in space, it taught them to ensure their tools were tied down or tethered. Everyone, from senior management on down, learned new ways of doing things, observing how things went well or not so well. Often an astronaut came up with their own ideas for procedures, tool design, and techniques, and all worked hard in finding efficiency in the timeline, to extract the maximum from every EVA.

**Al Vernacchio, managing operations and systems**

Over the 17 years that he worked on the program, Al Vernacchio's role on Hubble evolved. He served as HST Ground/Operations System Engineering Manager and most of his tasks were related to the service missions, serving as the Systems Engineering and Verification Manager. He joined the Engineering Directorate at Goddard straight after graduating from college and for a short time developed test systems to support the development of software for spacecraft computers.

Al Vernacchio, HST Ground/Operations Systems Engineering Manager. (Courtesy Goddard Space Flight Center)

"Around 1985," he explains, "I joined the newly formed branch that was responsible for maintaining flight software in computers for satellites. My job was to support Hubble before it was launched, so we were getting to a position to take over flight software. I helped set up and lead a team through the launch and activation period, all the way up to the first service mission and shortly after that. The team did a lot of software changes to deal with any issues that the telescope had, or new or additional capabilities the scientists wanted on the guidance and control computer."[2] His team was not directly involved with developing the corrective optics package to resolve the spherical aberration of the mirror; that was more on the flight operations side with Frank Cepollina's team. Vernacchio explained, "We were only a little involved in how to operate that thing from the ground, how to deploy those mirrors and get them in position and so forth, and the commanding required to achieve that. After I left the Engineering Directorate, I joined the Hubble Project in the operations and ground systems area. That is when I took the role of Ground Systems Engineering Manager, between 1994 and 1998 supporting science operations about the time of the SM-2 mission. I then took on a new role as Hubble Operations Service Missions Manager. It was a much more active role during the service missions, and immediately afterwards, to basically commission the new instruments for use by the science community."

The division of the Hubble Program Office at Goddard into two components of a "super project" saw one, headed by Frank Cepollina, deal with the on-orbit servicing, and the other, headed up initially by Preston Burch and later by Al Vernacchio, handle the operations and ground systems. Having primary responsibility for operating the telescope and its associated ground systems, Vernacchio managed the daily operations of the spacecraft. Another aspect of his job was to work with the Space Telescope Science Institute in Baltimore, Maryland. In addition, his team prepared all the ground systems for whatever changes had to be made for upcoming service missions, such as installing a new instrument, "not only where the camera was more capable from a physical perspective but also a computer perspective, with a much higher data rate. There was a lot of work done on the ground to prepare to operate the HST effectively and efficiently after a service mission. During a mission we were responsible for executing all the operations of the spacecraft, so we would maneuver the telescope into the proper orientation when the shuttle came to retrieve it, then power off its systems before the astronauts went out, and when they were done we would test those systems, make sure they were installed properly, and check them out."

Vernacchio's team was the link between Hubble science and engineering, and the flight controllers in Houston and the astronauts in space, ensuring a smooth transition to and from both sides. "We operated the satellite during the service missions, working very closely with the guys at JSC and Frank Cepollina's team to choreograph all that activity as efficiently as possible, particularly due to the expense of shuttle time. That takes a lot of preparation and planning on each service mission to be enabled to do that."

Asked about the time available to run simulations in preparation for the service missions, when there were other preparations going on for several shuttle missions, Vernacchio said it turned out to be a full time job for many on the team. "They would start the next mission as soon as the previous one was done—if not earlier in some areas. But they all built on each other, with the planning that went into first SM going into second, and so on. This included teleconferencing and quite a bit of traveling as well, I'd probably go to JSC each quarter to meet with the shuttle folks there, but there were more frequent telecons between teams with each discipline getting involved for planning purposes. And we would have our own internal simulations; we had our own simulator and we'd have [10 to 12] people play the roles of key shuttle interfaces that we were dealing with and conduct smaller sims broken up into days or couple of days in preparation for joint sims with JSC where computers ran the sims together as if in flight, it was pretty high-fidelity, and closer to the mission the astronaut crew would get involved in interacting with team as well."

The Operations Team usually involved the same engineers and operators who tended to work on telescope every day. Then a smaller team augmented the larger team, but focused entirely upon roughly ten discipline areas of the service mission, with at least two or three people involved in each of these areas. The overall systems team that supported a service mission involved between 30 and 50 people, if not more depending on the activities which were to be carried out. In an interview for this book, Vernacchio strongly emphasized how many people were involved with the service missions—ranging from the scientists to the engineers, designers, and subcontractors. Early in his career he was involved in supporting the science operations, flight software, operations, and engineering. Questions were always raised about how specific tasks should be tackled. For example, "What do we need to do to enable some of these activities which the astrophysicists that are operating that part of the telescope at the STScI want to do. So when unique things came around, like when Comet Shoemaker-Levy hit Jupiter, we had to do some things in terms of pointing [which were] a new activity for Hubble; we didn't *track* planets at that time so we had to do some software changes from a science perspective. Then, from a servicing perspective when a new camera comes along or a dramatic upgrade there is a lot of engineering that needs to be worked out, such as how do you double and triple the data coming out of the instrument and how do we get it to the ground, how do we get it to the science institute, and the folks over there would work out how they would process it. All of that is very tightly related to the development of that instrument, so it starts with the development of a new instrument and carries all the way through to installation on-orbit and checkout and operation. So we had teams focused upon each instrument, with a group of engineers who would determine how we would integrate it into the telescope, how we would operate it and make it available to the wider astrophysical community."

Budgeting was more the responsibility of Cepollina's team but Vernacchio's team would become involved if there were trade-offs to make, forming an integrated team that gave the systems people a say, whilst ensuring the instruments remained within their budgets.

Vernacchio says that his appointment as Hubble Operations Service Missions Manager relied on his biggest strength, namely a computer background and an intimate understanding of how the Hubble system worked. One former boss once told him, "The hardware is great, but software guys are the ones who really make it work." Vernacchio firmly believed this to be the case, "The hardware these days is not going to get very far without the software folks making it work." His former role also helped him to understand what was possible and what was not, amongst the suggestions and plans presented to him from scientists to engineers for what they wanted done in a particular way or to a capacity.

**The ART of Hubble management**

Vernacchio explained the role of the Anomaly Response Team (ART), which was based at both Goddard and JSC during a Hubble service mission. "You have a ground team at JSC as well as at GSFC, running the well-choreographed operation for the astronauts to do whatever repairs they were doing. In the Goddard control center we had all our disciplines monitoring those activities, anticipating many of them; when you're doing that you are really 'tied to the seat', you can't just get up and leave, and what was recognized very early was the need of a backroom team that could go off and work issues and bring answers back to the online team. So that's what was called the Anomaly Response Team.

"When a problem popped up on-orbit or on the ground, we had a group of engineers that were divorced from the real-time operations and could go off and work problems. There was a manager, because there was never just a single problem and we had to keep track of how to spend time, when to have the answers, how to deploy the team, and so on, primarily during a service mission. There was also a much smaller operations team, more hands-off in between the servicing. On a service mission, in particular with a crew outside, often you would need answers immediately, you have to be able to respond with the right answer quickly. So that was the point for having the ART supporting the shuttle crew—so everything they needed, they got as soon as possible. Cepollina's team was mostly deployed at JSC, as they wanted to be nearer the shuttle guys, [but] most of us wanted to be back where we had our data and where we were operating the telescope from. How we managed requirements over resources was always a challenge from a management perspective; especially if not directly involved with the hardware—how many detectors are required, how advanced should the camera be, and what could we afford. One of our challenges was how do you keep the systems modern, when operating a telescope for 30 years. You don't want to be operating equipment from the 1980s, so we did several upgrades along the way. That was always a balance between what we could afford and what was the technology at that time, plus the further challenge of how to integrate the changes into an operational system, how do you roll in a major upgrade into an operational environment without disturbing that.

"There were a lot of little things that failed on Hubble, but being the category of mission it is and its visibility as a national asset, we put a lot of resources into planning and preparation before we did anything. Of course, you are aware of surprises that were encountered like the spherical aberration at launch—I discount those, but along the way there are times you make changes and there are unexpected responses. There were times we changed the software and didn't realize it was going to have the reaction that it had or the unexpected effect that it had. We always prepared for those eventualities by having procedures to pull it back out quickly. Then as regards the observatory itself, it had been designed to be pretty robust in terms of its capacity to deal with failures and problems. We always had a sense that the spacecraft would take care of itself; it had to."

## Joyce King, managing the engineering

Goddard engineer Joyce King served as the Deputy Operations Manager for HST between July 2003 and April 2005 before assuming the role of Systems and Engineering Manager for HST Operations. Prior to that she was a mechanical engineer at the Kennedy Space Center in Florida, where she worked from 1989 to 1994 on the mechanical integration of payloads. As it happened, one of the payloads that she worked on during the early months of 1990 was the Hubble Space Telescope. Her responsibility was to ensure it was placed safely and securely in the payload canister for its trip out to the launch pad.[3] In 1994 King moved to Goddard, where she was assigned to the Flight Operations Directorate (Code 440) and detailed to the preparation of hardware (Code 442) for the second service mission, eventually being made SM-2 Lead for mechanical integration, managing the technicians and engineers building the ORU carrier for the mission and preparing the Flight Support Structure. Code 440 consisted of the Operations Team, Code 441, which controlled the telescope, and Code 442 supplying the hardware for each service mission with the support of the mechanical engineers, thermal engineers and so on of the Code 500 engineering team. As King explains, "I stayed with the Flight Operations Directorate, working in '441' until my assignment as Deputy Operations Manager for Hubble."

Joyce King, HST Mission Operations/Deputy Operations Manager. (Courtesy Joyce King)

As Hubble's Deputy Operations Manager, King reported to the Operations Manager, who at the time was Chris Wilkins. She was responsible for working the day-to-day operations of the flight operations team. This involved scheduling up-links and down-links, ground system activities, organizing the schedule of flight software releases designed to upgrade the onboard software and update the ground system security patches. This involved a *lot* of testing, and as Deputy it was her role to keep track of all the flight software that was needed to support each service mission. In fact the only Hubble mission that she didn't work on in some manner was STS-61/SM-1 in 1993.

The flight controllers and managers at JSC looked after the shuttle and its crew, while the Flight Operations Team at Goddard monitored the health and safety of the telescope during a servicing phase. Hubble was not normally switched off, it was still reporting on its condition although no science was being conducted. What remained operating depended upon the EVA plan. Usually only the instrument on which the astronauts were working would be switched off. The ground team isolated each of the items to be worked on in order to make them safe for the astronauts, and then powered them up again to verify the servicing had not disrupted the basic system. Therefore Hubble remained "alive and talking to the ground, charging the spacecraft's systems and keeping the data management system up and running".

When in March 2002 the STS-109/SM-3B astronauts changed out the power control unit, dubbed "the heart of Hubble", the entire spacecraft was powered down for the first time since its launch. There were a few stressful moments when a delay in starting the EVA meant that the telescope had to be powered up and then powered down again once the EVA finally got going. According to Joyce King, "A lot of training on the ground in sims was accomplished for this task. We had a training facility in another building and the Vehicle Electrical Support Structure (VESS) that simulated the whole power bus. [By the development of software] we were able to come up with 'super prox', allowing the team to send up one large command all at once, instead of command after command, [and] to turn power back on quickly. This was worked out in the build up to the service mission and then patched live, in real time, on the mission. We had completed a Service Commission Ground Test so any command to be sent during a service mission had been tested on the ground first." King was the Test Conductor for these evaluations. Therefore when a change of instrument was required, or where power had to be turned down, her team had already worked out the powering down sequencing in simulations to verify that it would work correctly. Then after the astronauts had changed the instrument, it was her team that powered the system up again in the proper order to return it to the desired temperature prior to sending any operational commands. This was followed by the "aliveness test" to verify the system was receiving power and sending signals. Later, the functional test was conducted while the shuttle crew was asleep. Doing the tests "overnight" allowed time to plan any necessary remedial action to be undertaken by the crew when they awoke.

For STS-125/SM-4, the final service mission, King's team was on duty at Goddard in support of the Planning (Orbit 3) Shift. This was normally a quiet shift that allowed time to review work completed the previous flight day. Unfortunately for King and her team, as this was the final visit to the telescope the astronauts had continued to complete a lot more work during their EVAs than had been scheduled. This "was fabulous," she acknowledges, "but my shift ended up doing a lot more work in their on-duty time. Each day, the planning shift would pick up where the command plan left off, and follow on, but there had been frequent re-plans so we'd re-work the schedule for the next day and that [evening] for the rest of the mission. As they got ahead, we fixed the schedule, so I was put in charge of the Re-Plan Conferences, completing an internal review at the control center in Goddard."

She was in charge of about 125 people, including scientists, management, systems, thermal, communications, hardware, mechanical engineers, etc. When necessary, they called upon expertise in a particular field. As the planning shift had already rehearsed the tests during earlier simulations, she worked directly with Flight Operations Control and the shift supervisor and the person in charge of Flight Control in Houston. The protocol was that Mission Control would send commands to Goddard for verification, then turn over the Hubble command panel to Goddard to perform direct communications with Hubble.

During a planning shift, the team could evaluate what had been done and plan for future actions based upon what was expected to have been done, but which still remained an open issue, and the priority tasks which needed to be done next day. As this was usually ahead of the schedule, the planning shift was also constantly seeking to take on even more get-ahead tasks to ensure that the EVA astronauts always had something to do. For instance, on SM-4 the shuttle had to adopt a given position in order to orientate the solar arrays to recharge the batteries of the telescope during a day-side pass. This obliged Joyce King to coordinate with Houston regarding when this could be done without conflicting with the activities JSC had planned. As the Systems Manager, she had about 30 engineers from different subsystems—electrical power, communications, optical telescope, safing, science instruments, and data management, etc. In her managerial role she had to be aware of all technical aspects of the telescope, such as pointing and control, thermal concerns, the instruments, etc. She had to know Hubble inside out, know how to solve problems and know the implications of actions, "and if I didn't, I knew someone who was going to help me". Fortunately for King, systems on the telescope in space were quite close to the simulations, and the match improved with experience gained in implementing changes during on-orbit servicing. "We began to know the differences between simulations and what was on the flight model. An example was the WFPC. It had gone through its thermal vacuum test in the simulated environment facility, which was supposed to simulate the temperatures it would be exposed to in space, on-orbit. But we had a problem with the WFPC-3 cold plate and variable conditions on its heat pipes, so they had a little problem making sure the instrument didn't get warm and was kept cold. The anomaly team discussed what we needed to do, and the engineers got together." Being the Mission Operation Manager (MOM) on the Planning Shift, they briefed her with, "This is what is going on, this is why it's happening and we need to change this set point so that we can get it to the right temperature." King would then get on the loop to Houston, explaining, "This is our problem and this is what we need to do." Houston then changed the command plan, and after some testing they used the "operations request" to document the revisions to the command plan, signed off, and gave Goddard the command panel to issue the required and agreed commands to Hubble, then the panel was returned to Houston's control. All of this occurred mostly while the astronaut slept.

**The Systems Management Ground Test**

The Systems Management Ground Test (SMGT) was applied to all new hardware going to Hubble in order to verify the planned command path in nominal and under "a big command plan". These could run for up to 24 hours, with anomalies being logged to be worked on in the future. A succession of such cycles would gradually eliminate the problems.

Joyce King chaired the SMGT, and her role was "to ensure these issues were addressed and then closed by identifying this was the problem, and this is what we did; some tasks were simple, some not so simple and required even more testing and analysis."

During the planning shift hours, separate investigative simulations were carried out by the Goddard Simulations Team, in which they reproduced the anomalies observed on the flight telescope and determined how to resolve them. The results were documented and tested, and if necessary carried over to the next mission. Post-servicing evaluation, including all systems anomalies, was an important element in planning the next service mission and compiling an archive library for future programs. The Post Service Mission Operations Team performed reliability analyses and studied potential failures, indicating how they should be addressed if they ever occurred. One off-nominal situation arose in the autumn of 2008 when one side of the SIC&DH failed. After 2 weeks of tests the decision was made to go to the "other side" of the spacecraft, a switch over which was possibly only because redundancy had been built into the design. In order to prevent a "single system failure" from occurring during the final service mission, management directed that a flight spare be refurbished, certified, and then installed by SM-4 in 2009. This added training on a new component to an already packed program, but the pressure was relieved by a postponement imposed by a hurricane striking the Cape. The team was able to conduct a lot of life extension initiatives, forward planning and cross-strapping. And despite one instrument having failed, they didn't opt to swap the whole side. This enabled the telescope to continue to function in a hybrid mode, based on extensive ground testing for exactly this situation, and controllers were able to adequately prepare for the situation and minimize the telescope's down time. After the SM-4 mission, the operation and staffing of the Post Servicing Team was downsized, with a reduction in electrical engineers, and the members who remained had to cross-train and increase their tasks. Consequently the former Thermal Branch and Mechanical Systems Branch at Goddard are now all part of the Power Systems Branch.

In her managerial role, Joyce King participated in standalone training at Goddard as well as joint integrated simulations with JSC, often with astronauts working in 1-g simulations or neutral buoyancy in the water tank. Each simulated EVA would be planned to last 6 hours, about the duration of a real EVA from the shuttle, but with a delay the crew would likely fall behind in their tasks. The anomalies included a delayed start to an EVA due to a (theoretical) leak in a suit. Any work "lost" would have to be reinserted, including replicating the activity of the Planning Shift and a Re-plan Integrated Meeting (by a '440' team) in order to see how the priority list would be revised. There were many technical discussions, each prioritized so that if an EVA required to be shortened the team would already know how to reprioritize its timeline. To add to the fidelity, the Program Manager would make the final decisions. King would talk to the management at Goddard and the Goddard controllers installed at JSC, then make her recommendations to JSC's management and flight controllers.

From her experiences on Hubble, Joyce King reflected, "It's important to pay attention to detail. Forward planning is essential, as is lots of contingency planning. If something went wrong we could quickly plug in because the systems had already been tested in contingency mode. Of course, time is critical in space flight. Training is essential. Basic core

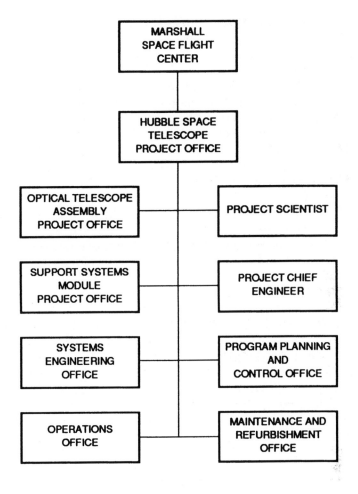

Marshall telescope organizational structure (1990).

training can handle likely problems on the ground by simulation so that when something occurs in flight they ought to be able to respond efficiently and effectively as a result of that training, focus, and determination."

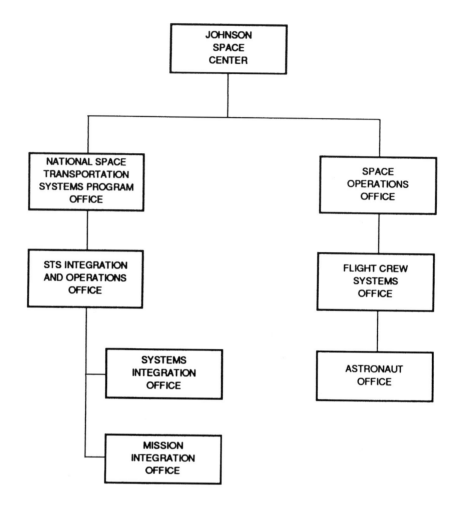

JSC telescope organizational structure (1990).

Hubble program office 299

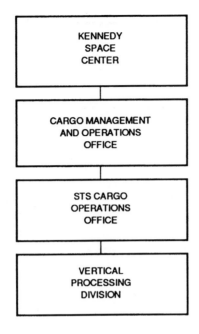

KSC telescope organizational structure (1990).

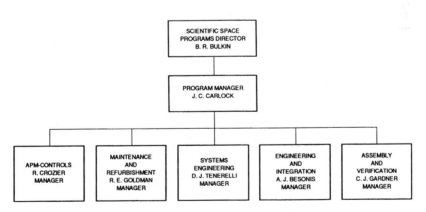

Lockheed telescope organizational structure (1990).

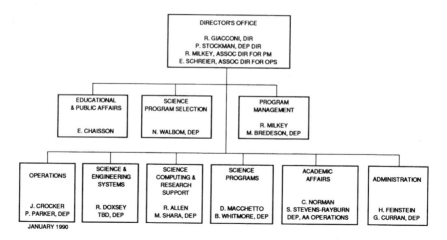

Hubble Space Telescope science organization structure (1990).

## OTHER SERVICE MISSION TEAM MEMBERS

It is always fascinating to learn of aspects of a space flight directly from a member of a flight crew. It is equally as pleasing to gain a wider appreciation of what goes into planning a space flight from those who manage, plan, control, and support missions, train crews, test hardware and prepare experiments and payloads. This, together with sifting the original documentation, gives a truer account of the missions or programs researched. With the growth of the internet over the past 25 years, NASA has expanded its coverage to include background information on those who fulfill support roles from management and engineering, to mission control and training, and many more. The final three Hubble missions benefited greatly from this new "outreach" and during SM-4 in particular a variety of team positions were featured on the web, showing how Hubble has been a focal point for most of their professional careers.[4]

Some of the key positions fulfilled during that final service mission reflect the strong infrastructure required, not just for servicing Hubble, but for all missions.

### NASA Headquarters

The senior positions at NASA Headquarters in Washington DC involved in the Hubble program included the following:

- The *Director of the Astrophysics Division* was responsible for the senior management of over 20 different flight projects, including Hubble.
- The *HST Program Scientist* was responsible for monitoring and maintaining the Hubble science program, whilst ensuring that the mission remained "true to NASA scientific objectives".
- The *HST Program Executive* headed all program activities for Hubble at NASA HQ (except those dealing with science content) to ensure that the program was carried out on a daily basis in accordance with NASA guidelines and to assess its performance against technical schedule and budget requirements.

## Marshall Space Flight Center

The Marshall Space Flight Center in Huntsville, Alabama, was responsible for total project management as the HST Development Lead Center. Its responsibilities included developing the Optical Telescope Assembly and the Support System Module, as well as the integration and verification of the telescope prior to launch and the initial orbital verification operations and maintenance and refurbishment planning prior to handing it over to Goddard to run the science program and manage the service missions.

Up to the time of the telescope's launch, the HST Project Office at Marshall included:

- Optical Telescope Assembly Project Office
- Project Scientist
- Support System Module Project Office
- Project Chief Engineer
- Systems Engineering Office
- Program Planning and Control Office
- Operations Office
- Maintenance and Refurbishment Office.

Lockheed Missile and Space Company engineers at their consoles in the Mission Operations Room of the Space Telescope Operations Control Center (STOCC) at Goddard, conducting a simulation prior to the launch of the telescope in 1990. (Courtesy LM&SC)

## Goddard Space Flight Center

The Goddard Space Flight Center in Greenbelt, Maryland, played a key role in the Hubble program, with the following senior positions:

- The *Deputy Program Manager, HST* was responsible for the overall management of telescope operations and science operations. This position was also responsible for service mission development, and headed the Management Team at Goddard which oversaw the EVAs of the final servicing.
- The *Lead Mission Systems Engineer for HST* led the development cycle for the HST systems engineering team. This covered defining any items that were to be added to the telescope from its initial requirements through its design, assembly, test, integration, launch, and orbital operations.
- The *Deputy Associate Director/Technical for HST* led the technical development of all program activities, including servicing, operations, and advanced studies.
- The *Deputy Project Manager for the HST Development Project* was responsible for the development of flight hardware for the service missions.
- The *Deputy Manager of HST Operations Project* managed the technical activities of the project team, managed and answered for the budget, ensured that work remained on schedule in preparing for a service mission, and then managed the post-servicing orbital verification. During a service mission, this person led the team that deployed to JSC and at Goddard was the primary interface on HST operations for each 12 hour shift and for supporting the EVAs.
- The *HST Senior Project Scientist* provided the scientific leadership for the program, including managing the science program and the telescope, and the development of new science instruments and the periods of on-orbit servicing. This was a broad all-encompassing role which was designed to support the scientific investigations of the observatory and ensure that the science obtained was both productive and successful over a long period of time. As a member of the senior management team, the Senior Project Scientist helped to create the planning and guidance of each service mission, and during a mission they would be at JSC to monitor its progress and be available should a contingency situation arise that required a decision by senior management.
- The *HST Deputy Senior Project Scientist* worked with the Senior Project Scientist to advise upon the preparation and execution of each service mission. This role ranged from science operations, the development of new instruments, overseeing anomalies that occurred on-orbit, and a number of public relations and outreach activities.
- The *HST Operations Project Scientist* performed an important liaison role during a service mission by being the point of contact between the STOCC Planning Shift of the project management and the various science teams which analyzed the data from the functional tests conducted on each new or repaired instrument, then reporting the outcome to senior project management. The Operations Project Scientist would also provide advice on the scientific priorities and any necessary trade-offs, and provide advice, guidance and oversight to the Operations Project Manager. The job required monitoring the performance of instruments and systems, assessing anomalies and the actions taken, supporting the execution of each service mission, and overseeing both current science operations and future instrument development.

- The *HST Development Project Scientist* was responsible for ensuring that any new hardware installed during service missions met its scientific goals.
- The *HST Observatory Manager* headed a team that designed, developed, fabricated, assembled, and tested the systems and components to be added to the telescope by a service mission. This team handled all the spacecraft systems, science instruments, support equipment, tools and EVA aids, and ensured that all hardware was brought together for integration, testing, and verification prior to the launch.
- The *HST Instrument Development Office Manager*. Before the Observatory Team could prepare an instrument for flight, first it had to be developed and that role was fulfilled by a team led by the Instrument Development Office Manager. This team was responsible for the design, fabrication, and testing of new items of hardware. It was a collaborative effort involving a great many individuals and facilities located across the United States, as well as personnel at Goddard.
- The *HST Carrier Development Manager* was responsible for four hardware carriers: the Super Lightweight Interchangeable Carrier (SLIC), the Orbital Replacement Unit Carrier (ORUC), the Multi-Use Lightweight Equipment (MULE) Carrier, and of course the Flight Support Structure (FSS). These were part of the hardware which made up the "payload" of a service mission. The goal was to safely deliver the new instruments, equipment and tools to orbit. To achieve this the Carrier Team ensured all hardware items remained secure for launch and landing and were maintained at their optimum temperature on-orbit, and that items to be returned to Earth were secured in suitable containers. The team also developed support structures that carried the instruments, avionics, and tools for a service mission. In addition they designed, fabricated, and tested the new hardware that was integrated on the carriers when preparing a shuttle for launch.
- The *Instrument Manager* headed the Instrument Team, which was responsible for ensuring that an instrument was built on time and within budget, and then met its technical requirements. To achieve this, the manager directed a team of engineers, technicians, and scientists in ten organizations distributed across the United States.
- The *HST Flight Servicing Project Thermal Systems Lead*. The HST must work in a harsh environment, so understanding the thermal characteristics of that environment was important in ensuring all elements of hardware, including scientific instruments, were capable of working throughout the telescope's lifetime. Members of this team were responsible for the thermal design, implementation, testing, and installation of hardware on the telescope. It had three main areas of responsibility: the new ORUs, the flight support equipment and carriers, and the crew aids and tools.
- The *EVA Activity Office Manager* headed a team involved in the development of the EVA procedures, techniques and tools for the service missions. It was responsible for training the astronauts at Goddard with engineering units, high-fidelity mockups, and flight hardware. The manager of the team also coordinated and managed all the HST hardware that the astronauts would rehearse with in water tanks, and also provided a team of scuba divers to support the astronauts during underwater training.
- The *Service Mission Operations Manager (MOM)* was located at the STOCC and was responsible for a team of about 90 engineers who ensured the correct procedures were built into commanding the telescope. During a service mission this manager worked the

MOM console on a 12 hour stint on the Orbit shift, alternating with the MOM on the Planning shift. They were on duty during the rendezvous, capture and berthing of Hubble, each EVA, and the eventual release of the telescope. Joyce King enjoyed her time as "Hubble's MOM" and recalled her experiences with pride and affection.

Space Telescope Operation Control Center (circa 1990).

## Space Telescope Operation Control Center

Located at Goddard Space Flight Center, the Space Telescope Operations Control Center (STOCC) has been the focal point of Hubble operations since its launch in April 1990 and has been upgraded several times. All commands are sent from STOCC to the telescope, and the data from the science instruments and onboard systems arrives at STOCC.[5] The science data is forwarded to the Space Telescope Science Institute (STScI) in Baltimore, Maryland.

The STOCC is located in the Mission Operations Room (MOR) and is used to control and monitor all HST flight operations, engineering, and science activities. The MOR also has the operational workstations and displays needed to monitor the health and safety of Hubble. All commands to the telescope originate from the MOR, which also monitors all engineering and science activities on a daily basis. This activity was carried out under the Mission Operations System Engineering and Software (MOSES) contract. This contract was implemented by the HST Operations and Ground Systems Project which managed the project.

During a service mission, the *STOCC Operations (STOCC OPS)* maintained a direct link to the JSC Payload Operations position at MCC in Houston and was also responsible for the operational application of the Command Plan (CP), which was part of the Service Mission Integrated Timeline (SMIT), and coordinated all STOCC operations and mission scheduled events and/or activities. The *Shift Supervisor (SS)* served as the lead controller for the flight operations team that commanded the telescope and was responsible for the configuration of the ground system. This role included implementing the planned activities, maintaining the health of the telescope, controlling STOCC commanding, and coordinating the transfer of data from MCC in Houston to STOCC during servicing activities.

STOCC OPS consisted of six main console engineer (CE) positions plus a CCS Support desk:

- *Control Center System Support* was basically a Help Desk to provide help in solving problems and advice on all the computers and programs used by the controllers and engineers in STOCC.
- *Data Management Subsystem/Instrumentation and Communication* (DMS/I&C) was responsible for monitoring the performance of the Hubble computers and managing the onboard data recorders, the flow of commands to the telescope, and defining the pertinent data mode, telemetry rate, format, and receiver and transmitter settings to support these functions. The controller also nominated the mode and both-way links between Hubble and the Space Network or Ground Network to maintain the correct pointing of the high-gain antenna of the telescope, and monitored the performance and status of the solar arrays and batteries.
- *Mission Support Analyst* (MSA) was the "timeline expert" at STOCC responsible for ensuring that operations followed the SMIT and the Command Plan. As a member of the Planning Team, they would update these plans during a service mission on at least a daily basis and more often as required.
- *Pointing Control Systems Engineer* (PCS) was responsible for rotating and stabilizing Hubble to accurately aim the telescope at a target. This person managed the telescope during the post-servicing checkout, prior to a resumption of the science program.
- *Pointing Control/Science Instruments/Mechanisms Controller* (PCS) was responsible for the health and safety of the scientific instruments, the SIC&DH, and the pointing control subsystem during the science program. This person monitored the status of the onboard computers, the telescope's slew and target acquisitions, and the gyroscopes, reaction wheels, and safing system.
- *Electrical Power Systems* (EPS) had responsibility for the solar arrays and batteries, and routed power across the telescope.
- *Senior Analysis and Calibration* (SAC) ran the computer program which was used to calibrate the pointing control system hardware, including the Fine Guidance Sensors, gyroscopes, and reaction wheels. The SAC also produced the computer loads for the telescope.
- *Ground Systems Manager (GSM)* was the focal point for all activities involving the ground systems, and supported the servicing by exploiting their experiences with the Control Center Systems (CCS), data flows, networking, and overall HST/JSC/orbiter operations.

STOCC Hubble's mission control network (circa 1999). (Courtesy Goddard Space Flight Center)

During the simulations and the missions themselves the GSM was assigned to assist in the ground systems and network troubleshooting efforts specific to both the Mission Operations Room (MOR) and the adjacent Service Mission Operations Room (SMOR) used to support the service missions, including the preparation, testing, and simulations conducted for each servicing flight while the MOR continued the daily activities of the telescope. The SMOR consisted of the following positions:

- *Mission Operations Manager* (MOM) was "controlling authority" for all STOCC operations, coordinating with both the Service Mission Manager and the Systems Manager for all nominal and contingency operations as well as for the Command Plan (CP) and Service Mission Integrated Timeline (SMIT) re-planning activities. The MOM informed the Service Mission Manager of operational status and coordinated all Go/No-Go calls.
- *HST Systems* served as the interface with JSC Payload Systems, providing systems engineering and analysis support directly to the Systems Manager, and coordinating all operational activities between STOCC, HST Systems Support, HST Engineering Analysis Support, and EVA specialists and Space Support Equipment Systems.
- *HST Systems Engineering Specialists:* This rotational console position included the following functions:

    - *Data Management Systems* (DMS) was responsible for the Data Management Systems.
    - *Electrical Power Systems* (EPS) was responsible for the solar arrays and batteries, and for routing power across the telescope. During a service mission the shuttle was the primary source of power for Hubble.

- *Instrumentation and Communications* (I&C) managed the communications onboard the telescope.
- *Mechanical Systems* (MS) worked in direct support of the MOM, verifying all HST mechanical activities, including the rotation of the solar arrays, and analyzing mechanical activity.
- *Pointing Control Subsystem* (PCS) was responsible for engineering and analysis support of the Pointing Control System/Attitude Control System. This included the Fine Guidance Sensors, gyroscopes, and star trackers.
- *Orbital Replacement Unit System Engineers* were present as appropriate for the ORU that was being installed on the telescope. This position included the *Optical Telescope Assembly (OTA) Engineer* who oversaw the main optics and the Fine Guidance Sensors, and the *Safing Engineer* who was responsible for analysis of the onboard safing systems.

- *Payload Operations Control Center* (POCC) provided support for all HST activities by coordinating communication and data acquisition with MOR personnel and NASA Network Control Center (NCC) at Goddard. POCC housed all processor computers, data receiving and recording subsystems, telemetry and commanding equipment, and the communications equipment necessary to enable Hubble to achieve its objectives.
- *Mission Support Room* (MSR) ran the day-to-day flight operations and engineering activities, with the support of off-line mission planning and data processing. MSR worked with the NCC to generate the science and engineering loads for the onboard computers, and performed science and engineering data processing. It was also the location where updates were issued to maintain navigational integrity and pointing accuracy of the telescope.
- *System Engineering and Evaluation Room* (SEER) supported mission operations by providing in-depth subsystem analysis and trending in near real-time and then post-analysis support using the engineering data obtained from the spacecraft. The room had the same capabilities as the MOR and could serve as backup operations control room.

In addition to the main MOR and SMOR, there were a number of smaller "back room" locations used in support of the Hubble operations:

- *Thermal Subsystem engineers* monitored the temperatures of the main telescope and new replacement hardware.
- *Engineer Support Systems* (ESS) had access to all historical data about Hubble since April 1990 and could plot voltages, temperature, currents, speeds, torques, changing switch positions, and other values of interest.
- *Anomaly Response Manager* (ARM) was described above.
- *Service Mission Planning and Re-planning Tool* (SM PART) prepared and revised the timeline and command plan that detailed every element of a service mission. It did so by coordinating the activities of hundreds of engineers and controllers, the shuttle, the telescope, the fleet of TDRS relay satellites.
- *Simulation Team* (SIM TEAM) trained the operations teams at STOCC for a year in advance of each service mission.

- *Electronic Data Control Center* (EDOCS) created and maintained an intranet site to provide reference resources for real-time technical documentation, tactical data, and plans to prepare and execute a service mission.

**European Space Agency**

The agreement by ESA to partner with NASA on Hubble had a provision that allowed up to 15 Europeans to join the Space Telescope Science Institute in support of work on the solar arrays, the Faint Object Camera, etc.[6] As Lothar Gerlach notes, "Each service mission to Hubble had different objectives. Sometimes we wanted to replace a camera, sometimes we had a gyroscope problem. So every time there were slight neuroses when NASA asked us to revisit all of our flight rules and contingency activities. And, of course, during a mission we were monitoring, constantly listening basically to three voice channels in parallel, trying to pick up something, anything, a decision being made or whatever, something that is outside the flight rules or the contingency work so that we could raise as an alarm. Everybody had a selection of the three channels which he wanted to listen to. For example, we supported the final service mission with just four people because there was no need for more, so we had two shifts of 12 hours with one guy of each shift concentrating on the electrical part and the other on the mechanical and structural part. So we always had two electrical engineers and two mechanical engineers. However, for an event like SM-1 where we replaced the original solar generators, or SM-3B where we retrieved the second set of solar generators, we had a huge team, I would say 10 to 15 roughly; so it depended on the task involving our hardware.

"We had a box in front of us and had 20 different voice channels to listen to. I listened to communications between the astronauts on-orbit and the Capcom on Earth. I also listened to the engineering management channel where decisions were being made. And because I am more of an electrical engineer than a mechanical engineer in my involvement with Hubble, I listened to the power system channel. Those were the three channels that I listened to. When monitoring, you try to pick up anything relevant to your point of view. For example, when the astronauts went out into the shuttle cargo bay on SM-2, air from the airlock was pushed towards one of the solar generators and rotated it out of its alignment, overcoming the brake and pushing the solar generator to its end stops. This was something we never expected and we had not considered in our contingency plan; this was the first time it happened. Then, at night the Planning Shift came up with the corrective actions or additional information which must be prepared. So we worked with the Planning Shift that night, asking what happend here? Did we damage the solar generator? We did a lot of analysis, and concluded that there was absolutely no problem on the performance of the solar generator. It had not been damaged, because all the forces were below the qualification limits. But the astronauts go out of the airlock every day and we don't want this blast of air to occur every day. So as a corrective action we wrote them a procedure that the release of air must not be so sudden, but have a certain profile. The corrective action was written in the Planning Shift, to be sent up to the crew when they awoke, so that the next time we reach that same point this problem won't happen again."

The HST ESA Team at KSC May 2009: *left* to *right* Manfred Schmid, Michael Eiden, Udo Rapp and Lothar Gerlach. (Courtesy ESA)

## Kennedy Space Center

At the Kennedy Space Center in Florida, a separate team accepted and prepared the hardware for flight: the orbiter, twin SRBs, External Tank, and all items that comprised the "payload" and its associated carriers.[7] This team comprised:

- *Pre-Flight Mission Management Team Chairman* was a shared position with the In-Flight Mission Management Team Chairman, assisting with the overall management, integration, and shuttle program operations. This assignment reported directly to the Shuttle Program Manager.

- The *Shuttle Launch Director* was responsible for making the final Go/No-Go for a launch after polling the Payload Manager, Shuttle Engineering Director, the Launch Weather Officer, and the Director of Safety to reach a consensus and verifying that the Test Director and Mission Management Director were in agreement.
- The *Test Director* led the shuttle test team and was responsible for the integration of the shuttle vehicle and ground support testing throughout the countdown phase, plus the safety of all personnel, including those at the pad following the loading of the ET. This assignment reported to the Launch Director.
- The *Mission Manager* led a team of engineers and technicians that assembled and tested the payload designated to fly aboard the Hubble service mission. This was the primary interface with the payload customers, responsible for solving any technical issues that occurred.
- *Shuttle Orbiter Flow Director*. Each Orbiter was assigned a Flow Director during its pre-mission processing to deal with all the payload and hardware integration, launch scheduling and various processing operations through the final countdown leading to launch.
- *Launch Weather Officer*. The weather conditions for launch were monitored from the Morrell Operations Center at the Cape Canaveral Air Force Station, just south of the Kennedy Space Center. Weather information for the launch and RTLS landing strip, as well as for the various abort sites across the Atlantic, were constantly updated to determine whether conditions matched the launch-commit criteria for the shuttle. It was in contact with the Launch Director and the Landing Recovery Director during the countdown, and participated in the final poll prior to launch.
- The *Assistant Test Director* supported the Test Director and was responsible for the management of the launch team and launch procedures for the terminal countdown. This role included the management of the countdown clock with the ground launch sequencer operator and for the "special crews" that were sent out to the pad to deal with major issues or emergencies.
- *Director of Safety and Mission Assurance*. This was an advisory role to the Mission Management Team with responsibility to review the rationale in the decision-making process and particularly to ensure that any dissenting opinions were encouraged, not overlooked. It could call in additional resources to address any issue that required a rapid response.

## "Crip" Crippen, Deputy Director Shuttle Operations

Former astronaut Bob Crippen, who flew four early shuttle missions, was also involved in the management side of the shuttle program during the period following the loss of Challenger in 1986 through to the deployment of Hubble and preparations for its first service mission. He used his experiences and skills as Deputy Director Shuttle Operations at KSC between July 1987 and December 1989 and at NASA Headquarters from January 1990 to January 1992.[8]

Crippen says there was a great deal of focus devoted to returning the fleet to operational flying after the loss of Challenger, with a lot of people working hard for long hours not only on Hubble but a multitude of other missions. "We wanted to fly, and fly safe. The first task

The April 10, STS-31 launch scrub at Launch Control Center, KSC. At *left* is former astronaut Bob Crippen, Director Space Shuttle, NASA HQ. Next to him is another former astronaut, Bill Lenoir, Associate Administrator for Space Flight.

was to look at the mission manifests. The sequence of launches was made in Headquarters and Houston." The first flight would be a "simple" mission, and those which would follow "fell out naturally". Hence the Tracking and Data Relay Satellites (STS-26, STS-29), DOD payloads (STS-27, STS-28, STS-33, STS-36), planetary probes (STS-30, STS-34), retrieval of LDEF (STS-32) and deployment of Hubble (STS-31).

Crippen says his time in Washington was more challenging than rewarding, and certainly had its frustrations. His responsibilities essentially dictated that "if something went wrong it was going to be my fault". It was the suggestion of the Shuttle Program Director that, after the loss of Challenger, veteran astronauts should become more involved in the management structure and decision-making process; experience being considered useful in that process. A group was set up to devise the new structure that would ensure that the field centers worked together, but their rivalries persist today. Crippen's primary role was with the shuttle vehicle and in particular what was being done with the orbiter. There was a comprehensive move to eliminate problems with the shuttle. He was not that involved with the day-to-day issues of every mission, he was more of an overseer if things were going well. During preparations for the first service mission, STS-61, he visited Goddard several times to review progress and to satisfy himself that all was going well. As he points out, STS-61 had a higher priority than the other 21 manifested missions at that time "because NASA was extremely embarrassed about the problems with the telescope's mirror in the wake of the tragedy of the Challenger accident that shook the space agency to the core. The desire was to make all the customers happy—the DOD, commercial customers, scientists—but it was almost impossible to make everyone happy."

## ASTRONAUT OFFICE SUPPORT

In addition to the named flight crew and any backup crewmembers who were assigned to a shuttle mission, there was a cadre of experienced and rookie astronauts staffing a variety of support roles. Some of these roles, such as Capcom at Mission Control or Launch Support at the Cape, were part of a "tour" that involved covering several missions in the same support position. This expertise in certain fields added to an individual's technical skills and was an important stepping stone for rookies eager to receive their first flight assignment, as well as for veterans in between flights.

These support roles could include:

- The *Capcom* (CAPsule COMmunicator) was the Astronaut Office representative on each flight control team and served as the point of contact between the flight crew in space and the controllers on the ground. At least one astronaut was assigned to each flight control team: Ascent/Entry/Orbit 1, 2, 3, and where necessary Orbit 4. The role of Capcom had its origins in the Mercury program and was one of the most visible of the mission support tasks which astronauts undertook. During EVAs, the Capcom on console was usually either EVA experienced personally or had worked with the crew during their EVA training and was familiar with the planned tasks and procedures.
- *SPAN* (SPacecraft ANalysis) was operated throughout a mission in 12 hour shifts by a group of astronauts and other support personnel assigned to each mission or perhaps to a particular flight day. This group supported the mission by reviewing analysis data returned from the orbiter, offering support in real time or in the event of malfunctions and equipment failures.
- *SAIL Support* (Shuttle Avionics and Integration Laboratory) was the only facility in which the hardware and flight software of the shuttle could be integrated and tested in simulated flight environments. The avionics mockup was designated OV-095 (a non-flight "vehicle") that was a skeleton of the orbiter flight deck. The support team could accurately reproduce flight stations and procedures using identical electronics to those on the flight vehicle.
- *SMS Support* (Shuttle Mission Simulator). Building 5 at JSC housed two shuttle simulators. In addition to the fixed-base simulator that consisted of a high-fidelity mockup of the orbiter flight deck, with computer-aided visuals out of the forward, aft and overhead windows, there was a low-fidelity mockup of the middeck for training. Building 35 housed the Guidance and Navigation Simulator (GNS). This motion-based simulator offered a six-axis motion system but with computer-aided visuals out of the forward windows only. Support astronauts used these simulators during a mission to replicate flight issues and to explore contingency and workaround tasks to assist the crew on-orbit.
- *KSC Launch Support* (known as the Cape Crusaders or $C^2$) was a team of astronauts who supported shuttle activities at Launch Complex 39, such as checking the status and

preparedness of the crew cabin and setting switch positions prior to launch. The designated Astronaut Support Person assisted the final checks and strap-in activities with the flight crew inside the vehicle prior to hatch closing.
- The *Weather Coordinator* was the Astronaut Office link to the Mission Management Team, the Flight Director at Mission Control, and the Launch Director at the Cape.
- *TAL (Trans-Atlantic Landing) Support.* Two or three astronauts would be sent to the various launch abort sites, mostly on the west coast of northern Africa or in Spain, to assist in abort situations. They monitored the ascent and stood by as Capcom and NASA coordinator should a TAL situation be initiated.
- *Contingency Action Centers.* If a contingency situation occurred, the field centers at KSC, JSC, and Dryden (in California) activated an Action Center to coordinate any required activities in real time. The Astronaut Office would normally be represented by a veteran astronaut or a member of Astronaut Management, although not always.
- *STA Weather Pilot.* Shuttle Training Aircraft weather pilots (code name WX, derived from the Morse Code telegraph designation) were senior astronauts who flew the Shuttle Training Aircraft from KSC or Edwards (in California) to report on weather conditions at altitude immediately prior to a shuttle launch or landing. They supplied information to the Mission Management Team to steer a decision either to proceed or to call a delay. WX pilots played a similar role in certifying whether to permit or wave off a landing at the primary or alternative sites.
- *EOM (End of Mission) Exchange Crew.* This was formed from members of the Cape Crusaders to provide an "astronaut crew" for de-configuring the orbiter after its flight crew had left the vehicle. The ground closeout crew would then transfer the orbiter to the OPF if it landed on the Shuttle Landing Facility in Florida or prepare it for return to the Cape if it had landed elsewhere.
- *EVA Support.* A team of normally EVA-experienced astronauts and support divers available for simulating EVA operations underwater to rehearse real-time situations and contingencies.
- *Mishap Representative.* Usually a senior astronaut was available in the event of a serious mishap during a mission.
- *Family Escort.* A member of the Astronaut Office was assigned to each prime crew member's immediate family or dependents to provide advice, guidance, and support during preparations for the mission, launch and period in space, as well as to protect the privacy of the family in the event of a major incident.
- *Extended Family Escort.* Several astronauts were also assigned to support each prime crew member's extended family group.

## MISSION CONTROL

Traditionally, shuttle missions were "controlled" from the Mission Operations and Control Room (MOCR) at the Johnson Space Center in Houston, Texas, but the management and mission operations structures for the Hubble missions also required a close working relationship with the teams at Goddard Space Flight Center.[9]

Once the twin SRBs had ignited and the shuttle cleared the tower at Launch Complex 39, all responsibilities for conducting the mission transferred from Kennedy Launch Control to JSC—whose involvement had started about 5 hours earlier and would continue around the clock through the immediate post-landing activities.

Commonly known by the radio call sign "Houston", the Mission Control Center at JSC became active in June 1965 during the Gemini 4 mission. For the next 30 years, the Flight Control Rooms (FCR) were located on the second floor of Building 30. Then in 1995 new flight control rooms came on line in an enlarged five-story block called Building 30 South (30-S), dedicated to shuttle and space station operations. This area also housed the Payload Operations Control Center (POCC), the Mission Operations and Integration Room (MOIR) which would become the Shuttle Mission Evaluation Room (MER), several support rooms, and a miscellany of new state-of-the-art electronic apparatus. After alternating flight control with the old control rooms for seven shuttle missions, the new control rooms became fully operational in 1998. On April 14, 2011 the original Building 30, which had been designated as a national monument, was renamed the Christopher C. Kraft, Jr. Mission Control Center but retained its old call sign. Chris Kraft, a retired NASA engineer and manager, had been instrumental establishing flight operations for Project Mercury and in creating the current facility.

Flight Control Room 1 (FCR-1, pronounced "Flicker") was converted from the original Mission Operations Control Room 1 (MOCR-1, "Moeker"), and was used for most of the unclassified shuttle missions through 1996, including the deployment of Hubble (STS-31) and its first service mission (STS-61). After sharing flight operations with FCR-2 on seven non-classified missions (STS-70 through STS-76) FCR-1 handled all the subsequent shuttle missions through to the conclusion of the program. The new control rooms were designated Red, White, and Blue. White was used for the shuttle missions from 1996 to 2011, including Hubble SM-3A, SM-3B and SM-4; Blue was formerly used for ISS operations; and Red has always been used for training flight controllers.

Table 11  Flight Directors and Capcoms for Hubble related shuttle missions.

Each assigned Flight Director adopts a personal 'Flight' name which lasts for their career. The numbers denote their sequence on-console

| Mission | Position | Flight Director | Number | Flight name | Capcom |
|---|---|---|---|---|---|
| STS-31 | Ascent | Ronald D. Dittemore | 27 | Phoenix | Steve Oswald |
| | Orbit 1/Lead | William D. Reeves | 23 | Alpha | Story Musgrave (Prime) Don McMonagle (BUp) |
| | Orbit 2 | J. Milton. Heflin | 25 | Sirius | James Voss (Prime) Ken Bowersox (BUp) |
| | Planning/Orbit 3 | A. Lee Briscoe | 20 | Aquila | Kathy Thornton |
| | Entry | N. Wayne Hale | 28 | Turquoise | Mike Baker |
| | MOD | B. Randy Stone | 18 | Amber | N/A |

(continued)

**Table 11** (continued)

Each assigned Flight Director adopts a personal 'Flight' name which lasts for their career. The numbers denote their sequence on-console

| Mission | Position | Flight Director | Number | Flight name | Capcom |
|---|---|---|---|---|---|
| STS-61 | Ascent/Entry | Richard D. Jackson | 34 | Burgundy | Ken Cockrell Charlie Precourt/Curt Brown (Wx) |
| | Orbit 1 Rendezvous and deploy | Robert E. Castle | 29 | Antares | Susan Helms |
| | Orbit 2/Lead EVA | J. Milton Heflin | 25 | Sirius | Greg Harbaugh |
| | Orbit 2 Orbiter ops and systems | Jeffrey W. Bantle | 32 | Aurora | N/A |
| | Planning/Orbit 3 | John F. Muratore | 35 | Kitty Hawk | Carl Meade |
| | MOD | B. Randy Stone | 18 | Amber | N/A |
| STS-82 | Ascent/Entry | N. Wayne Hale | 28 | Turquoise | Kevin Kregel Dom Gorie (Wx) |
| | Orbit 1/Lead | Jeffrey W. Bantle | 32 | Aurora | Marc Garneau |
| | Orbit 2 | Bryan P. Austin | 37 | Perseus | Kathryn Hire |
| | Planning/Orbit 3 | Charles W. Shaw | 24 | Altair | Chris Hadfield |
| | MOD | A. Lee Briscoe | 20 | Aquila | N/A |
| STS-103 | Ascent/Entry | N. Wayne Hale | 28 | Turquoise | Scott Altman Rick Sturckow/Joe Edwards (Wx) |
| | Orbit 1/Lead | Linda J. (Hautzinger) Ham | 33 | Corona | Steve Robinson |
| | Orbit 2 | Bryan P. Austin | 37 | Perseus | Ellen Ochoa |
| | Planning/Orbit 3 | Jeffrey M. Hanley | 41 | Ares | Chris Hadfield |
| | Orbit 4 | N. Wayne Hale | 28 | Turquoise | N/A |
| | MOD | Jeffrey W. Bantle | 32 | Aurora | N/A |
| STS-109 | Ascent/Entry | John P. Shannon | 38 | Midnight | Mark Polansky Charles Hobaugh (Wx) |
| | Orbit 1 /Lead | Bryan P. Austin | 37 | Perseus | Mario Runco |
| | Orbit 2 | Anthony J. Ceccacci | 57 | Intrepid | Steve MacLean |
| | Planning/Orbit 3 | Jeffrey M. Hanley | 41 | Ares | Dan Burbank |
| | MOD | N. Wayne Hale | 28 | Turquoise | N/A |
| STS-125 | Ascent/Entry | Norman D. Knight | 51 | Amethyst | Greg Johnson Eric Boe (Wx) |
| | Orbit 1 /Lead | Anthony J. Ceccacci | 57 | Intrepid | Dan Burbank (Lead) |
| | Orbit 2 | Richard E. LaBrode | 46 | Pegasus | Alan Poindexter |
| | Planning/Orbit 3 | Paul F. Dye | 36 | Iron | Janice Voss |
| | Orbit 4 | Bryan C. Lunney* | 54 | Onyx | N/A |
| | MOD | John A. Mccullough | 50 | Eagle | N/A |

*Second generation Flight Director, son of Glynn Lunney, #4 Black Flight (Class of 1963, retired as Flight Director in 1974)

**For Hubble missions the control rooms used were:**

- Deployment/STS-31: FCR-1
- SM-1/STS-61: FCR-1
- SM-2/STS-82: MCC White FCR
- SM-3A/STS-103: MCC White FCR
- SM-3B/STS-109: MCC White FCR
- SM-4/STS-125: MCC White FCR.

Teams of flight controllers alternated shifts in the control center and the nearby analysis and support facilities. The handover of control teams usually took about 1 hour, to permit each flight controller to brief his or her oncoming colleague on the course of events over the previous two shifts. There was generally a press conference by the off-going Flight Director within an hour of a shift handover being completed.

The shuttle flight control teams were Launch (Ascent) and Landing (Entry), Orbit 1, 2 or 3, and if necessary (see below) Orbit 4. Some positions were staffed by the same people for Ascent and Entry activities and for Orbit 1 operations. Others were alternated by specialists in launch and landing activities, or orbital operations as necessary. Orbit 1 staff were usually responsible for the deployment (on STS-31) and retrieval of Hubble. Orbit 2 would handle the EVAs. Orbit 3 coincided with the crew sleep period and was known as the "night shift", "planning shift" or, more affectionately, the "graveyard shift".

The STS-31 mission was the first since the early days of the program where the vehicle's electrical and environmental systems were split over two console positions called EECOM and EGIL. However, the position of EECOM continued to have responsibilities for the life support systems of the orbiter as well as cabin pressure, active thermal control systems (e.g. the Flash Evaporator System) and management of the supply and waste water tanks.

The FCR-1 console positions circa 1990–1993 were:

- The *Flight Director* (FD or "Flight") had overall responsibility for the conduct of the mission.
- The *Capcom* was by tradition an astronaut responsible for all voice communications with the flight crew. The name derived from when the astronaut in the control center "communicated" with the lone astronaut aboard the Mercury "capsule". The original pilot-astronauts disliked the engineers' term, preferring instead "spacecraft" because, as they put it, you "take a capsule" for medicine but "fly a spacecraft". Nevertheless, the name of this console became firmly embedded in the vernacular and is in use for ISS operations.
- *Flight Activities Officer* (FAO). In addition to being responsible for procedures and crew timelines, this console offered expertise on flight documentation and checklists, and prepared messages and maintained all teleprinter (early in the program), text and graphics systems, and (finally) e-mail traffic to the orbiter.
- *Phase Specialist* (PROCEDURES). A specialist position which sometimes occupied the FAO console and offered expertise to the FD in the specific procedures required during a complex operation.

STS-125 Mission Control Houston Teams: Ascent and Entry (*upper*) and Orbit 1.

STS-125 Mission Control Houston Teams: Orbit 2 (*upper*) and Orbit 3 Planning.

- The *Integrated Communications Officer* (INCO) was responsible for all orbiter data, voice and video communication systems, monitoring the telemetry links between the vehicle and the ground, and overseeing the uplink command and control processes.
- The *Flight Dynamics Officer* (FDO or "FIDO") monitored the shuttle's performance during ascent and assessed the abort modes as required, calculated orbital maneuvers and resulting trajectories, and then monitored the vehicle's flight profile and energy levels during re-entry.
- The *Guidance Procedures Officer* (GPO) was responsible for navigational software aboard the shuttle and maintained its "state vector".
- The *Trajectory Officer* (TRAJECTORY or "TRAJ") aided FDO during the dynamic phases of the flight and was responsible for maintaining the trajectory processors in MCC and for trajectory inputs made to the mission operations computer.
- The *Environmental Engineer and Consumables Manager* (EECOM) was responsible for all life support systems, cabin pressure, thermal control, supply and waste water management, and the consumption of oxygen and nitrogen.
- The *Electrical Generation and Illumination Officer* (EGIL or "EAGLE") managed the fuel cells and power distribution system, the vehicle's lighting, and its Master Caution and Warning System.
- The *Payloads Office* (PAYLOADS) was responsible for coordinating all the payload activities, and served as principal interface with remote payload operations facilities; for Hubble this was mainly Goddard and STOCC.
- The *Data Processing Systems Engineer* (DPS) was responsible for all onboard mass memory and data processing hardware, monitoring the flight software (both primary and backup), and managing the operating routines and multi-computer configurations.
- The *Propulsion Engineer* (PROP) was responsible for the OMS and RCS thrusters in all phases of flight, monitoring the use of fuel and storage tank status and calculating the optimal sequences for thruster firings. For the Hubble missions this was a critical position, as the orbiter flew close to the limit of its capability. During rendezvous and proximity operations, as well as re-boosting the telescope, onboard consumables had to be carefully managed to ensure that there would be sufficient propellant available for the de-orbit burn.
- *Booster Systems Engineer* (BOOSTER). This was another "carryover" from the early days of the space age when a launch vehicle was known as a "booster". The console was responsible for monitoring the performance of the solid rocket boosters and main engines of the orbiter during the ascent phase.
- The *Guidance, Navigation and Control Systems Engineer* (GNC or "Guido") was responsible for all the internal navigational systems hardware on the orbiter, such as the star trackers, radar altimeters, and the inertial measurement units. This console also monitored radio navigation and digital autopilot hardware systems.
- The *Ground Controller* (GC) coordinated the operation of ground and other elements in the world-wide space tracking and data network, including the geostationary TDRS satellites. This controller was also responsible for the computer support and displays in MCC.
- *Maintenance, Mechanical, Arm and Crew Systems* (MMACS), formerly known as RMU (Remote Manipulator Unit), was responsible for the RMS system, making it

another important role during Hubble service missions. This console also monitored the auxiliary power units and hydraulic systems of the orbiter, as well as managing the payload bay and vent door operations.
- The *Rendezvous Guidance and Procedures Officer* (RENDEZVOUS) monitored the onboard navigation of the shuttle during a rendezvous operation and advised the FD of the developing profile vis-à-vis the target. It was another key console for Hubble missions. Had a revisit to the telescope been required during STS-31 to attend to an issue with the telescope, this console would have been reactivated in support of that decision.
- The *Extravehicular Activities System Engineer* assisted in planning the spacewalks to be performed during a mission, worked with the astronauts during their training, and assisted in the development of tools and techniques, then monitored the pressure suits as the astronauts went EVA on-orbit.
- The *Flight Surgeon* (SURGEON) was the nemesis of the astronauts, monitoring their health and providing procedures and guidance on all health-related matters. Over the years some astronauts, whilst complying with the medical requirements, experiment protocols and flight rules, were opposed to what they regarded as excessive intrusion into their private medical conditions not only for pre-flight and post-flight activities, but also in-flight. Due to the nature of their primary task, Hubble missions were not overly burdened with additional medical objectives. Nevertheless, the status of the crew were monitored as the mission progressed, especially the levels of stress and concentration to operate the RMS and the physical exertion of the EVA astronauts.
- The *Public Affairs Office* (PAO) was a member of JSC's Public Affair Office who gave real-time explanations of mission activities to supplement and explain air-to-ground commentary during all phases of a flight from launch to post-landing.

When flight control moved to the new rooms, there were changes made to the console positions to reflect improvements in technology and reviews of the control team's role in supporting the remaining shuttle missions. The White FCR console positions circa 1997–2009 were:

- *TRAJ/FDO—Trajectory Officer/Flight Dynamics Officer*: Planned maneuvers and monitored trajectory in conjunction with the Guidance Officer.
- *GPR—Guidance Procedures Officer:* Ensured the onboard navigation and guidance computer software executed the required tasks to accomplish the objectives of the mission.
- *GC—Ground Controller:* Directed maintenance and operational activities affecting Mission Control hardware, software, and supporting facilities, and also coordinated the space tracking and data network and TDRS in conjunction with Goddard Space Flight Center.
- *PROP—Propulsion Officer:* Monitored and evaluated the OMS and RCS thrusters in all phases of the flight, and managed propellant and other consumables available for maneuvers. This was a critical console position for Hubble missions.
- *GNC—Guidance, Navigation and Control:* Monitored all guidance, navigation and control systems on the orbiter, advised the flight crew regarding malfunctions of the

guidance system, and notified the Flight Director and flight crew of impending abort situations.
- *MMACS—Maintenance, Mechanical and Crew Systems Engineer:* Monitored the structure and mechanical systems of the orbiter, the use of onboard crew hardware, and any in-flight equipment maintenance.
- *EGIL—Electrical Generation and Integrated Lighting Systems Engineer:* Monitored the cryogenic reactants for the fuel cells, electrical power generation and distribution systems, and all vehicle lighting.
- *DPS—Data Processing System Engineer:* Monitored the status of the orbiter's data processing system, including the five general purpose computers, flight-critical and launch data lines, the displays, and the onboard mass memory and software.
- *PAYLOADS—Payloads Officer:* Coordinated onboard and ground system interfaces between the flight control team, the payload users, and any interfaces with payloads.
- *FAO—Flight Activities Officer:* Planned and supported crew activities, checklists, procedures and schedules, developed a timeline for the attitude of the orbiter which would optimize mission activities, and provided daily updates to the Crew Activity Plan (CAP).
- *EECOM (DF)—Emergency, Environmental and Consumable Systems Engineer:* Monitored the avionics and cabin cooling systems aboard the orbiter, as well as the cabin pressure control systems.
- *INCO (DF)—Instrumentation and Communication System Engineer:* Planned and monitored in-flight communications and the configuration of the instrumentation system.
- *FLIGHT—Flight Director:* The final authority in Mission Control over all decisions regarding a safe and expedient mission, including the performance of the payloads.
- *CAPCOM (Spacecraft Communicator)*—Primary communicator between the flight controllers and the flight crew.
- *PDRS (Payloads Deploy and Retrieval Systems Engineer)*—Primarily monitored the status and operation of the Remote Manipulator Systems (RMS).
- *PAO—Public Affairs Officer:* Issued real-time explanations to the media and general public to supplement air-to-ground commentary and explain flight control procedures and actions during all phases of a flight from launch to post-landing.
- *MOD—Mission Operations Directorate:* Provided the link between the flight control team and upper levels of mission managers and NASA administrators.
- *EVA—Extravehicular Activity Systems Engineer:* Monitored the astronauts' pressure suits during EVA operations.
- *BOOSTER—Booster Systems Engineer:* Monitored and evaluated the performance of the three shuttle main engines, twin SRBs and ET during pre-launch and ascent, then monitored the helium pressure in the Main Propulsion System for entry. The EVA console was used for this activity because there was never a scheduling conflict.
- *Surgeon—Flight Surgeon:* Monitored the activities of the flight crew, coordinated medical operations of the flight control team, offered crew consultation services for private medical matters, and advised the Flight Director on the status of the crew's health.

## Multi-purpose Support Rooms

The Flight Control Rooms have been the most visible element of MCC-Houston for over 50 years, but there were numerous support rooms within the building which were used to assist the controllers on-console. Referred to as Multi-Purpose Support Rooms (MPSR) or "back rooms", it was there that data from the shuttle and payloads were analyzed. The rooms also provided locations for real-time simulations which give valuable advice and information to the duty flight controllers.

## THE ROLE OF A HUBBLE FLIGHT DIRECTOR/MISSION DIRECTOR

The head of each flight control team (or shift) was the Flight Director (referred to simply as "Flight" on the voice loops). A senior member of that esteemed group was appointed as the Lead Flight Director for each mission and assumed overall leadership for the duration of the mission. It was a fine balance of supporting their teams, following mission rules, answering to higher management in addition to monitoring and guiding the crew through the execution of their mission.[10] For Hubble, six flight directors took the lead with a further 22 shift flight director positions being filled from the ranks of the Flight Director Office (Code DF).

At various times Charles "Chuck" Shaw served as a Hubble mission Shift Flight Director, a Lead Flight Director and a Mission Director. He offers a fascinating insight into the role of the Flight Director, management of the flight control teams, and the unique role created for the Hubble service missions, namely that of Mission Director.[11]

### "Altair" Flight, Charles "Chuck" Shaw

Chuck Shaw was assigned by the Air Force to NASA in 1980, initially as a shuttle systems instructor, then simulation supervisor. In 1983 he was selected as the first non-NASA Flight Director for both NASA and DOD shuttle missions. As the 24th Flight Director, he selected the call sign "Altair". Shaw had joined the Air Force in 1969 and worked on the Minuteman missile program as a launch crew commander, then a training manager, and finally a launch director. He retired from the military as a lieutenant colonel in 1989 but remained a shuttle Flight Director until 2003. In all, he served as a Flight Director for 31 missions; 10 of them in the leading role.

For the Hubble deployment by STS-31, Shaw was Flight Director for Team 4. But as he explains, "Team 4 didn't normally pull any shifts during the mission unless something went wrong and the on-console flight control team needed either off-line help to solve a technical issue, or needed us to sit on-console while the original team stepped off to work the issue.

"For SM-2, I was the Planning Team Flight Director. Generally, for an HST service mission, many of the failure scenarios had been thought out to the point that if special tools were needed, they were developed and practiced with, and the timeline for performing the failure response or workaround had been developed. The Planning Team pulled together the re-planned timelines to incorporate the recovery activities and confirm that the assumptions which went into the recovery activities were valid for the situation being faced for the next day's activities."

# The role of a Hubble flight director/mission director

Shuttle Flight Director/Mission Director Charles "Chuck" Shaw. (Courtesy, Chuck Shaw)

"For SM-4, I was the Mission Director. The Mission Director was a special management position defined for HST servicing, starting with SM-1, and was directly responsible to the Associate Director for Manned Space Flight and to NASA Headquarters. There were three roles to the responsibility. Since HST missions could absorb a lot of resources of the Space Shuttle Program Office, the Mission Operations Directorate, the HST Program Office, the Johnson Space Center, the Kennedy Space Center, and the Goddard Space Flight Center, the first role was to ensure that all those organizations were working together and to address any disconnects. The second role was to perform the function of the 'Flight Manager' position in the Space Shuttle Program Office. This included overseeing all of the cargo engineering work and the schedules for integrating the cargo components into the shuttle orbiter. The third role was in real time during the mission, to act as the coordinator between the HST Management Team and the Shuttle Program Management Team."

Of course, flight controllers work on other missions as well as the primary mission but, as Shaw explained, this was dependent upon their role in the primary mission. "For the Flight Director assignment for the deployment mission (STS-31) and for SM-2, I was also working on other missions. The preparations for shuttle missions varied in length depending upon the complexity of the mission and your role in it. The Lead Flight Director and his flight control team and the prime flight crew carried the biggest burden of designing and preparing for the mission. As a result, they had the least time available to work on other missions during those preparations. The other Orbit Operations Teams and the Planning Team would participate in the training and any special aspects of the preparations unique to their shift's responsibilities but they would have time to work on other missions in parallel—until the integrated training simulations start about 4 to 5 months before the mission, when everyone that is assigned to it is dedicated to getting ready for their role in that mission."

To transfer experience from one mission to the next, several astronauts flew more than one Hubble mission; usually, although not always, flying consecutive missions. The flight control team tended to follow this pattern. "Most 'generic' shuttle mission operations did not require any special HST experience, but the coordination, partnerships, and working relationships of the many organizations involved in bringing about and performing the unique servicing tasks for Hubble profited by having as many HST-experienced team members as possible." And as Shaw pointed out, having several shuttle flights in planning on the manifest at the same time demanded that operations be done as explained above, where the Lead Flight Director's team bore the brunt of preparing for a specific mission while the remaining teams would work on other missions until the commencement of integrated simulations. All of this was dependent upon the complexity of the mission and the personal involvement of each flight controller.

The rotation and assignments of the various flight control teams for the Hubble missions was also addressed by Shaw. "For the most part, the shift positions and rotations were along the lines of the other shuttle missions. There were generic guidelines for the number of days in a row for 'planned' work and the handover times were tailored to the nominal timeline in order to allow the teams to focus on their planned tasks during training."

The rendezvous, Shaw says, whilst *never* a simple or straightforward process, was pretty well understood, especially after several flights to Mir and the ISS. "The EVA timeline was also nothing out of the ordinary for preparing and operating the suits. However, Hubble put extreme demands on everyone during the large amount of time spent on EVA, especially on the crew. EVA is a *very* physically demanding activity for the astronauts, and jamming daily EVAs back-to-back pushed everyone to the limit—that is a very success-orientated timeline, and demanded a *lot* of very close orchestration, coordination, and preparation, especially if it was flexible enough to address any issues that could happen in real time and still accomplish the objectives."

## Mission direction and management

The role of Mission Director, as noted above, was specially created for Hubble, and Shaw offered an expanded description of the role and its place between the flight controllers and the Mission Management Team for the service missions. "There were two facets to real-time operations for a shuttle mission: the 'Operations' Team (i.e. Crew and Flight Control Team) and the 'Management' Team. The Operations Team was led by the Flight Director, and was responsible for the real-time execution of the mission. You can't fly an airplane or conduct real-time shuttle operations by committee. The Flight Director was responsible for 'making it happen', and had the authority to do anything (within the rules, or not) required for mission safety and anything within the Flight Rules to make the mission successful. The Operations Team did not own the hardware, but they were responsible for operating it as per the plans that they pulled together pre-mission and then submitted to

Management for approval. As long as the mission was going according to plan, Management watched as the Operations Team executed." Shaw provided an analogy to operations in commercial airlines, where an airline pilot is totally responsible for the execution in real time for the flight. However they do not own the plane or make the decision for where it flies to or from, or what and who is carried; that is down to airline management.

"The shuttle was the same," he said. "The Shuttle Program Office owned the shuttle. The HST Project Office owned Hubble. The Shuttle Program Office contracted with the payload customer, in this case the HST Projects Office, to perform the repair mission. That contract was the Payload Integration Plan, or PIP. The Flight Manager pulled that contract together. The Shuttle Program also contracts with the Mission Operations Directorate or MOD at the Johnson Space Center to plan and conduct the mission that is contracted for in the PIP. The parameters which described how the mission was planned, and would be executed in real time in order to meet the contractual obligations in the PIP were captured and agreed in the Mission Specific Flight Data File—these were the timelines and check lists the crew and mission control used—and the Mission Specific Flight Rules Annex to the General Flight Rules Document. So the Shuttle Program, the HST Project, and the Mission Operations Directorate all jointly agreed to those documents."

**Structure and responsibility**

The Management Team was also referred to as the Mission Management Team (MMT). It was chaired by the Shuttle Program Manager or their representative. The Payload Program Office for whatever payloads were aboard the shuttle (for HST this was the HST Program Office at Goddard) was part of the Customer Management Team (CMT), also referred to as the Customer Support Team (CST). The Mission Director chaired the CST and represented the payload community to the MMT.

A NASA Mission Management Team meeting. In this case it was during STS-114, but the concept was the same—to review, debate and discuss prior to making a decision.

Adjacent to the main flight control rooms in Houston there were areas set aside to enable the payload customers to bring in their own operational and managerial support teams, or to interface with remote payload teams back at the contractors and subcontractors, or both. As Shaw says, the HST series of missions always had a very large contingent of operations and management at Mission Control at JSC, and any of the 'secondary payloads' manifested on the Hubble missions were all carefully chosen and planned in order not to interfere with the primary mission of servicing the telescope. Secondary payloads could be small experiments performed on the middeck, either with or without astronaut participation, or small test and development experiments or investigations with implications for new technology or for the development of procedures.

The Payload Officer and the EVA Officer were the primary points of contact between the shuttle Flight Control Team and the Payload Operation Team(s). During a service mission, the HST Operations Team could call upon the huge technical support capability at Goddard in real time to assist in solving any technical issues on which the Operations Team members required help. As Shaw pointed out, for shuttle orbiter issues, a similar type of engineering support was available to Mission Control in the area referred to as the Mission Engineering Room (MER).

To achieve mission success, the management team would deliberate upon any additional operations and risks that were outside the boundaries defined for the Operations Team in the pre-mission planning. "For any operation which HST or the HST service operations had to consider that was outside the payload planned activities and risks, the payload management team had to authorize such action, and that approval was channeled up and discussed at the MMT. The Mission Director was responsible for making that process work, and for making sure the Payload Team wasn't asking to do something that the shuttle MMT wouldn't agree to. In the same vein, if the shuttle had a failure that put the HST or its repair operation at risk, the Mission Director would represent the Shuttle Program to the HST management to resolve the issue." If an issue should arise that was time sensitive and need coordination going either or both ways between the shuttle and the HST, it was the Mission Director's responsibility to ensure that happened. "The day-to-day in-flight management level oversight of the mission was done by the Mission Management Team and chaired by the Shuttle Program Manager. The MMT had representatives for all the interested and responsible parties participating in the mission. At the MMT meetings, the Mission Director/Flight Manager represented the Customer Support Team."

According to Shaw, the purpose of the Mission Management Team was "to keep a clearly defined and authorized operating 'space' available to the Operations Team, and be proactive rather than reactive." This was particularly necessary for such closely orchestrated missions such as Hubble service missions, since there was essentially no "down time" until after the telescope was released. Most shuttle missions included periods of intense activity involving time-critical actions, but otherwise they were planned to be "paced" wherever possible, and to "run like a marathon not a sprint". Hubble missions, however, "were a sprint from launch to Hubble release, near the end of the mission"— which was about 10 days into the flight on average. "To do that required an extraordinary amount of planning, teamwork and practice," explained Shaw. It was also planned that most of the primary activities of the mission were front-loaded, meaning that the items that were considered essential to a successful mission were prioritized towards the first few days of the flight, so that should anything happen that required the mission to be shortened then the major objectives had a better chance of being completed than if they were schedule later in the flight.

STS-125 Flight Day 3 Flight Director Tony Ceccacci (*left*) and Capcom Dan Burbank (*right*) monitor progress from Mission Control.

**Flight controller training**

When asked how the flight controller training was integrated into the system, Shaw pointed out, "The term 'training' can be somewhat of a misnomer when referring to the simulations leading up to a mission, since the preparations and development of the tasks to be performed provide excellent training in and of themselves. However, that being said, the training teams at GSFC and JSC jointly provide an environment designed to find any weaknesses in flight rules, procedures, timing, and malfunction troubleshooting, recognition, plans and strategies, and most importantly communications between the team elements. The environment is a no holds barred 'real' environment. The sophistication and fidelity of the simulator modeling is state of the art, and all the systems on the ground that are involved are fed a command and telemetry stream that for the most part cannot be discerned from that seen in flight. Also the crew is subjected to 1-g in the simulators unless in the NBL, but often even that is forgotten about in the heat of the moment! The training teams look for whether the teams can resolve tricky problems in a time-pressure environment, and whether the synergy of two (or more) failures or problems that combine into something of much greater impact (and is also much more subtle) to follow-on activities that are time-critical, can be dealt with. There is an old saying that the meanest simulation supervisors that have ever lived are Mother Nature and Lady Luck. The stakes were high for missions the shuttle flew, no matter what the payloads, for many reasons, including the lives of the crews involved. HST SM-4 seemed a bit larger than life in this respect, however, so even when facing limited resources for supporting both ISS and SM-4, there was really never a moment when SM-4 did not carry the same weight, and sometimes more when the pressure of the SM-4 timeline was taken into account. That same spirit of cooperation ensured we all were able to 'make it happen'."

**The 'magic' of the service missions**

Shaw also explained how the team came together for the mission, with different contractors and support teams on-site at JSC and remote links from their own locations, underlining the huge infrastructure and teamwork that made the Hubble missions a success. "The GSFC is where the HST STOCC is located, and where the GSFC engineering support is situated. It has their systems verification and testing infrastructure that would be used during a service mission to verify new or modified procedures or failure signatures. Those parts of the HST Team that were most closely involved with the crew and the flight controllers were at JSC. They were the developers of the crew aids and tools and the developers of the procedures used to operate the telescope's systems that the crew was working on or actually operated. The engineers that built the carriers in the shuttle's payload bay were also in Houston, with support back at GSFC.

"The timeline that the STOCC used at GSFC actually was a composite of the activities to be done in the STOCC—the commands to be sent and telemetry to be received and verified, etc. These were lined up with the crew's timeline and detailed procedures, and the timeline that MCC in Houston was performing in managing the orbiter's systems, communications, tracking, and planning for future maneuvers, etc. This was the 'SMIT' (Service Mission Integrated Timeline). Coordination and clear communications between all the parts of the system was essential to keeping on the timeline and regaining it after something happened. Because there were so many facets of the activity to keep up with, and so many groups of people responsible for each part, the organizational chart look a bit complicated. One of the key aspects of a Joint Integrated Simulation was to exercise the channels of communication and of command and control in real-time operations. The founding fathers of manned space flight operations developed the system of command and control of team operations used in Mission Control, and the teams supporting payload operations are a part of that. SM-4 was the fifth service mission and for the most part the HST team that supported all those service missions had many of the same people, so the continuity was maintained. It was part of the magic of HST missions."

**REFERENCES**

1. AIS interview with Preston Burch November 14, 2013
2. AIS Interview with Al Vernacchio October 24, 2013
3. AIS interview with Joyce King, December 2, 2013

4. www.nasa.gov/mission_pages/hubble/servicing/SM4/main_Team.html dated November 12, 2008. Last accessed September 14, 2012
5. Space Telescope Operations Control Center (STOCC) Overview, thanks to Phil Newman, Kenneth Carpenter, J.D. Myers, NASA Goddard, November 16, 2001 http://asd.gsfc.nasa.gov/archive.sm3a/stocc-desc.html~smpart Last accessed: December 22, 2014
6. AIS Interview with Lothar Gerlach February 15, 2013
7. Meet the STS-125 Launch Team, http://www.nasa.gov/mission_pages/shuttle/shuttlemissions/sts125/launch/launch_team_prt.htm Last accessed 25 May 2009
8. AIS Interview with Robert L. Crippen January 25, 2013
9. Information provided (over several years) by Jeff Carr, Eileen Hawley, James Hartsfield and Barbara Schwartz former Public Affairs Officers (PAO) at NASA JSC, and Chuck Shaw, former Shuttle Flight Director, JSC. acknowledgments also to Mike Scott, NASA
10. Actor Ed Harris immortalized the role when playing legendary Apollo flight director Gene Kranz in the 1995 movie *Apollo 13*.
11. Various emails from Charles Shaw to D. Shayler, 2013–2014

# 8

# Service Mission 1

> STS-61 is the most extensive and perhaps the most difficult service mission NASA has ever attempted. Over 7 days of this 11 day flight, the astronauts will rendezvous with the Hubble Space Telescope; capture and berth the bus-size, 13 ton spacecraft in Endeavour's cargo bay; and perform five spacewalks totaling more than 30 hours, more than on any previous mission, to repair and service the telescope so that it can continue its 15 year mission.
>
> Rockwell International, December 1993

The press release that contained this statement was issued by Rockwell International shortly before the STS-61 mission flew in December 1993, and it was no exaggeration. After all the hoop-la of getting the telescope authorized, funded, designed, built, tested, and launched, the discovery that its optics were flawed made what had been promoted as the "telescope of the century" appear to fail before it had even begun its examination of the universe. This was, to say the least, extremely embarrassing for NASA.

## THE HUBBLE COMEBACK

The agency was already struggling to fund at least something of the highly over-budget and complicated Freedom Space Station, nowadays known as the International Space Station. If there ever was a mission which NASA wanted, indeed *required* to succeed and succeed well, it was STS-61. The pressure was not only on the agency but also on the planners, controllers, and astronauts. In fact the entire Hubble and shuttle team needed to 'make good' in the eyes of the science community, the politicians, and the public.

## Hubble humor

In the early 1990s, before the service mission, Hubble was the favorite topic of jokes on late night talk shows; it was an embarrassment for NASA and the astronomical community, and was rated fair game, with pictures of a stricken telescope alongside the greatest disasters and flops in history. The pressure was on to repair the telescope and prove the value of humans in space, not only to Congress and the American people but also to the astronomical community and the wider world. In his 2009 oral history, Jeff Hoffman, who would be a spacewalker on the STS-61 mission, recalled NASA Administrator Dan Goldin telling the astronauts that the agency's future was literally in their hands. With the space station struggling to get funding and increasing concern about the likely amount of EVA activity required to assembly it, the spotlight fell on STS-61 and its record-breaking five back-to-back EVAs. If the crew could not demonstrate their ability to service and fix Hubble, then that would undermine support and confidence in the even more complicated and demanding EVAs for the station.

There was pressure on the training, on organizing the flight plan, and in being confident that so much work could be achieved on a single flight. It was an ambitious mission, and at one point it was thought that splitting it in two (as was done for the third service mission) might be a better option. It may have alleviated the pressure on the first mission somewhat, but as Hoffman noted, "How do you know that you've not left the items that were going to fail," thereby causing more problems for the next crew. As fellow STS-61 spacewalker Story Musgrave wrote in his book, "At times during the preparation the only peace I could find was in the dentist chair."

The emblem for Service Mission 1 details servicing the solar arrays. (Courtesy, Joachim Becker, Space Facts, Germany)

## Learning to deal with the press

In 2013 HST Program Manager Preston Burch explained the mood at Goddard Space Flight Center during the lead up to the first service mission to overcome the spherical aberration problem, especially in dealing with the press as "being careful not to put your foot in your mouth". Burch acknowledged that because this was the first service mission there were lots of risks, and that the team had to deal with being the butt of late night talk show comedians. In the comedy feature film *Naked Gun 2.5* the character played by Leslie Nielson goes to the Blue Note Café, a bar for losers where there was a line of photos from historic accidents and disasters and in between a picture of the Hindenburg and a sinking Titanic there was Hubble. "We could look and we could laugh, we thought it was hilarious, [but] we were so sure that we would succeed."[1] Being constantly under the spotlight, it was decided that the Goddard team would take control during the mission. To enable the team to focus on the job in hand, no press would be allowed in the control center; though a webcam would operate. Key staff received several days of professional media training, mock interviews were videotaped and then critiqued. Staff learned to express themselves with greater clarity and to provide better explanations. It had been frustrating when their comments were reported out of context, so they learned how to minimize that. They were "put on spot" to answer awkward questions, and subjected to "dirty questions" designed to embarrass. In addition, a list of "Frequently Asked Questions" was drawn up for Goddard management. NASA Headquarters and other field centers and institutes asked for copies. The plan was to ensure that everyone involved "spoke with one voice". The team prepared fact sheets for the major aspects of the service mission. With each successive mission, the team improved its outreach skills. Of course, it helped that by the time of the final mission in 2009 the press, having been won over by the success of previous service missions and the skills of the whole team, was more friendly.

To achieve the unprecedentedly demanding STS-61 mission required selecting the right crew and devising a very aggressive training program. If it succeeded in restoring Hubble's vision, it would be a vindication of the role of humans in space; if not, then it might be the end of NASA.

## CHOOSING A CREW

For the first service mission it was important to choose a crew with experience. It was the first time that anything like this had been attempted, and it was imperative to make sure the crew was adequately prepared for what lay ahead. By choosing experienced astronauts, their adaptability to space flight was known, and they had knowledge of what to expect, and how to use the equipment. Experience in controlling the orbiter, preparing for and making EVAs, using the RMS, and detailed knowledge of Hubble would all be beneficial in the preparation and execution of the mission.

On January 25, 1990, 3 months before the launch of the telescope, NASA had created the role of Payload Commander (PC) as a senior Mission Specialist to provide long range leadership in the development and planning of payload crew activities. The PC had overall crew responsibility for the planning, integration, and on-orbit coordination to ensure mission success and flight safety. A Payload Commander was assigned prior to the remainder of the crew to identify and resolve training issues and operational constraints in advance of formal crew training. Although mainly focused upon missions that had largely scientific objectives, and serving as a foundation for the development of the Space Station Commander concept, the role of PC would also be applied to missions with significant EVA activities such as the Hubble service missions and the early stages of the spacewalking program at the space station.

With the launch of Hubble in April 1990 and the subsequent problems with the spherical aberration and jittering solar panels, there were already substantial additional tasks added to the first service mission, in addition to the scheduled routine maintenance. Therefore a long lead time was envisaged to aid in crew training. On March 16, 1992, veteran astronaut Story Musgrave was named Payload Commander and MS4/EV2 for STS-61, the manifested first service mission.[2] Musgrave had been selected as a scientist-astronaut in August 1967, and had conducted extensive training and support work in the development of EVA techniques. He served as backup Science Pilot for Skylab 2, then flew on STS-6 in 1983, on which he made the first shuttle-based EVA, STS-51F in 1985, STS-33 in 1989, and STS-44 in 1991.

The STS-61 crew on the flight deck. *Left* to *right*: Story Musgrave, Dick Covey (*rear*), Claude Nicollier, Jeff Hoffman, Ken Bowersox, Kathy Thornton and Tom Akers.

It was not until August 25, 1992 that Tom Akers (MS5/EV4), Jeff Hoffman (MS3/EV1), and Kathy Thornton (MS1/EV3) were named as the other three EVA crewmembers, along with Musgrave.[3] Professional astronomer Jeff Hoffman had been an astronaut since January 1978. He had made an EVA on his first mission on STS-51D in 1985, flown the

ASTRO-1 astronomy mission in December 1990, and returned from his third mission on STS-46 only 17 days prior to the Hubble announcement. Kathy Thornton became an astronaut in May 1984 and in addition to flying with Musgrave on STS-33 she made an EVA with Tom Akers on STS-49 to test space station construction techniques. Akers joined NASA in June 1987 and had flown on STS-41 in October 1990 and on STS-49 in May 1992. It was during his second mission that, in addition to carrying out an EVA with Thornton, Akers had gone out to assist Pierre Thuot and Rick Hieb in manually retrieving an Intelsat satellite, thereby making the first and so far only three-person EVA.

Being named to this make-or-break mission for the agency was expected to cause a few problems in the Astronaut Office. Thornton reckoned everyone would hate them for it, but Hoffman said this was not the case. Akers once joked that he checked the tires of his car in the parking lot every day that they were in training, sure they'd be slashed.

The 'EVA crew' were well into their training program when the members of the 'orbiter crew' were announced in December 1992.[4] Dick Covey was named as Commander. He was selected as an astronaut candidate in 1978, along with Hoffman. He had flown in the pilot's seat on STS-51I in 1985 and STS-26, the Return to Flight mission in September 1988, then commanded STS-38 in November 1990. Ken Bowersox joined NASA in June 1987, along with Akers. He flew in the pilot's seat on STS-50 in June 1992, the longest mission to date. Rounding out the crew was ESA astronaut Claude Nicollier of Switzerland as MS2/FE and primary RMS operator. Nicollier had been selected by ESA in 1978 as a candidate payload specialist for the Spacelab 1 mission, and then under a special agreement between the two space agencies he underwent NASA MS training with the Group 9 intake in 1980 and 1981. He was scheduled to fly in 1986 but this was canceled in the wake of the loss of Challenger and he didn't make it into orbit until 1992 aboard STS-46, flying with Hoffman.

The complexity and importance of this mission was reflected in the first official NASA astronaut backup assignment since STS-3 in 1982. On March 9, 1993, Greg Harbaugh was assigned as BUp EVA astronaut for SM-1.[5] He would be fully trained and ready to go if any of the EV prime crewmembers became disqualified. Prior to being selected as an astronaut candidate in 1987, Harbaugh had worked at the Johnson Space Center for almost a decade. In 1991 he flew on STS-39, and some of his technical assignments in the Astronaut Office had included work on HST EVA development at JSC and Marshall Space Flight Center.

**Telling the wife**

In his 2009 oral history,[6] Jeff Hoffman revealed that he had hoped for a flight to the Russian space station Mir, but was told that he was too tall to be qualified for the Soyuz spacecraft or the Orlan spacesuit, but in August 1992, when in quarantine for STS-46, he was asked what he would like to do next. Although a Spacelab flight was attractive, he pointed out, "What I would really love of course, being an astronomer, is to go on the Hubble mission." Most of the Astronaut Office had also requested that mission, but Hoffman was successful. He later learned that the prior experience criteria for Hubble were someone who had already been a commander or pilot, four people who had completed at least one EVA and somebody with extensive RMS work. It may also have been a factor that crewmembers had flown together before: Musgrave and Thornton on STS-33, Thornton and Akers on STS-49, Hoffman and Nicollier on STS-46. In addition, Covey and Hoffman had been selected in the same group, as had Akers and Bowersox, so they had gone through the Ascan training together.

Hoffman had completed three missions and was considering his future options. His wife thought they would be moving out of Houston, and when told that he was to be on the crew of STS-61 he wanted his wife to hear of it from him, rather from the radio. She was upset at first but accepted it. Apparently at an astronaut get-together, other crewmembers were told how fortunate they were to be assigned to the service mission, but Hoffman's wife had their sympathies because they knew she really wanted Jeff to be moving on after nearly 15 years with NASA. As an astronomer he was really excited with the challenge.

Tom Akers recalls being told of his assignment to the first service mission several weeks after returning from STS-49.[7] On that flight he participated in two spacewalks, including the world's first three-person EVA to retrieve a satellite and a space station assembly simulation with Kathy Thornton. Akers was aware of work to develop methods to fix the problems with the Hubble mirror but he had no expectation of flying that mission. Shortly after learning of his assignment, he accompanied Musgrave on preliminary EVA work at the immersion pool at Marshall, which was large enough to handle an almost complete Hubble mockup. Whilst there, Akers learned that Hoffman and Thornton would be joining the first servicing crew. The official announcement was made a short time later. He acknowledged that his previous experience, together with working with Thornton in an EVA team, probably assisted in his assignment.

**Primary tasks**

Three tasks were assigned the highest priority on the mission.[8] Firstly, to restore the planned scientific capability of Hubble; secondly to restore the reliability of its systems; and finally to demonstrate on-orbit servicing, something that had been vigorously promoted for years. This appeared straightforward, but that was far from the truth. The complexity of the first service mission is made clear by the individual servicing tasks:

- Both of the original electricity generating solar arrays had developed a "jitter" due to the thermal stress as the spacecraft crossed the terminator between the cold darkness of the Earth's shadow and the heat of daylight, twice per orbit. This was impairing the ability of the telescope to point at targets accurately. The arrays were to be returned to Earth for extensive evaluation and be replaced by new units designed to eliminate the jitter problem.
- The original Wide Field/Planetary Camera (WFPC) was to be replaced by a new one capable of compensating for the spherical aberration of Hubble's primary mirror.
- The astronauts were to install P16 fuse plugs designed to correct sizing and wiring discrepancies. These protected the telescope's electrical circuitry.
- With three of the six rate sensor gyro units having failed after only 3 years in space, the first in December 1990, the second in June 1991 and the third in November 1992, the astronauts were to install two new units (RSU 2 and RSU 3, each containing a pair of gyroscopes) to assist in pointing and tracking the telescope.

- One Electrical Control Unit (ECU-3) was to be replaced for increased reliability and support for the other two "electronic brains" onboard the facility.
- An improved Magnetic Sensing Systems (MSS-1) was to replace an earlier unit, for better measurements of the spacecraft's position relative to Earth's magnetic field.
- A failed Solar Array Drive Electronic (SADE) was to be replaced. This was one of a pair that transmitted positioning commands to the array assembly.
- To compensate for the spherical aberration of the main mirror, the astronauts were to remove the High Speed Photometer and replace it with the Corrective Optics Space Telescope Axial Replacement (COSTAR). Naturally this task had received most of the press attention, as it was this package that was intended to clear Hubble's vision problem. This task was likened to placing "spectacles" on the telescope—a concept which certainly gained the attention of the both the media and the general public.

If this were not enough, there was also a shopping list of additional activities that were to be attempted on a time-available basis with as little disruption as possible to the flight plan. These included:

- Installing the DF-224 co-processor which was designed not only to improve the degraded memory redundancy on the telescope but also to improve its memory capacity and speed.
- Replacing P15 fuse plugs to improve the wiring discrepancy.
- Installing a second ECU (ECU-1) unit and the second MSS unit (MSS-2).
- And if there was still time available, repairing the High Resolution Spectrograph, one of whose detectors systems was not working correctly as a result of a power supply problem.

To undertake these servicing activities, the four highly experienced EVA crew were split into pairs, with Jeff Hoffman (EV1) and Story Musgrave (EV2) forming one team and Kathy Thornton (EV3) and Tom Akers (EV4) the other. The astronauts were to conduct alternating EVAs, to allow each pair a period of rest between excursions. To make the most of back-to-back EVAs during the servicing period, Hoffman and Musgrave trained to conduct the first, third and fifth EVAs, and Thornton and Akers tackled the second and fourth. Half-jokingly, Hoffman and Musgrave were known as the "Odd Team" because they would undertake the odd-numbered spacewalks and Thornton and Akers were the "Even Team".

STS-61 was the first shuttle mission to fly with five cryogenic storage tanks in order to provide sufficient consumables not only to support the planned five EVAs but, if required, two unscheduled spacewalks, and still retain sufficient margin for a contingency EVA (for example, to close the payload bay doors at the end of the mission).

The minimum criteria for "mission success" was listed as installing at least three reliable gyroscopes and either an operational WFPC or COSTAR. The "full success" would require accomplishing all seven primary payload tasks. If the mission were only partially successful then a second service mission would be requested, the timing of which would depend upon which items had not been installed and the operational state of the telescope.[9]

## Hubble's spectacles

During a hearing into the problems on Hubble, it was feared that the entire project might be canceled, with either Congress or NASA simply writing off the loss. But the euphoria that emerged from these dark days spurred on the team to ensure the first service mission would do as much as possible to achieve full success. Replacements for existing instruments were already being planned, and from studying the optical characteristics of the mirror it became apparent that the flaw could be "corrected". Numerous astronomers and engineers offered a variety of suggestions for how to achieve this. Retired astronaut Bruce McCandless became an advisor for the proposals.

The ideas ranged from mechanically or thermally deforming the mirror in some way, or over coating it in order to modify its shape, or installing a full-aperture gas-filled corrective optics in front of the main mirror, or replacing the secondary mirror. Many of the proposals were lacking in practicality and downright dangerous to the telescope and/or the astronauts. As there was no way to make a major modification to the primary mirror while Hubble was in space and ground servicing had been eliminated years before, the only possibility was to figure out what could be achieved on the first service mission, due in about 3 years.[10]

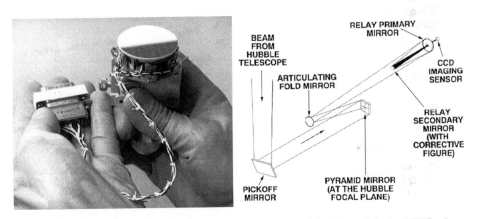

The tiny mirrors of the COSTAR package for Hubble (*left*) and the light path in the WFPC using the corrective optics.

## Inspiration from a dentist mirror and a shower head

Fortunately, a possible fix was available. A decade earlier, workers at JPL had started work on a successor for the main imaging camera, the Wide Field/Planetary Camera, and it was suggested by Murk Bottema of Ball Aerospace that the errors in the primary mirror could be compensated for by the insertion of small mirrors which were planned for WFPC-2 in order to adjust light for the other axial instruments. The challenge was how to install these mirrors precisely "in the light bundle" behind the main mirror. The small mirrors were no larger than those used by a dentist to look into a patient's mouth, and would have to be shaped to "undo the erosion of the reflected light caused by the faults on the main mirror".

As often happens, the solution came by a chance observation. Electrical engineer James Crocker of the STScI was at a Strategy Panel Meeting at the Space Telescope European Coordinating Facility in Garching, Germany. Taking a shower one evening in his hotel, he observed that the shower head was mounted on articulated telescoping arms, and as he adjusted them he realized this technique could be the answer to extending the small mirrors in front of the primary mirror on the telescope.

The carrier for this unit was already available. In the early 1980s, the Space Telescope Axial Replacement (STAR) was developed as an empty replacement for one of the axial scientific instruments, just in case one wasn't ready to fly or had to be removed, with no replacement available to fill the void. In that case this mechanically and thermally benign structure would be inserted so that Hubble could function without the missing instrument. The solution was to incorporate the mirror and articulated arms to intercept and correct the light reflected by the main mirror, then feed it to the three remaining axial instruments. As there were a total of five entrance apertures to the instruments, the unit would require ten mirrors. It was not just the mirrors and telescopic arms, the now named Corrective Optics Space Telescope Axial Replacement (COSTAR) featured mechanical assemblies and the electronics to control them from the ground. In all, it had 5300 parts. The tiny mirrors that were supplied by Tinsley Optics in California were reportedly an optician's nightmare to develop, being handmade "anamorphic fourth-order aspheres on toroidal blanks".

It was decided to remove the High Speed Photometer, which was causing the least of the telescope's problems but also unfortunately the least used instrument, and to replace it with COSTAR. After design and proposal work, in January 1991 Ball Aerospace was contracted to build COSTAR. Although its task was to interpose ten tiny mirrors into the optical paths from the main mirror to the three remaining coaxial instruments, the finished photo-booth sized package weighed in at a hefty 487 pounds (220.9 kg).

**Crew training**

Tom Akers recalled that even though he and Kathy Thornton had worked together and made a joint EVA, they were still required to go through the training process. "It wasn't like they deleted any training because Kathy and I worked together before—we still did all the same training for the Hubble repair mission as if we hadn't done that. Training is good, you never have too much training, you know it's something that when you are trained and ready to go. You know you'll get rusty during a mission delay of several months, which happens on some missions. You still have to keep training to stay up to speed. So I would say that Kathy and I had a good working relationship from flying together on STS-49. I don't think that when we went on the STS-61 mission Kathy and I were any better at it than Story and Jeff, who hadn't worked and trained together." But they were familiar with working together, and on STS-49 they assisted Rick Hieb and Pierre Thuot into and out of their suits. They were very familiar with progressing through the pre- and post-EVA procedures on-orbit, which is a completely different experience from rehearsing on the ground.

Of course, the training for the STS-61 spacewalks attracted the attention of the media and various onlookers monitoring their progress, adding to the pressure on them and underlining the importance of the mission. As Akers explained, "We had, I think over 20 different review teams looking over our shoulder as we were training for our mission, because it was the first time we had attempted to do five spacewalks in a row, four suits and four crewmembers. We had done four spacewalks on STS-49 but we hadn't done them in a row, so it was something we hadn't done before. But we were comfortable with doing that.

"As a crew, we were actually surprised by so much scrutiny from the outside and that we were obviously, like any mission is, worried to the point that we had made sure that we had thought of everything, but it wasn't like that was a big major background. When you were in the pool training there were a lot of people watching you, and probably being critical of what you were doing, but you get used to it, you didn't think or worry about that. I remember the first sim that we did. It was kind of in the back of our minds, and we had said that we would have to be careful what we said because of these people watching, but you just totally forget that after a few minutes. You say what you think and what needs to be done, and press on.

"I think our Commander, Dick Covey, probably behind the scenes, unknown to us, did a great job of keeping all that type of pressure away from the EVA crewmembers as we were training. I don't think it really impacted on us; now it may have impacted the folks trying to train us, having to deal with the outside people there, all having to have headsets and listen and all that, but I remember that last review team, they actually had some good suggestions, many of the review teams had some good suggestions that we incorporated. We were really in trying to save time, so we were open to any and all suggestions for how we could make a particular task easier and more efficient to do."

In his 2009 oral history, Jeff Hoffman remembers about 13 different committees that became involved, but the appointment of former Flight Director Randy Brinkley as Mission Manager helped keep these committees away from the crew, to allow them to focus upon the job in hand. Hoffman did say that some of the committees were quite useful. In one example there were too many tools outside and it would take time to assemble them at the start of the EVA. Time during an EVA is a limited commodity, as was the oxygen in the backpacks, so suggesting that the tools should be relocated inside to save about half an hour in time alone would help. It was still several months from launch, but the weight and balance calculations had been performed. "Just because the crew asks for it doesn't mean you're going to get it," Hoffman said, "but this was Hubble, and sure enough the tools were inside."

Claude Nicollier reflected the European interest in the mission. "This mission is important for Europe because of our overall participation in the Hubble Space Telescope Program. A lot of people in Europe have this telescope in their hearts, and we want to serve them well."[11] As primary RMS operator, Nicollier was crucial to the success of the spacewalks on the mission, and spent many hours working with the EVA team on simulating each task. Kathy Thornton once mentioned that he could make the arm go anywhere, "It can almost wrap itself around the telescope in the payload bay, and Claude gets us where we need to go. He almost knows before we tell him."

### Cross training

For STS-61 there was so much to train for that it was essential early on to establish a system of cross-task training. To back each other up as a precaution against someone not being able to attempt a task that they had specifically trained for, or as a result of serious suit failure or crew illness, Akers explained, "We trained equally on all of the tasks until the last month or two; from there we trained just primarily on our planned tasks. Early on in the training, the planners didn't know in what order they were going to do things, and what was going to be done on what day; they had a general idea, but for example the magnetometer test came in late so there had to be some reshuffling. So yes, we were all very well trained and could've gone and done any one of the tasks as a team. In fact, if something had happened to one of the team members, if I'd been sick on one of my spacewalks, Jeff or Story would've taken my place and gone and done that. Within a spacewalking day, you kind of specialize. For example, on the solar arrays KT [Kathy Thornton] and I kind of had our own tasks that we did and had done more training for. Kathy was usually on the end of the remote arm during our spacewalks and I was the free floater, so if I had been sick, Jeff or Story whichever one had trained for those specific tasks, would have taken my place; and if it had been KT that was sick then the other one would have."

Scheduling the EVA tasks, Akers explained, had not only to include flexibility but also priority in case not all tasks or EVAs could be accomplished due to something requiring the shortening of the mission, and this was organized by the assorted management and training teams. "Because you never know that when you go up to do five spacewalks, you may only get one of them done for whatever reason—you could have suit failures, or you could have space shuttle issues that make you come home early—so you plan for the worst case: if we can only have one spacewalk, here is what we want to get done. Now you would think that they would do the highest priority on day one, but you also have to have some optimism that you are going to get them all done. So you start off planning that you *are* going to get them all done, but if things turn sour they always have a backup plan. For example, on the second EVA that Kathy and I did, when we went out to changeout the HSP for the COSTAR, that obviously was high priority because that was going to correct a lot of optics problems. If we knew that we weren't going to get this spacewalk, we might have done something different, like changing out the co-processor—we would have just gone straight into just doing that. You have to be flexible even when you're out doing a spacewalk, that things may change."

Despite all the earlier rivalry between Marshall and JSC and Goddard, by the time of the first service mission everyone was pulling together and determined to get Hubble fixed no matter what; it became almost a personal crusade to get the job done.

**Frostbitten fingers**

On June 4, 1993, it was reported that Story Musgrave had been slightly injured performing simulated EVA activities for SM-1 in an altitude chamber at JSC.[12] He suffered a mild case of frostbite on his fingers through the EVA gloves. The report went on to point out that his rapid recovery meant the incident would have no impact on his crew assignments. Sixteen years later in his oral history, Jeff Hoffman expanded on that incident. In 1985 Hoffman had experienced the cold of space during a lull in his EVA on STS-51D, and was worried about such an incident occurring during the Hubble mission. He knew that there would be a lot of intricate work on the EVAs, some of it with Hubble's servicing doors open and in shade as much as possible so that the ultraviolet optics would not become contaminated or damaged. The plan was to fly the shuttle with its belly facing the Sun and the payload bay, containing Hubble, facing Earth. Hubble would be in shadow, while radiated heat from

Earth kept the astronauts warm. However, it was estimated that during the night pass the temperatures on some of the metal components would drop to minus 150 degrees Fahrenheit and Hoffman was concerned about his fingers.

Once again the power of Hubble came though and a very expensive altitude chamber test was organized at JSC. In May 1993 the human thermal vacuum tests began. Hoffman went into Chamber B first, nitrogen was pumped through the walls and the temperature dropped. He found that trying to remove the tools from the toolbox was difficult and the little pip-pin used to lock the devices was frozen solid. "It would've been a disaster," he explained. "We would've been out there and we couldn't have gotten half the tools we needed. So that got people's attention. The engineers went to work and took out all the residual grease. I guess they filed it off, to increase some of the clearances." So now Story Musgrave went into the chamber and tried to repeat the tasks. Sure enough what the engineers had done worked, but he complained that his hands were really cold, which considering he was working with very cold metal was not a surprise, so he carried on. When the flight doctor in attendance checked upon his status, Musgrave said his hands must have warmed up, because he could no longer feel them. However, this was a bad sign that was overlooked. The 1998 NASA book *From Engineering Science to Big Science* cited a March 1995 interview with Musgrave.[13] In that interview Musgrave called vacuum testing "the world's worst hell. That's the hardest day you're ever going to have as an astronaut." In the black chamber in deep vacuum, dragging his 480 pound (217.72 kg) inflated pressure suit around was like being "a plough horse". After several hours at minus 170 degrees, his fingers felt numb from manipulating the tools, and despite being a professionally trained medical doctor he did not recognize the signs of "going from pain to injury". He tried to pull his hands inside the sleeves of the pressure suit or place them close to lamps that simulated sunlight in an effort to warm them up, but to no avail. When Musgrave emerged from the altitude chamber and doffed his gloves, he found that his fingers had turned purple and black and eight fingers had suffered severe frostbite.

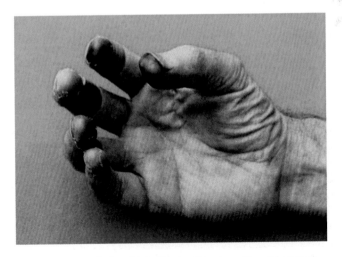

Story Musgrave's frost bitten fingers. (Courtesy, Story Musgrave)

"That was serious," Hoffman explained, "because he might no longer be able to fly. So they sent him up to Alaska, where they have the world's frostbite experts." For a while it looked like Musgrave would have to be grounded and replaced on the mission by the EVA backup astronaut, Greg Harbaugh. But the medics managed to save his fingers and he was certified as fit to fly. This experience prompted a review of the entire EVA operations. The result was a new approach to keeping the astronauts warm whilst outside the vehicle.

This test got the attention of management and resulted in a request to change the thermal profile. The problem was that the "fix" required the shuttle to make two attitude maneuvers per orbit, further eroding the already tight propellant margin due to the altitude of Hubble's operating orbit. To overcome this, engineers devised a new technique for working the RCS jets to use far less propellant in changing the attitude of the vehicle.

**Incorrect drawings, again**

The crew conducted some of their training at Goddard Space Flight Center, where the real flight hardware was in a clean room. A problem was revealed when WFPC-2 was placed in the mechanical simulator suspended by a crane for a simulation. The crew had brought their helmets and lights, and rigged the room to reproduce exactly what they would see during an orbital day or night. Desiring not to break anything on the flight, they wanted to ensure they had a good view of inserting and removing the instrument. But it became hung up half way into the high-fidelity simulator. "We pulled it out, and it turned out that the thermal shielding on the outside had been installed in such a way that there was a right angled piece [that had] interfered with the ledge that it went on. There was no way it was going to go in," Hoffman explained.

Goddard contacted JPL, who had assembled the original WFPC on Hubble, and was told by technicians there that they'd had the same problems on WFPC-1 before launch "because the drawings were wrong". JPL had modified the thermal covering on its first unit to insert it into Hubble before launch but had not changed the drawings. As a result WFPC-2 was made the same way. Luckily this had been found out prior to trying to slip it into the telescope on-orbit, which could have been a serious and expensive embarrassment.

A short time later, a training session with COSTAR revealed that it would not slide in the final 2 inches (50.8 mm). As Hoffman pointed out, "This doesn't build up a lot of confidence when neither of your two main instruments go in properly." On examining the unit, there was a bolt extending from the bottom. As COSTAR was to occupy the location initially assigned to the smaller High Speed Photometer, there was no guarantee that it would fit. Checking the drawing, the bolt appeared flush, but on COSTAR it wasn't. They couldn't know whether it was meant to be flush or protrude, because the HSP was in Hubble on-orbit and it could have been installed with the bolt protruding, but COSTAR, being larger, could not. The astronauts would have to wait and see which position was required. When they got into space, Thornton and Akers were armed with contingency tools such as hacksaws and vice grip pliers, in case these were needed. When the HSP was pulled out of the telescope, there was the bolt, flush. So they installed COSTAR with little difficulty and a flush bolt. As had been noted in water tank simulations years before, there was no guarantee that drawings would match the flight hardware.

## Virtual Reality, a new training tool

The STS-61 crew was the first to use Virtual Reality in training. In his 2009 oral history, Jeff Hoffman explained that water tank training was very expensive with all the charges for divers and facility support, so "it's silly to go in there just to figure out if I need to stand on the right side or left side of a door in order to get access to the compartment." He also pointed out that unlike earlier shuttle EVAs, where everything was done in the payload

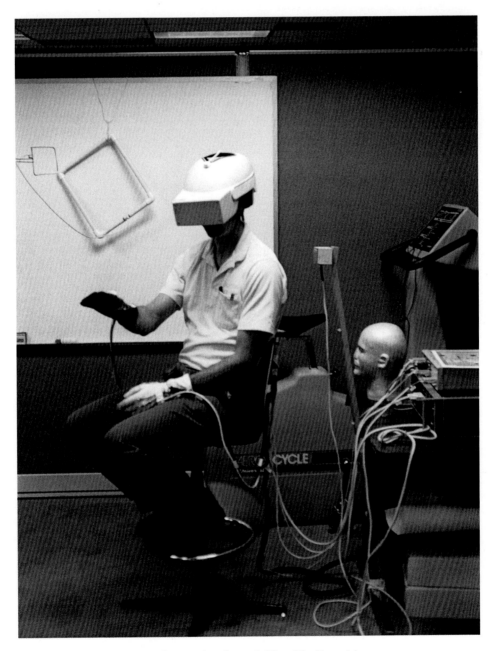

Jeff Hoffman undergoing early Virtual Reality training.

bay, due to Hubble's large size it was not possible to fit an entire vehicle into the tank, and this made simulating a task at the top of the telescope with a high-fidelity RMS difficult. Hoffman mentioned a trip to the National Air and Space Museum to view the former Structural Dynamic Test Vehicle that was on display there. They used a cherry picker hydraulic crane to reach the level of the magnetometers, but that was not geometrically accurate.

The Virtual Reality facility could depict a full extension on the RMS and the area which they were to work in, to enable an RMS operator to work out the angles of the arm joints. It was new, and it was slow, but it worked and Hoffman found it useful. Although it was not a formal training tool at that time, it was capable of rapid upgrades and changes outside of the bureaucracy usually associated with instigating a change.

With the endorsement of the STS-61 crew, the Virtual Reality Laboratory was developed considerably and in addition to assisting the later Hubble service missions, it proved itself in training for International Space Station.

## PLANNING THE MISSION

Work on planning the service missions began in August 1988, and over the next few years in preparation for STS-61 a lot of attention was devoted to the crew training program and EVA activities. The extent of this work is revealed by the documents for EVA simulations prior to the deployment of Hubble by STS-31 and in preparing the SM-1 mission. The data for EVA verification included:

- Crew aids fit-checked to HST (91 tools)
- ORU installed and removed by 9 astronauts
- EVA simulations practiced in water tanks (738 hours)
- Contingence spacewalk training (94 hours)
- Manned thermal vacuum chamber tests (20 hours)
- Hardware spare fit-checked (32 of 44)
- EVA procedures document for use on STS-61
- Six long-duration mission simulations planned as part of the STS-61 Joint Integrated Simulations:
  - 1. Rendezvous/EVA#1 (39 hours)
  - 2. EVA#2/Solar Arrays (36 hours)
  - 3. Hubble Space Telescope deploy (10 hours)
  - 4. EVA#3 (12 hours)
  - 5. Post insertion/Flight Day 2 (39 hours)
  - 6. EVA#4/EVA#5/HST deploy (59 hours).

Initially SM-1 was only to changeout some of the planned replacements for the original instruments, but following the development of COSTAR and problems with the solar arrays more tasks were added. Although Hubble could perform some useful science with spherical aberration and jittery solar arrays, the pressure put on the first service mission increased as time went on. By the end of 1990 the problem with the jitter that occurred as Hubble passed in and out of the Earth's shadow twice per orbit was understood but software improvements designed to damp out the vibrations took half an orbit, leaving no time before the telescope passed over the next terminator. And the software occupied far too much of the memory on the vehicle. So the replacement of the solar arrays with ones that would not jitter was added to the service mission. Other problems were interrupting the science program. In particular, the Fine Guidance Sensors which accurately pointed the telescope, had started to act up. So attending to these units was added to the list of tasks. In May 1991, after barely 1 year in space, the memory unit on the main computer failed, placing Hubble into a deep safe mode that was likened to a coma. By the next month the spacecraft was down to a single backup gyro unit; two had failed and three were needed to determine attitude and position. It is said that things come in three's and this was true for Hubble in that summer, because in July the Goddard High Resolution Spectrograph suffered a problem with its power supply which cut its capacity by half. By the turn of the year it was evident that SM-1 would be a much more ambitious and challenging mission than originally envisaged.

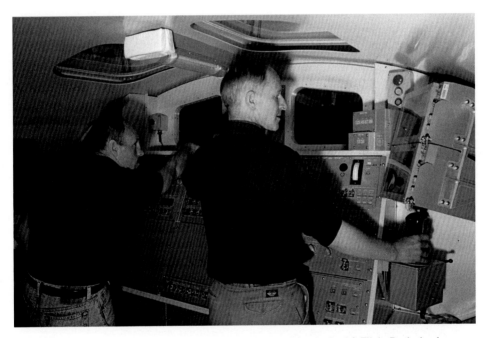

Ken Bowersox and Claude Nicollier participate in RMS training in the Aft Flight Deck simulator.

NASA insisted that STS-61 was not a rescue mission, but a planned servicing flight with some additional tasks which, although not originally scheduled, were one of the purposes of the series of service missions: to respond to unforeseen incidents and occurrences. When the cost of the servicing was calculated, it came out at about $500 million.

The construction of the telescope was estimated at $1.5 billion, so the service mission became about one-third of the initial outlay. When the cost of supporting the mission and analyzing its data was added in, the analogy was made that if Hubble had been an automobile it would have been written off.[14]

By September 1992 the Faint Object Camera had also developed power supply problems that denied the instrument half of its capability. Then another gyro failed, leaving just three. If another gyro were to fail, Hubble would have to halt its science program and adopt a safe mode until the service mission. Then, as if there was not enough for the service mission to deal with, the memory on a second computer failed.

With an overburdened EVA schedule, it was a scramble to organize the relative priorities of the failed items. Originally planned for three EVAs, this conservative approach had to be amended in April 1993 to five EVAs. This unprecedented amount of spacewalking gave rise to further reviews.

**Reviews**

In developing the first service mission, there were numerous review boards established to look at satellite rescue and repair, the space station, and Hubble. When the STS-49 mission unexpectedly required three EVAs to grab hold of the Intelsat satellite, with the successful effort involving an unprecedented three astronauts, the headlines praised the drama and the capabilities of the astronauts, but behind the scenes NASA ordered a review of the training methods. The agency's new Administrator, Dan Goldin, publicly praised his team for their "daring satellite rescue" and then privately made cutting comments about "those cowboys!" He created an independent review team to determine why an *improvisation* during the EVA had been required to achieve success.

One of the team members was former astronaut Tom Stafford who, in his 2002 book with author Mike Cassutt, *We Have Capture*, outlined his involvement in the STS-49 and Hubble Service Mission Review. "What we found was a mess. The Intelsat retrieval [concept] had not been properly simulated and the documentation did not match the hardware. [However,] the biggest problem was the lack of a single authority over the mission." He recommended a return to the Apollo style of management, by appointing a Mission Director specifically for the forthcoming Hubble mission. The person who filled that role was the former Marine test pilot Randy Brinkley, then working at McDonnell Douglas in St. Louis. Stafford was asked by Goldin to set up an independent oversight committee known as the HST Service Mission Review Team specifically for SM-1. This included former astronaut Joe Engle and Dr. Joe Rothenberg, at that time the Hubble program manager at Goddard. "The bureaucrats at JSC and Marshall fought Brinkley tooth and nail," notes Stafford, but he did an outstanding job with Hubble and progressed to head up the new ISS program at Houston.[15]

Then in June 1992 there was the JSC Space Shuttle Program Office's own HST Review Team headed by Richard Fitts. The following month NASA Headquarters created the HST SM-1 Program Review Team headed by Michael Greenford. Furthermore, former astronaut Joseph Allen reviewed the EVA aspects of the mission for the Headquarters Office of Space Science, and former Chief Astronaut John Young was asked to review the mission plan for STS-61.

By the Flight Readiness Review of November 17, 1993, only 27 of 195 recommendations made by 12 review teams remained open. It was time to put all the planning, training and evaluation into practice.

**Processing milestones**

Work to prepare the hardware for the STS-61 mission began at the Cape months in advance of the planned December 1993 launch.[16] The planning for a shuttle mission could take years, with the final steps towards that moment of launch commencing months earlier. The intensity and depth of the effort involved in preparing a mission like this was lost in the general press coverage. It required a variety of processes, items, and events to converge at the right time to enable the vehicle to lift off from the pad. The 'journey' of the hardware for this first service mission was typical of a shuttle flight.

On July 1, 1993, Endeavour completed its 10 day STS-57 mission by landing on Runway 33 of the SLF at KSC. Within hours it had been towed into Orbiter Processing Facility (OPF) Bay 1 for post-mission de-servicing. Preliminary inspections established it to be in excellent condition. By July 13, both the Spacehab augmentation module and the recovered EURECA satellite had been removed. Post-flight processing activity continued with de-configuring the payload bay that had supported the STS-57 cargo manifest, servicing the crew compartment of the orbiter, removing the three main engines, post-flight hypergolic servicing by emptying the propellant tanks, and inspecting the thermal protection system. Normal preparation for its launch as STS-61 some 5 months (about 20 weeks) later was expected to be tight.

On July 24 a report was issued by Richard U. Perry, Director of the Space Flight Safety and Mission Assurance Division at NASA Headquarters and Chair of the Shuttle Processing Review Team which had been formed at the request of KSC Director Robert Crippen. Throughout NASA, similar teams periodically carried out studies to ensure that the relevant work was being conducted in a safe, efficient, and cost-effective manner. In its report, this Review Team stated, "The Space Shuttle Processing is the best that it has ever been and is continually being evaluated for improvement." This was indeed good news, especially for the teams preparing for the critical Hubble service mission, but on that very same day an event demonstrated that processing a shuttle to meet a tight manifest was always prone to delays, slips, and mishaps. Following on from the failed first launch attempt of STS-51 on July 17, the second attempt a week later pushed the start of that mission into August. This would have a knock-on influence on the remaining missions of that year, including STS-61. In a similar review of quality control problems at KSC also issued by Richard Perry but this time ordered by NASA Administrator Daniel Goldin, the Review Team said, "There is a fear [with staff] that noted mistakes will lead to loss of employment. Therefore there is a tendency to not report problems, close calls and incidents because of the fear of reprisal. There is no evidence that reporting problems led to any serious punishments but the fear was attributed to recent layoffs at the center and media reports of shuttle mishaps, which would not help the confidence and morale of the workers."[17] These types of incidents were not good for NASA on top of the situation with Hubble, adding to the pressure in preparing SM-1 for launch.

By the beginning of August, the fifth set of cryogenic tanks were installed in Endeavour to increase the consumables for the mission and, despite delays to other vehicles in the manifest, work continued on target. In parallel, elements of payload hardware required for the mission began to arrive at the Cape. On August 6, following checkout at Goddard, the Space Support Equipment (SSE) was loaded in containers previously used for the Gamma Ray Observatory and the Long Duration Exposure Facility and shipped to the Cape on a barge which normally delivered External Tanks. The SSE would house some of the replacement components inside the payload bay during the flight. The SSE consisted of the Orbital Replacement Unit Carrier which was a dedicated Spacelab pallet adapted for the maintenance and repair role, the Solar Array Carrier which would support the two replacement solar arrays and carry the retrieved arrays back to Earth, and the Flight Support Structure which would hold and orient Hubble whilst it was in the payload bay for servicing (this was the same hardware as had held Solar Max in Challenger during the 1984 STS-41C mission). Arriving at KSC, the hardware was offloaded from the barge and taken to the Payload Hazardous Servicing Facility (PHSF) in the Industrial Area to undergo further preparations prior to loading aboard Endeavour.

Less than a week later, on August 12, a third delay in launching STS-51 further impacted the manifest for the rest of the year, and particularly STS-61. As a result of STS-51 not being able to launch, the Hubble service mission entered an indefinite delay pending a review of its options. Nevertheless, work continued on processing Endeavour and its payload. On August 19, the WFPC-2 instrument arrived at Hangar AE, a NASA Spacecraft Checkout Facility at the Cape Canaveral Air Force Station immediately south of KSC. Over the next fortnight a convoy of 17 trucks ferried flight hardware from Goddard to KSC. This hardware included COSTAR, two replacement solar panels, the High Resolution Spectrograph redundancy kit, rate gyro sensor and electronic control units, and two magnetometers. For several months it had been undergoing integration and testing in a clean room at Goddard.

On August 30, Mission Director Randy Brinkley said that the team was trying not to let the launch of STS-61 slip into the Christmas and New Year period, but if it slipped beyond December 10 there was a real possibility that the mission would be postponed into the new year. A series of recent events, especially involving STS-51, made this a real possibly. The successful launch of STS-51 on September 12 was a great relief to the Hubble team. On the other hand, technical issues with Columbia, which was being processed for STS-58, raised doubts about getting STS-61 off the ground in 1993.

As stacking of the twin SRBs (designated BI-063) on Mobile Launcher Platform 2 got underway in Vehicle Assembly Building High Bay 1, work continued on Endeavour in the OPF, including testing the RMS (serial number 303) that would grapple the telescope and support the planned EVAs. At the Payload Hazardous Servicing Facility, work progressed with the intended payload, including the arrival on September 10 of the twin solar arrays. During the second week of the month, both COSTAR and WFPC-2, followed by the solar arrays, were installed in their protective enclosures on their respective payload carriers. By September 24 closeout activities were complete and the hardware was ready to be taken to the Vertical Processing Facility (VPF) using a Payload Canister that had undergone a series of tests prior to being loaded with the Flight Support Structure, then the solar array carrier, and finally the ORU carrier holding WFPC-2 and COSTAR. With all the payloads aboard, the canister was rotated from horizontal to vertical for a final check before its move to the pad on October 28. But then on October 3, after a decision based on an independent optics panel assessment, WFPC-2 was returned to the PHSF for additional tests to re-validate its focal point. There was some risk involved in these tests, but carrying them out would not impact on this instrument being ready for launch. Meanwhile in the OPF, by October 4 all three SSMEs had been installed on Endeavour, with engine 2019 in the upper (number 1) position, engine 2033 in the lower left (number 2) location, and engine 2017 in the lower right (number 3) position.

Over the weekend of October 2–3 the flight crew visited KSC for the Crew Equipment Interface Test (CEIT) and other checks, and over the next few days the pace of processing picked up. In the VPF the testing of the payload using the Cargo Integrated Test Equipment (CITE) began on October 8. This was an essential milestone in preparing and qualifying the hardware for flight, except for WFPC-2 and COSTAR which were not required for this test. The program started with the electrical Interface Verification Test (IVT) which verified the readiness and compatibility of the HST's systems with those on Endeavour, particularly the command paths and associated circuitry that connected the flight deck to the FSS. Though their control would be limited, the crew could issue commands to direct power to activate various latches and heater circuits, as well as monitor telemetry coming from the telescope while it was on the support structure. The IVT was followed the next day by an 8 hour end-to-end test in which the Merritt Island Launch Area (MILA) tracking station at KSC was connected by satellite to the communication switching and distribution facility at Goddard before the signals were forwarded to JSC in Houston. As this test was being completed, the Solar Array Drive Electronics (SADE) arrived at the VPF and was installed in the East Test Cell for testing.

By early October, the stacking of the SRBs in the VAB had been completed, along with the mating of the External Tank (ET-60). October 7 was therefore a significant milestone in readying the flight hardware for launch on time. Despite problems in getting STS-58 off the ground, work on the Hubble service mission hardware continued smoothly—so well in fact that discussions were held by the NASA management team as to the possibility of bringing the launch of STS-61 forward from the planned December 2 to November 30. The deciding factor in this suggestion was the planned November 28 launch of an Air Force satellite, but the general feeling was that they should try and get Endeavour back into the OPF before the Christmas break. According to Loren Shriver, who was now the Manager of Shuttle Launch Integration at KSC but had commanded the STS-31 Hubble deployment mission, a decision would be made within a week. The DSCS-III type of military communications satellite was launched as planned by Atlas II on November 28, and in the final analysis it was decided to target December 1 as the optimal date on which to launch STS-61.

Endeavour was towed over to the VAB on October 21 and the following day was hoisted vertically for mating with its ET. After a week of combined tests, the orbiter was powered down in readiness for transferring the stack to LC 39A. This occurred on October 28. With the protective Rotating Service Structure wrapped around the vehicle the pre-launch checks and other preparations continued. On November 2, following the installation of the IMAX Cargo Bay Camera (ICBC) into the payload bay, further payload installation was suddenly placed on hold due to "contamination" found in the Payload Changeout Room at

Veteran spacewalker Jerry Ross at the Capcom Console during an STS-61 integrated simulation, a position he served during the flight. Along with fellow astronaut Susan Helms, he is monitoring neutral buoyancy EVA simulations at MSFC.

Susan Rainwater monitors an EVA simulation from the EVA console at Mission Control in Houston during a joint integrated simulation for STS-61 in which astronauts were rehearsing EVA tasks in the neutral buoyancy tank at the Marshall Space Flight Center in Alabama.

the pad—very fine sand particles blown by an exceptionally windy weather system that had recently passed through central Florida had penetrated the room and, despite an earlier cleaning, an inspection found small amounts of sand in other parts of the room. Four teams were formed to investigate the problem and introduce any changes to clean the area now and prevent a similar incident occurring again.

As this work progressed, managers continued to evaluate their options to undertake the launch of STS-61 from either LC 39A or 39B without impacting the planned December 1 launch date. Meanwhile work aboard Endeavour continued with the Terminal Countdown Demonstration Test. The flight crew had arrived on November 4 to participate in the final parts of this test, which ended the next day. After an investigation of the contamination, on November 5 the HST payload was removed from the pad and taken back to the PHSF for a cleanliness inspection which was expected to last 10 days. The same day, the decision was made to transfer the STS-61 stack from LC 39A to LC 39B. This meant that all the built-in contingency time in launch processing, apart from several days around Thanksgiving, had been used, but NASA was confident of certifying Pad B by parallel processing designed to meet the target launch date. Routine vehicle processing would continue at Pad A, including the loading of hypergolic propellants, until Pad B was ready. Even further pressure was put on the launch teams on November 10, when inspections of Endeavour's three main engine high pressure fuel turbo pumps were ordered in response to a report of minor discoloration and minuscule cracks being found in a test pump at the Rocketdyne manufacturing plant in California. Officials wanted to verify that there were no imperfections in the pumps aboard Endeavour that might interfere with the final countdown and launch. An inspection cleared the pumps fit for flight. On November 15, the stack was moved to LC 39B in an operation which lasted 5 hours, and the HST payload was transported out to the pad later on the same day.

On November 17, the Flight Readiness Review reflected on all the challenges that had influenced the decision to move the shuttle to the second pad and preparing that to support the launch. After reviewing the status of the stack and the cleanliness of the payload it was decided to set the launch for December 1. Meanwhile at the pad, final pre-launch tests and checks were completed in readiness for loading the payload aboard Endeavour.

Then just 2 days later, on November 19 a problem with a small sensor threatened the launch. This was one of four small sensors which measured hydraulic fluid pressure in the actuator that moved the elevon on the right-hand wing of the orbiter. Though technically a safety issue, even if all four sensors failed, the crew would be able to determine from other apparatus whether the elevon was working properly. The suspect sensor was disconnected, retested and passed, but still wouldn't work. To replace the sensor would require returning the stack to the VAB, certainly delaying the mission into January at the earliest. In Houston, flight controllers evaluated the options and consequences of not flying versus flying with the faulty sensor. They decided that redundancy would be maintained with just three sensors and reported this to mission management, who decided to waive the launch rule and proceed with the reduced redundancy of only three instead of four sensors. Another niggling problem that was being worked around at this time was excess noise on the radios in two of the EVA suits being installed in the orbiter. Issues such as this were common during the final days prior to launch, and illustrated the vast number of small items, circuits, and processes that required to be tested, checked, and passed for flight, all of which contributed to the point where it was decided to put all the elements of a shuttle stack to the ultimate test and launch.

The flight crew arrived at the Cape on November 27, just ahead of Thanksgiving, when many activities were shut down at the Cape. Mission commander Dick Covey reported, "We look forward to putting on quite a show. We have trained hard, and feel confident that two weeks from [today] we're going to be back… after a very successful mission." The next day an Air Force rocket successfully launched the military satellite, despite a computer error that delayed is departure by 31 minutes—had the Atlas needed a greater delay, this would have required STS-61 to be postponed by at least 24 hours. Shuttle Test Director Mike Leinbach said, "The shuttle is in good shape. Endeavour is a good ship, and we hope to get her off on the first attempt to get on with this very exciting mission for America and the world."[18] The schedule allowed Endeavour until December 6 to launch and complete a full mission, but if any last minute delays pushed its launch beyond the 48 hour turnaround, which at the latest meant a launch on December 9, then the mission would have to be delayed into the coming year in order to accommodate the seasonal holidays. The countdown for STS-61 started on November 28, and as the days slipped by workers, onlookers, and those close to the mission prepared to witness the launch that would, they hoped and prayed, restore the Hubble Space Telescope to full operational service.

So heartfelt was the upcoming mission, an astronomer and professor at a technical college in nearby Melbourne was quoted in *Florida Today* as having students and "several colleagues whose jobs depended" on its success. Such was the anticipation, interest, and expectations for Endeavour and its seven astronauts.

**December 1, 1993 launch attempt**

The one thing that could not be controlled was the weather, and on December 1 higher than permissible cross winds, along with scattered showers within 20 miles of the Shuttle Landing Facility (SLF), meant the countdown that day had to be scrubbed. It was decided to try again the following day, when conditions were predicted to be better, so the flight crew returned to their quarters to wait out the 24 hour delay. Mission commander Covey, mindful of the spotlight on this mission, and with the knowledge there was only enough propellant on Endeavour for one rendezvous due to the altitude of the telescope, had earlier acknowledged, "We have to do it right first time."

**December 2, launch**

Endeavour blasted off the pad exactly on schedule, and just 8 minutes later entered into its initial orbit of 308 by 214 nautical miles (354.4 by 246.3 statute miles, 570.4 by 396.3 km). Normally MS1 supported the ascent and entry, mainly as a backup to read checklists during emergencies. It had been decided that Covey, Bowersox and Nicollier would handle ascent and entry, to enable MS1 Thornton to concentrate on her EVA training along with her three colleagues. Shortly after entering orbit with no reported problems and all systems working well, the crew were told to proceed with orbital operations. As Covey prepared for the first firing of the OMS to give chase to Hubble he observed, "It's a beautiful sunrise."

The first day on-orbit was one of adjustment to weightlessness, but the experienced crew had no difficulties and were able to work through the many tasks involved in converting the shuttle for orbital flight.

On the ground, Covey's former astronaut classmate Loren Shriver said, "This mission is higher profile than most. I have big confidence we will get this one done." It was now down to the seven astronauts to apply all that they had learned in preparing for the Hubble service mission.[19]

## FIRST HOUSE CALL AT HUBBLE

With the successful launch, Endeavour was now heading to perform the first "house call" at the telescope. Following years of planning, simulation and preparation, the skills of satellite servicing proposed in the 1970s and pioneered in the 1980s were about to face the toughest challenge at Hubble.

### December 2–3, chasing Hubble

The first full day on-orbit saw the four EVA astronauts check out the pressure garments and EVA equipment on the middeck to ensure their life support and communications units were working correctly. Up on the flight deck, ESA astronaut Claude Nicollier checked out the RMS which he was to use to grasp Hubble and berth it in the bay, support the five planned EVAs, and then redeploy the telescope at the end of the servicing period. He also switched on the numerous TV cameras to verify that the system was ready to support the forthcoming operations. The RMS installed for this mission was equipped with a new generation color CCTV camera at the elbow position, and several planned surveys were revised during the mission to take full advantage of the much higher quality image provided by this new camera. So good were the images that sometimes the color elbow camera was used instead of the black-and-white one on the wrist of the arm.[20] In preparation for the extensive EVA program, the crew started to depressurize the cabin atmosphere from 14.7 psi to 10.2 psi in order to reduce the amount of time that they would have to pre-breathe oxygen prior to an EVA.

As Endeavour closed in on Hubble at a rate of 60 nautical miles (69.04 miles, 111.12 km) per revolution of Earth, the telescope was closing out scientific observations with the WFPC and HSP. The Space Telescope Operations Control Center (STOCC) at Goddard then closed the aperture door and commanded the vehicle to adopt the correct solar-inertial attitude for the shuttle's arrival the following day.

### A tiny point of light

"It's an exciting process when you're doing a rendezvous, and you first pick up the object as a tiny point of light out in the distance, and then it gradually gets brighter," pointed out Jeff Hoffman. As they closed in he used binoculars to inspect Hubble and announced that one of the solar arrays was bent. The reflection off the thermal coating on the telescope tube was so bright that he had to don sunglasses. It was evident that Nicollier would have to do likewise when using the RMS to grapple the telescope.

Hubble is placed back in the payload bay of a shuttle for its first "house call".

## December 3–4, capture and EVA 1

The shuttle's third day in space initiated one of the most important and busiest times of the mission. In preparation for the rendezvous, the Hubble controllers at Goddard commanded the twin high gain antennas to fold gently against the side of the body of the telescope, but there were telemetry indications that two latches on one antenna and one latch on the other were not showing the expected "ready to latch" signal. Since both appeared to be in a stable condition it was decided not to close their latches. The failure of the latches to lock was not expected to influence the rendezvous, grapple, or servicing of the telescope. However, as a precaution to fully assess the situation, Goddard requested additional information from the planned camera survey once the crew had captured the telescope.

When Endeavour was 40 miles (64.36 km) behind the telescope, in a lower orbit, it fired its RCS for a change in velocity of 4.6 feet (1.40 meters) per second which would increase the apogee of its orbit. This was followed by the NC-3 maneuver, where the OMS adjusted the velocity by an additional 12.4 feet (3.78 meters) per second in order to yield a catch-up rate of approximately 16 nautical miles (18.41 miles, 29.63 km) per revolution of the Earth. Two revolutions later, the separation was 8 nautical miles (9.20 statute miles, 14.81 km). A third burn, named NPC, of just 1.8 feet (0.54 meters) per second then aligned the planes of the two orbits so that they followed precisely the same ground track. The terminal initiation (TI) burn was achieved by firing multi-axis RCS burns. This put Endeavour on an intercept course with Hubble, and would allow Covey to complete the final stages of the rendezvous manually. The telescope had traveled over 530 million miles (853 million km), circling the world 19,695 times since its deployment on STS-31 just less than 44 months earlier.

As Covey eased Endeavour to within 30 feet (9.14 meters) of Hubble, Nicollier used the RMS to gently grapple the observatory at 2.48 am EST on December 4, as the combination flew over the Pacific Ocean, not far from the eastern coast of Australia. As the snare of the RMS end effector closed on the grapple fixture, Covey announced, "Endeavour has a firm handshake with Mr. Hubble's telescope." Thirty-eight minutes later, Nicollier successfully berthed the telescope in the payload bay.

During the rendezvous and grapple sequence, Akers had taken photos and made himself useful. "We did that—on a crew usually everyone is assigned to do something, so if you're not doing something, you grab a camera and try to take pictures. We also had something we called 'Icky Bicky' in the back of the orbiter, an IMAX camera, and I was operator for that and I'm pretty sure that we got some IMAX photos as we approached Hubble."

Visual surveys and observations of the telescope once it was installed in the payload bay showed a kink and twisting of the outer bi-stem of one of the solar arrays, but after a review the managers decided to follow the original plan of rolling up and retracting the arrays at the end of the first EVA period. Meanwhile, the crew had already begun their final preparations for that first spacewalk.

Each "flight day" aboard Endeavour spanned the late afternoon of one day and the early morning hours of the next day for controllers at the control centers in Johnson and Goddard, so the crew were awakened at 5.57 pm CST. Reflecting the herculean task in front of them, the wake-up music was Aaron Copeland's 'Fanfare for the Common Man'. Preparations for the first EVA went ahead smoothly, enabling Story Musgrave and Jeff Hoffman to initiate their planned 6 hour EVA an hour ahead of schedule.

As the astronauts on Endeavour and flight controllers in Houston prepared for the EVA, the controllers at Goddard were hard at work preparing Hubble for its servicing, powering down both Rate Sensor Units 2 and 3 as well as disabling the associated heaters.

With five back-to-back and extremely ambitious EVAs planned, it was easy to overlook the fact that this also equated to a whole working week on Earth. As such, the first priority when Musgrave and Hoffman left the airlock was to prepare Hubble for the work ahead by installing protective covers on the low gain antenna and the exposed voltage bearing connecter.

The initial EVA at the telescope lasted for 7 hours 54 minutes, with the astronauts' prior experience clearly showing in the way that they meticulously progressed though their tasks. Their motto was "Do useful work." The corollary was that they must not waste time. They successfully changed out the rate sensing units and the electronics control unit and replaced eight fuse plugs in the telescope's electrical circuitry, restoring the six gyros to full capacity. In training, Musgrave and Hoffman had devised a new procedure of positioning Musgrave underneath the gyroscope inside the telescope; it worked nicely and saved about an hour in EVA time.

As the astronauts carried out their tasks, the Goddard controllers managed the systems of the telescope. For example, when the crew were ready to replace fuse plugs, the elements of Hubble that were powered by those fuses were turned off, and once the fuse plugs had been replaced those elements were powered up again. After the astronauts had replaced both rate sensor units, Goddard performed an "aliveness test" on the new units. Two and a half hours later, following the completion of the EVA and with the crew asleep, a functional test of all three telescope gyros was conducted, together with a similar test of the new fuse plugs.

Not everything went according to plan. The two astronauts encountered great difficulty in securing the latches on the door to the gyro compartment after the new gyros were in place; two of the four bolts on the door failed to reset. They had opened and closed the doors over 100 times in training without difficulty, but the actual doors were warped and closing them took a combined effort of both men with helpful support from controllers on the ground. It was essential that the doors be closed "overnight" to stop light leaking in and changing the thermal levels inside the telescope.

The astronauts tried various hand holds and even pushed their helmets again the doors in an effort to make them close, but they would not. Then after talking over their options with the engineers, it was decided to employ a ratchet tool to tighten a payload retention device, similar to webbing, across the door handles. Hoffman recalled "a pretty spirited discussion" with the ground, which was concerned that too much tension would be applied in tightening the straps. It was Flight Director Milt Heflin who gave the go-ahead, telling the ground team that the crew had been selected for their skills, they were trained, and in the end they should be trusted.

Musgrave used a payload retention device to anchor himself against the base of the doors and thus apply body forces against the doors. Meanwhile Hoffman, riding the RMS, worked the top of the doors and was finally able to force them shut and engage the four bolts. It was later suggested by the telescope's engineering team that the length of time during which the doors had been open to allow the astronauts to work inside might have been sufficient for a temperature change to expand or contract the doors, distorting them so that they would not align properly when an attempt was made to close them. "That was something we hadn't planned for," Akers recalled. "We didn't anticipate any problems, and probably could not have because no one knew what the problem was going to be until it occurred. With Story being there and Jeff with his eyes on it and talking to the ground team, they came up with a plan fairly quickly to attempt to close the doors and it worked. And in fact that was the only significant problem out of all five spacewalks in terms of something that cost a bit of time." After STS-61, a door-closing simulator was installed at JSC to test a number of techniques for closing the warped doors.

According to Hoffman this delay took the best part of an hour to solve, but thanks to the time gained by the procedural change on the gyros earlier in the EVA it all turned out well.

Jeff Hoffman, high above the payload bay, works on changing Hubble's fuse plugs during the first EVA.

Had they not tried the new procedure, Hoffman was told, then they would not have had the time to complete all the tasks planned for that first EVA. So they were still able to perform "get-ahead tasks" for the second EVA to be conducted by Kathy Thornton and Tom Akers, who were to replace the solar arrays. "We always tried to do things and get ahead if we had time," explained Akers. "For example, on the first spacewalk, Jeff and Story were out there, and while Jeff was putting up the MFR for the day on the end of the remote arm, Story went around and released some locking bolts on the solar arrays on the carrier to help Kathy and I the next day. The ground team were also really good at always trying to get ahead, so there was never a spare moment during a spacewalk to just lie back and look at the Earth passing by below." Although these were small tasks, the time-saving for the next crew might prove critical in a tight EVA timeline. As much as they could, the space-walking pairs helped and supported their counterparts on the other shift. It was a team effort involving the rest of the crew and the controllers on the ground.

Akers explained his role as the Intra Vehicular (IV) crewmember for the mission. "I was the primary IV crewmember during Story and Jeff's EVAs. Kathy and I did the majority of readying the suits and tools for their EVAs. We were IV when they were out there and they were IV when Kathy and I were out there, with Story the IV on the mike. I don't remember having hardly any spare time once we started those EVAs but still it worked like clockwork. There wasn't a scramble to get ready, we had a good well-oiled team by then and everyone knew what had to be done and they got it done. The way EVA works, is you do not have a checklist of instructions such as, let's say, 'Okay I'm going to set my torque on the wrench by this amount for this bolt and it is going to be five counter-clockwise turns to loosen it.' That's all inside, in the checklist, so the IV crew inside is basically giving the directions to make sure the guys outside are doing things in order and don't miss anything. They make sure everything is done properly. So the 'spacewalk director', if you will, is the IV person inside the shuttle." This choreographer paces the tasks of the EVA in real-time against the checklist and use of consumables, always mindful of safety. "Our checklist actually showed the expected time, so you knew if you were ahead or behind the planned timeline. So if the spacewalkers were getting behind or ahead—and it's just like most jobs you're going to do, sometimes you get a little behind and sometimes you get a little ahead—if we got very far behind, the ground, which of course is watching this a lot closer than we were, might start to modify the tasks that we would try to get done that day. But that never happened, most days we were a little bit ahead and got a little extra done. I remember Kathy and I, on our second spacewalk, after we had swapped out the co-processor and we were getting kind of ready to come in, we went and got some insulation material off one of the carriers that was to be put on the magnetometer on the fifth spacewalk."

As an IV crewmember watching what was going on outside from the flight deck, Akers explained that it was not always possible to use the CCTV and RMS cameras to assist with the viewing. "Generally it was all visual, you always have a visual because you never know when you aren't going to have cameras. The aft bulkhead cameras, for example, can't see a whole lot more than by looking out the window. And of course on the end of the arms is an end effector with a camera, but we had a spacewalker all the time on the MFR at the end of the arm so we didn't use that for a visual unless we were looking close

Story Musgrave during first EVA at Hubble.

at some area. I don't remember at all looking at a camera to check the spacewalkers, we just visually looked out the window."

When Musgrave and Hoffman returned to the airlock at the end of their EVA, theirs had been the second longest spacewalk in NASA history, second only to that on STS-49 when three astronauts, one of whom was their colleague Akers, had retrieved the Intelsat satellite. As Hoffman explains, mission commander Covey "was a very strict 'mom' and made us go to bed on time". Like many astronauts, Hoffman liked to look out the window whenever he could, and since he didn't sleep 8 hours on the ground he would spend a couple of hours looking out the window. But Covey was determined to have a well-rested crew, and placed blinds on the windows and turned the lights out. He knew he could not *make* the crew sleep but he wanted to ensure they didn't do anything else. Once they started the EVAs, everyone slept down on the middeck.

**December 5, EVA 2**

Hoffman said he felt like he could have gone out again the next day, as he was not fighting the suit in order to work, but the plan was for the other EVA crew to take center stage.

In advance of the second EVA, STOCC controllers had planned a 73 minute retraction of the two European-built solar arrays. ESA Project Manager, Derek Easton was confident that the arrays would retract, despite the deformity reported during the post-grapple survey of the telescope. However, when the right-hand array failed to fully retract due to the kink in its bi-stem framework the decision was made to detach and jettison it. As a consequence, the start of the EVA was advanced by 90 minutes. Meanwhile, controllers at Goddard powered down the solar array electronic control boxes and prepared them for the changeout operation. After the new arrays were installed and the astronauts were back inside Endeavour, the controllers would restore power and conduct a 23 minute aliveness test on the newly installed arrays.

As Thornton entered the payload bay she discovered that her suit could no longer receive transmissions from Endeavour. Akers could still communicate with her, so the only way she could communicate with her colleagues inside the orbiter was by Akers relaying messages. Despite this problem the EVA was allowed to continue. Some 3 hours 15 minutes later the radio link suddenly came alive, with no action taken. Towards the end of the EVA she lost communications again. Back in the airlock, she switched to the hardline circuit inside but was still unable to receive orbiter communications. (This problem would return during her next EVA, when once again Akers had to act as relay for her.)[21]

In an interview in 2013 Tom Akers, after having reviewed his flight notes, confirmed that Thornton had an intermittent communications problem, but recalled the dropouts differently. "She could not hear the ground call the shuttle on that loop, but she could hear me and the guys inside the shuttle. How the EVA communication works is the guys outside, doing the spacewalk, we never—I wouldn't say never, but the standard way it's done is spacewalkers never talk to the ground directly, spacewalkers talk to the IV astronaut inside the cabin and the IV crewmember communicates with the ground. Of course, the ground can hear all that communication between the spacewalkers and the IV crewmember, but Kathy just couldn't hear the ground when they talked. She could hear me. As I recall, she could hear the folks inside the shuttle. I know that she could hear me. And everybody could hear her. Kathy's situation did not impact the spacewalk at all." In most circumstances the IV served as EVA Capcom, and relayed between the EVA crew and the ground, allowing the spacewalkers to focus on their work—unless they really needed to communicate directly with the ground to clarify a point or explain a situation. Akers again, "Generally spacewalkers speak to the IV crewmembers, but if you ever had problem out there, where you were looking at something and wanted to explain it, like when Story and Jeff had trouble closing the doors after working on the gyros, Story talked directly to the ground because he had the eyes on what the problem was and there was no point in him passing it to me and then me passing it to the ground."

This second EVA of the mission was planned to last 6 hours, including about 30 minutes to jettison the buckled solar array, but it actually ended up at 6 hours 36 minutes. Waking up to the sound of the Beatles song '*With a Little Help from My Friends*" seemed apt for such a challenging EVA that depended on the astronauts giving Hubble new solar panels, and then ground controllers ensuring that they worked. This operation had to be a coordinated effort between the EVA crew and the orbiter crew. Thornton positioned herself on a foot restraint locked onto the end of the robotic arm and attached a transfer handle to the right-hand solar array, then held on to that unit while Akers physically disconnected it from the body of the telescope and cut the electrical connections. With the array loose, Thornton swung it away from the telescope and the payload bay and released the structure. Nicollier, operating the RMS from inside Endeavour, then lowered Thornton back down into the bay while Covey and Bowersox gently moved Endeavour clear of the drifting array to ensure there would be no collision.

Part of the EVA training was to address the situation if the array had not rolled up, but it was not something that had involved any specific training. "We knew we had to roll the old solar arrays up to bring them home," Akers recalls. In training they discussed the options for if something didn't go as planned. "We played 'what if' scenarios. If the solar array doesn't roll up, what are you going to do. It was obvious that we couldn't put it back in the payload bay, so we would just discard it. We didn't actually train for this but once the array is off the telescope it makes no difference whether it's rolled up or not. If it isn't rolled up, the remote arm will take you up behind the payload bay and you'll release it. I don't think that we ever actually simulated doing that, because that was actually pretty straightforward."

Akers recalled the difficulty in adjusting when they had a bit of time on their hands, after tightly planning the EVA. "When Kathy got rid of the solar array that we couldn't re-stow, I think we waited 5 plus minutes while we were waiting for daylight because they didn't want her to jettison that in the dark. Remember your mind set is to get the job done and so you are totally focused on not wasting time when spacewalking. As there is so much to do, you never know when something's going to go wrong and will take up your time out there, time which is obviously limited. Even if you did have 2 or 3 minutes to look around, it's hard to get into the mind set of actually relaxing and doing that."

As the array drifted farther away, pulses from the shuttle's maneuvering engines imparted a "flutter" on the solar blankets, causing it, as Thornton commented, to flap "like a bird". The discarded +V2 array was expected to survive for up to a year before atmospheric drag pulled it out of orbit to burn up. It was given the international designation of 1990-037C and catalog number 22920, and actually remained in orbit for almost 5 years before re-entering

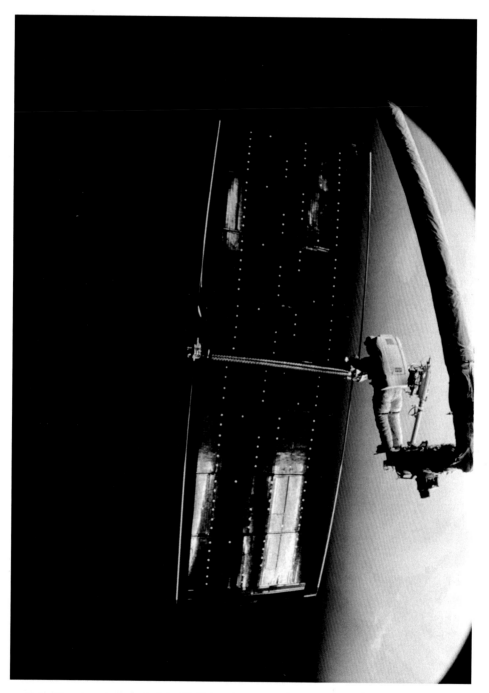

Kathy Thornton, on the end of the RMS, is about to discard one of the old first generation solar arrays.

The discarded solar array.

on October 28, 1998. The persistence of its orbit has been attributed to the difference between the high drag nature of the array and the low drag characteristic of the denser attachment mechanism. This would have resulted in an overall lower area-to-mass ratio which would slow the decay during a period of relatively low solar activity.[22]

The next task was to install one of the new arrays onto Hubble. Thornton, maneuvered on the arm by Nicollier, attached a transfer handle to the new folded array on its carrier and was then taken to the side of the telescope, where both astronauts installed the new unit. Next the pair manually folded the second array and stowed it in the carrier, with Akers strapping it in securely for the trip back to Earth. Engineers would have preferred to have both arrays back for post-flight evaluation after over three and a half years in space, but packing the first one would have been impossible in the time available. The experience of working together on a previous EVA showed as Thornton and Akers easily installed the second new array. Shortly thereafter Goddard verified that both new arrays were electrically alive, but they were not to be unfurled until after the fifth and final EVA.

The shuttle spent roughly half of every 90 minute orbit in darkness out of direct sunlight, so illumination had to be provided to allow the spacewalkers to continue to work.

"We didn't have floodlights," Akers recalled, "we had our little helmet lights. If you were spacewalking out there you would turn your helmet lights on, which is an interesting thing as your batteries are only good for so long obviously. You know to turn them on, because you cannot see very well without them if you're working in close. With the Sun coming up and going down, you know, coming up and going down once every 90 minutes, you forget to turn your lights off when it gets daylight; it was the job of someone on the flight deck to remind us to turn them off. We didn't have any floodlights, but what we did have was a big spotlight inside, it was kind of like one of those hand held floodlights and we just shone it out the back window and we did that on occasion if someone was working on a close area, such as when Kathy and I were changing out the co-processor, the guys inside lit up our area with that."

With the major task of the day finished, Thornton and Akers wrapped up their spacewalk and were informed by Capcom Greg Harbaugh that Akers now held the EVA record for the longest accumulated EVA time in shuttle history, his 22 hours 50 minutes exceeding fellow astronaut Jerry Ross's record by 1 minute. In reply, Akers predicted that his record would soon be beaten. During the re-pressurization in the airlock, Thornton reported suffering ear problems, so with both astronauts still in their suits the airlock was depressurized again and then re-pressurized at a much slower rate until it reached the required 10.2 psi and this time Thornton reported no ill effects.

Akers notes that there were few surprises in the accuracy of training or flight hardware to match the telescope, something which had troubled earlier EVAs and some of the water tank simulations. "The only thing I remember, was the Portable Foot Restraint location up on the side of the telescope that I used when I was de-mating the old solar arrays. It was not exactly like it was in the training sims, or perhaps it was *exactly* like it, I couldn't tell. I couldn't get into the position that I was used to doing in training when preparing to remove the little lock wire that I had to on the solar array. It could have just been the stiffness of the suit. The suits that we used in the pool are to the same psi, but they're more flexible because they are used more. The suits that we use in space don't get used a lot and they may have been stiffer. But nevertheless, I had to get *out* of the foot restraint and basically just wrap my legs around it in order to hold myself in place to remove that wire. Actually, Kathy had to come up and hold me in place. The foot restraint appeared different. It didn't feel exactly the same. It seems to me that the restraint was offset a little bit farther from the solar array than in training. I'm not sure. It could have just been the pressure of the suit. Overall, the training hardware was very high-fidelity and pretty much as we found it up there."

In the post-EVA press conference Joe Rothenberg, Hubble Space Telescope Flight Project Director, reflecting on the unplanned action of manually discarding one of the arrays, said, "I believe the first objective has been met." He then added, "We can handle on-orbit servicing, and we can handle contingencies."

For the crew, completing a shorter second EVA was a benefit that allowed them to catch up and enjoy a relaxing supper and then get some well-earned sleep.

Hoffman remembers some of the lighter moments of the mission. Knowing that on such long spacewalks they would drink all the water in the standard in-suit bags, the EVA crew had requested larger bags. In preparing for EVA 1, these bags were nowhere to be found so Hoffman asked the ground for assistance. An hour later the call came, "Well, Jeff, the good news is we have located the drink bags. The bad news is they're in Houston."

The bags had never been shipped. So the astronauts had to make do with the smaller bags. Then there was the case of the candy bar. Inside the helmet was a small candy bar in a retention sleeve that allowed it to pulled up by the teeth and bitten off, leaving enough in the sleeve for the next bite. Hoffman invariably knocked the bar back into the sleeve instead of leaving part of it accessible, and thus never did get to eat it. For Thornton, it was a different problem. When Hoffman helped Thornton doff her gloves at the end of EVA 2 her fingers were bright red. He presumed it to be blood, but it proved to be from the candy bar that happened to be red. The circulating air flow in the suit had carried some crumbs down the arm of the suit to the gloves, where they had been mushed up by her fingers, staining them bright red and giving Hoffman an initial shock when he saw the result.

**December 6, EVA 3**

The focus of the mission's third spacewalk was to be the replacement of the first generation Wide Field Planetary Camera with the upgraded version. The first WFPC had experienced difficulty focusing, but this new one had four precision-ground mirrors designed to remove the blurring of images by focusing stray light of the telescope's primary reflector. The time allocated to Hoffman and Musgrave to complete this exercise was 4 hours. The other main task was to install two new magnetometers, which was expected to take about an hour. The magnetometers measured the Earth's magnetic field in the same three directions as used by the momentum wheels in controlling the telescope's orientation.

Shortly before the mission, an astronomer friend of Hoffman had asked whether NASA thought they could pull off the double replacement on Hubble, and Hoffman said he hoped so. His friend said the astronomical community would be "deliriously happy" if the mission got just one replaced, and therefore Hoffman should not "feel that you will have failed, if it didn't work".

Prior to starting the EVA the Goddard controllers powered down the old WFPC and then reconfigured their apparatus to support the installation of the new unit. The EVA was begun over an hour ahead of schedule. While Hoffman mounted the foot restraint on the end of the robotic arm, which was again being operated by Nicollier, Musgrave stood in a portable foot restraint that had been anchored on Hubble near the location of the WFPC.

Hoffman calls Musgrave a perfectionist who had done a lot of work in figuring out where to position the foot restraint to comfortably handle the WFPC. As Musgrave eloquently puts it, "You are a ballerina, everything has been worked out [in advance]. Every move is worked out, every finger, and every toe. The movements have been worked out. Maybe you get some help, that's absolutely fine, looking at your back which you cannot see, but you must not rub things, rubbing is contamination and that's a disaster."[23]

Working about midway up the side of the telescope, the first task was to remove the old WFPC from the telescope. Hoffman grasped the handles to pull the unit out, with Musgrave stabilizing it as it emerged. As it slid along its guide rails a short stance, Hoffman paused to allow Nicollier to reposition him on the arm into a better position before extracting the unit almost completely out of its recess. However, prior to sliding it completely out, he shoved it back in to rehearse installing its replacement. Satisfied, Hoffman drew the old instrument all the way out and, maneuvered on the end of the arm by Nicollier, relocated it to a temporary position in the payload bay. Meanwhile, Musgrave inspected the cavity in the telescope and then went to prepare the new camera for installation.

Jeff Hoffman exchanging the Wide Field/Planetary Camera instrument.

A transfer handle was attached to the new 620 pound (281.23 kg) unit to extract it from the storage location. As he did so, the astronomer in Hoffman was impressed by its pristine condition and prospect of new science. "Oh, look at that baby, it's a beautiful spanking new WFPC, we'll see some nice pictures with that," he noted, adding, "I hope we have a lot of scientists eager to use this beautiful thing." Prior to inserting the new camera into Hubble, Musgrave removed the protective mirror covers on the camera. Less than 40 minutes later, again using the guide rails, the two astronauts gently slid the new camera into the telescope. The installation had been completed in record time, much less than the 4 hour slot allotted.

Within 35 minutes of WFPC-2 being inserted, Goddard announced that it had passed its electrical aliveness test. This was followed by functional tests and, much later, after Hubble had been released, a science data dump of test images which were processed on the ground within 30 minutes and confirmed that the new camera was much better than its predecessor.

Sixteen years later, Hoffman received some grief for over-torqueing the bolts holding in WFPC-2, when it proved difficult to remove the unit in exchange for WFPC-3 during SM-4. "I tightened it to the torque they told us to," he says. "On STS-125 Drew Feustel used what was thought to be the same torque but with a different tool to release the bolt but it wouldn't come out. He finally did it, but risked breaking the bolt." Apparently there was a wide range of calibration on the torque; Hoffman was toward the upper end of his torque whilst Feustal was at the lower end of his.

After the EVA and the success of installing the two important instruments, Hoffman was excited but Musgrave was concerned that the placement of the foot restraint was not exactly where he planned, and he could not reach where he wanted. Despite the satisfaction of a job well done, the perfectionist in Musgrave was frustrated that his plan had not worked out the way it should have.

It was during this EVA that the astronauts had difficulties with the JSC-supplied power tool, resulting in their switching over to the Goddard-supplied units. As the official NASA Mission Report says, "One of the two JSC-supplied HST power tools failed (flight problem STS-61-F-07). Indications are that a switch problem developed in the tool and caused the failure. The other power tool also had a speed setting failure, but [it] remained usable." The Problem Tracking List says, "During the Fine Guidance Sensor Bay closure, the HST power tool abruptly stopped working. Changing the batteries [and then] cycling the switches failed to resolve the problem."[24]

With the main objectives completed early, the pair of EVA astronauts moved to restraints on the end of the RMS to be moved to the magnetometer work area. As this was near the top of the telescope, Hubble had to be tilted forward on its support structure to enable the 50 foot (15.24 meter) arm to reach that section. The replacement of the magnetometers was done in two stages. After Goddard had prepared one magnetometer, the astronauts removed that unit and installed its replacement, then Goddard performed functional tests on the new unit. Only when that unit was confirmed satisfactory was the second magnetometer dealt with likewise. It was during this part of their EVA, while awaiting confirmation that the first magnetometer had been installed correctly and was working, that the astronauts reported parts of the unit's thermal insulation shell had become detached. This discovery led to plans for Thornton and Akers to devote part of their next EVA to preparatory work for the later installation of new insulation around two of the older magnetometers.

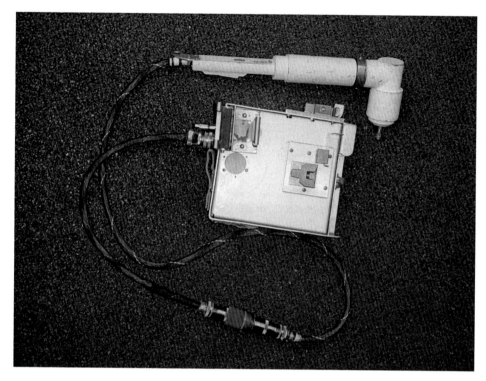

A close up of the ratchet tool used by STS-61.

This third excursion had taken 6 hours 47 minutes, and by smoothly completing all of the assigned tasks it demonstrated the wisdom and foresight in selecting and training four EVA-experienced astronauts for this important mission. It was over 10 years since Musgrave had performed the first EVA of the shuttle program in April 1983, along with Don Peterson, but his work on EVA systems, procedures, and hardware during a 25 year career as an astronaut was evident in the way that he went about the tasks. It was also over 8 years since Hoffman had performed his first EVA on STS-51D in April 1985, but the training program to prepare the EVA team for servicing Hubble had clearly paid off.

**December 7, EVA 4**

The most important task for this EVA was installation of the much-reported and awkwardly named Corrective Optics Space Telescope Axial Replacement—referred to by its acronym COSTAR. Thornton and Akers were also to replace a computer processor and, if there was time, collect some of the aluminized Kapton and Dacron mesh multi-layer insulation which the final EVA would install as impromptu protection for the two older magnetometers.

As these tasks had been likened to performing "brain and eye surgery" on Hubble, it was perhaps inevitable, if a little premature, that the wake-up call for that day should be

Johnny Nash's '*I Can See Clearly Now*'. Spirits were high as the two spacewalkers headed out into the payload bay.

To install COSTAR meant sacrificing the High Speed Photometer. Before this could be removed, the Goddard controllers turned off its power supply. Riding on the foot restraint at the end of the arm, Thornton used a power ratchet tool to open the large access door to allow Akers to move inside to disconnect the HSP. Then he assisted her to slide out the unit, which at 7×3 feet (2.13×0.91 meters) was the size of a telephone booth. With Thornton gripping the two transfer handles firmly, Nicollier maneuvered her clear of the telescope and down to a temporary stowage fixture in the payload bay. Next, Nicollier positioned Thornton above COSTAR, upon its payload bay carrier, to enable her to grasp its transfer handles. Thornton and the new unit were then returned to the installation area, where Akers helped to guide it into the telescope. The next task for Akers was to tighten fasteners to secure COSTAR and reconnect the electrical cables. In pre-flight planning this exercise was expected to take the astronauts 3 hours 10 minutes, but in reality they accomplished it in only 35 minutes. This reaffirmed that prior experience of EVA was beneficial in getting a job done promptly and smoothly. This was especially so if the two spacewalkers had worked together, as Thornton and Akers had on STS-49 just over 2 years earlier. As was becoming routine for a Hubble servicing task, as soon as the astronauts were clear Goddard carried out an aliveness test to verify the new unit's systems, communications, electronics, and telemetry.

COSTAR is removed from its storage in the payload bay prior to being installed in the telescope during the fourth spacewalk of STS-61.

A close view of Kathy Thornton during an EVA.

The extracted HSP was stored for return to Earth in the same container in the payload bay that had held COSTAR. This caused some difficulty for the astronauts. As Akers explained, "When KT and I were putting the HSP back in the carrier to bring it home, I was driving the B latch and she was holding the instrument in place. Now as the plunger was going into the connection to hold it in the carrier, I couldn't see it. So I would drive it in a little way, then I would come up and look to see what it was looking like and it was looking good." As Akers drove the power tool to tightening a bolt into the hole, he knew how many turns were needed before it was tight. "It turned out that I was counting aloud while rotating it and if you listen to the spacewalk you'll hear when we are turning the power tool that we are counting turns. That paid off, because it got tight before it was supposed to. When I went up to have a look, sure enough it hadn't gone in the hole properly, so I had to back it off and had to re-do it. So that was something that was very fresh in my mind to tell the next crew about: when you put that instrument down in the carrier you have to be careful that you don't let it float around in the carrier while you are driving that bolt."

With one of the most important objectives of the mission achieved, Thornton and Akers then moved to the next challenge of installing a new co-processor, designed to process data much more rapidly and expand the memory capacity. Obviously the old computer had to be disconnected. It was an anxious time for the astronauts, controllers, and scientists,

Tom Akers working in tight confines inside the telescope during installation of COSTAR. Thornton is just visible on the end of the RMS.

knowing that the Great Observatory that they were working so hard to restore to its full potential was to have its central core turned off. The operation was equated to taking every precaution in performing major surgery on a human patient, but there was no guarantee. The installation went smoothly, and there was a collective sigh of relief when Goddard announced the new unit had been powered up successfully. Of course, a lot more testing would be required to confirm that the telescope was in full health.

During a post-EVA press briefing Ken Ledbetter, the HST Program Manager, reiterated "essentially what we were doing was brain surgery". Dr. Dave Leckrone, the Senior Project Scientist, noted, "We also conducted eye surgery on the telescope." Ledbetter added, "In a day or so, hopefully the patient will be ready to walk on its own." In fact, it would be 6 to 8 weeks before the initial checkout of the telescope was completed and the first images using COSTAR were received. The scheduled science program would not be able to resume until about 3 months after the service mission, upon the completion of a comprehensive checkout.

Another item of trivia occurred during this EVA. Akers surpassed the all-time American career EVA record of 24 hours 14 minutes set by Gene Cernan during Gemini 9 in June 1966 and Apollo 17 in December 1972; the latter accrued while moonwalking. On closing out his second spacewalk on STS-61 Akers had an accumulated career total of 29 hours 40 minutes.

**December 8, EVA 5**

Records were being broken and set almost every day on this mission. For the first time on a single shuttle flight a fifth EVA was to complete the remaining tasks on what was already a remarkable and highly successful Hubble servicing. Hoffman and Musgrave were to replace one of the Solar Array Drive Electronics (SADE) units that enabled the telescope to face its solar arrays towards the Sun. They were also to install multi-layer insulation around two of the older magnetometers as improvised thermal protection until a later mission could either install a better covering or replace the magnetometers. After the two new solar array panels had been deployed, the EVA program would be completed by installing the Goddard High Resolution Spectrograph Redundancy Kit.

Although functional tests were completed on the new co-processor without any problems being reported, there was a loss of downlinked telemetry data from the computer (DF-224) during the night. The problem was traced to imprecise pointing by the Ku-band antenna on the shuttle, which caused intermittent interruptions in telemetry and communications being relayed by the shuttle between the telescope and the STOCC at Goddard over the Tracking and Data Relay Satellite network. Contingency actions had been added to the EVA in case the unit needed replacing. Hoffman and Musgrave would have attended to the computer in between insulating the magnetometers and installing the GHRS. However, this extra work was not required.

## The great screw chase

Hoffman explained the difficulty in working on Hubble using tools. "The first problem was just getting access, there was a small screwdriver [but] that turned very fast, so it was very hard to get it on the screw and keep it there. The only tool that we had which turned slowly enough that we could control it was [a] big ratchet tool. That was like swatting a fly with a mallet. It was much too big for the job we needed to do with it." He explained that once the screws were removed the unit just floated on its wire. When he removed the connector, the screws resting in the connector started floating out. Musgrave collected and placed them in his trash bag. The problem was that the things in the bag floated and some would escape each time the bag was opened. For this operation Hoffman was free-floating, hanging onto the arm with one hand while Musgrave was in the MFR on the RMS. When one particular escaped screw floated out of reach towards the payload bay it could have caused a problem later in the mission, perhaps preventing the bay doors from closing or becoming trapped in some other mechanism. As Nicollier started to maneuver the arm to chase the screw, Ken Bowersox told the computer to override the software which limited the loaded arm motion rate, setting it to that of an unloaded mode in order to enable it to move rapidly enough for Hoffman to capture the wayward screw. After the mission, the trash bag was redesigned to prevent items from floating out.

After the SADE unit had been replaced and the Goddard High Resolution Spectrograph Redundancy Kit installed, the RMS swung the two astronauts up to the top of the telescope where they installed Mylar covers over the older pair of magnetometers, not only to protect the instruments from ultraviolet degradation but also to contain any contaminants or debris which might become detached from them before the next service mission, when the issue would be properly dealt with.

Hoffman recalled the experience of being at the very top of the telescope, high above the payload bay. "It was a spectacular view from up there, even more fun for me because I was free-floating that day. Of course, we're both attached to the arm by stainless steel tethers but every once in a while I would just let go." He could not float away but he broke the physical bond between himself and the shuttle and the tether was loose, not pulling on him. "It was a remarkable psychological transformation," he remembers. "Instead of being attached to the shuttle, I became a free-flying satellite. I grabbed on: 'now I am part of the shuttle'. I let go again: 'now I'm a satellite'. It was really exhilarating, particularly at night." At one time the astronomer-astronaut turned his back on the shuttle, "so it was just me floating in space. All the stars around, I felt like I was alone. It was a moving experience, I'll certainly never forget it."

Towards the end of the EVA, after the Primary Deployment Mechanism on the array units became stuck, Musgrave gently pushed each array to coax it to unfurl. Each array took about 5 minutes to fully deploy, with the astronauts in "front row seats" ready to assist if necessary. One array developed a slight twist, but this was expected and was thought to be the result of the manufacturing tolerances and residual stresses of flight. It was predicted the array would achieve its designed configuration after just a few orbits exposed to the warmth of sunlight.

As the EVA closed, news was relayed to the astronauts that the GHRS kit had received a good health check. Several hours later the SADE was also looking good. Hoffman observed that this would make some of his astronomer friends very happy. Musgrave praised the large training team which had prepared the crew for the demanding EVA program on the mission, "You're in our hearts, you're in our heads. What we've done, and what we're going to do, is a simple reflection of what you've given us."

About an hour after the end of the 7 hour 21 minute EVA, the controllers at STOCC were able to report that the twin high gain antennas had successfully deployed in readiness for the release of Hubble the next day.

Over the five EVAs of STS-61, the four astronauts had accumulated 35 hours 28 minutes in the payload bay, creating a new record for the shuttle. It was a busy but satisfying week's work. "Actually, it went pretty much as planned," recalled Akers. "We were all pleasantly surprised at the end of every day as to how much things went just as we planned, with few surprises." In addition to accomplishing all of their assignments, they had placed protective shielding on two magnetometers, which was not planned. It would be some time before the post-installation tests established whether the telescope's problems had been solved but the crew knew they had done all that had been asked of them, and so they gave each other hugs and celebrated.

In preparation for releasing Hubble, Endeavour made a small RCS burn to increase its orbit. At capture, 4 days earlier, the orbit had been 319 × 313 nautical miles (367.1 × 360.2 miles, 590.7 × 579.67 km). Prior to the maneuver it was 320 × 313 nautical miles (368.25 × 360.2 miles, 592.64 × 579.67 km). The "boost" circularized the orbit at 321 by 320 nautical miles (369.40 × 368.25 miles, 594.49 × 592.64 km), only marginally below its deployment altitude.

## December 9, releasing Hubble

As the crew slept, Goddard staff investigated conflicting data concerning the Data Interface Unit (DIU) which monitored the telemetry from the telescope's subsystems. As a result, the second orbital boost maneuver was canceled to allow additional time for the problem to be investigated. Meanwhile the new solar arrays recharged the telescope's batteries. The release of Hubble was delayed for three and a half hours to fully investigate the telemetry problem. The DIU under suspicion was one of four units which monitored the engineering telemetry and commands. It was experiencing drop-outs and reporting conflicting readings. This was not a new problem, and was not related to anything which had been serviced or replaced by the STS-61 crew. Each of the four DIU had a two-sided redundancy with an A side and a B side. The controllers at STOCC isolated errors which occurred only on Side A of the DIU-2 unit, though there was no problem associated with its command capability. It was therefore decided to use Side A only in a backup mode with only a small degradation in its capability. It was clearly a problem that would need revisiting during a subsequent service mission but was beyond the capabilities of the current mission, so the release was authorized.

Deploying Hubble was not a simple or quick activity; it required a coordinated effort by the crew on-orbit and controllers in Houston and Goddard. Once authorized to proceed, the controllers at STOCC loaded new navigation tables into the telescope's onboard computer, temporarily switched off its solar arrays and powered up the reaction wheel assemblies and magnetic torquers. About 10 minutes later, Nicollier used the RMS end effector to grip the telescope's side grapple fixture. Some 40 minutes later he severed the umbilical connection and released the berthing latches which had held Hubble.

At this point, Nicollier gently commanded the arm to raise Hubble high above the cargo bay towards its release position. Thirty minutes later, with Hubble still firmly on the RMS, the STOCC controllers commanded the opening of the aperture door of the telescope, an operation which took a further 33 minutes. Even after that activity, there remained another half an hour of checks.

At 4:27 am CST on Friday, December 10, the snares in the end effector of the RMS were relaxed and Hubble gently slipped its moorings. It had spent 6 days, 1 hour and 1 minute in the payload bay.

Unlike on the deployment mission, the STS-61 crew did not perform a fly around to take pictures of the telescope. Almost immediately, Endeavour executed two small maneuvers to withdraw at a rate of about 1 foot (0.30 meter) per second. Meanwhile Hubble's solar arrays had locked onto the Sun and the telescope had established communications directly through TDRSS.

With Hubble operating independently, the crew of Endeavour received a 15 minute phone call from the Oval Office of the White House in Washington DC. Both President Bill Clinton and Vice President Al Gore offered their congratulations, with the President telling the crew that their mission was "one of the most spectacular space missions in all our history. We are so proud of you… I want to thank each and every one of you for what you did. You made it look easy." He added that the first Hubble service mission had given "an immense boost to the space program in general and to America's continued venture in space".

It was a happy end to a very tiring week for the space "fix it team", who were allowed to sleep in the next morning and then have a "day off" with only light duties in preparation for landing. Because the duration of the flight was limited, they had had to work hard without a break. As Tom Akers recalled, "I guess it was tiring. I don't remember ever feeling tired—I don't think any of us did. We had a choreography going, with team members getting ready every day but we did not have any spare time. You got up in the morning and when you got done in the evening, you weren't—since you had to get the LiOH canisters out, the batteries needed charging, we had to change everything out, re-do the tools. Every day we had to put different tools on our mini-workstation for each EVA crewmember for whatever the coming tasks were going to be. So I remember going to bed for a couple or three nights that I didn't even have time to read my emails or send anything back home." Akers did not recall being more tired after the release of Hubble, more a sense of relief. "That was a big event and we didn't really get to relax until after it was gone. But no, I don't remember ever feeling tired. When you're up in space there is always work to do, you don't get a lot of time to look out the window, and you're used to that from your previous missions. Down here on Earth, if I had to work 6 hours in a row, without a rest, I'd be whipped."

Akers also suggested that there was not much difference between flying an experienced over a rookie crew on such an important mission. "I think that even an inexperienced crew would probably had done just as well, in my personal opinion. You can't give enough credit to the training team, and even the program office, coming up with choreography, so I think anybody in our office could have gone and done the job."

### December 9–11, the calm after the storm

Over the next few days, the cabin pressure on Endeavour was gradually returned to 14.7 psi in preparation for returning to Earth. The crew recorded IMAX imagery, stowed the RMS, checked and stowed their EVA garments and equipment, performed supply and waste water dumps, and participated in press conferences. They also took time out to conduct one of the most popular activities on every mission, simply looking out the window as Earth rolled by.

On December 11 it was reported that Covey had observed "the brightest morning star" he had ever seen. This of course was Hubble, its silvery surface reflecting sunlight, and orbiting some 1 nautical mile (1.15 miles, 1.85 km) above and 76 nautical miles (87.46 miles, 140.75 km) behind Endeavour, with the separation increasing at about 4 nautical miles (4.60 miles, 7.40 km) per revolution. The discarded solar array was 2200 miles (3539.8 km) ahead with the range opening by 45 miles (72.4 km) per revolution. The astronauts were pleased to hear that Hubble was continuing to work well.

Being at its highest altitude, the shuttle circled the Earth in about 95 minutes instead of the traditional 90 minutes; this small difference meant they got to see 15 sunrises and sunsets per day instead of 16. As they were also much closer to the region of enhance radiation known as the South Atlantic Anomaly, they had been issued special radiation monitors to wear during the mission. Hoffman said that the dose of one or two rads was not a health hazard, but their mission was equivalent to spending 6 months on the space station.

As they prepared the orbiter and themselves for landing, they were informed that owing to developing weather trends at KSC, it had been decided to perform the de-orbit maneuver one orbit earlier than previously planned.

### December 12, landing day

Late on December 12, during orbit 161, Endeavour completed a successful night landing on Runway 33 at the Shuttle Landing Facility (SLF). The flight duration was logged at 10 days, 19 hours, 59 minutes 26 seconds. Following safing and the exit of the crew, the orbiter was towed in the early hours of the 13th to the OPF. The initial post-flight inspection established that the thermal protection system had sustained a total of 120 debris hits, 13 of which had a dimension of 1 inch or more; both figures were less than average.

## PASSING THE BATON

One of the important activities conducted shortly after each mission are the debriefings that the crew have with the training teams and other astronauts. A second service mission was intended but the crew hadn't yet been assigned, so there would be a gap before the STS-61 crew could talk to the next Hubble crew. As Akers recalls, "Once that [STS-82] crew was assigned, we talked with them and passed on anything that we hadn't already told the world in terms of when we did debriefs. So yes, we met with them and discussed the tasks that were similar to what they were going to be doing, like opening and closing the doors. I recall that they were to changeout one of the axial instruments—like the HSP to COSTAR that Kathy and I did. I talked with them especially about what we call the B latch, and how to insert it without a problem."

STS-61 was a very successful mission from the point of view of the servicing operation, and was a clear vindication of the decisions taken almost 25 years earlier to ensure that the telescope was serviceable on-orbit, to utilize the shuttle as the base for that servicing, and to train specialized teams both on the ground and to fly in space to conduct those servicing missions. It also demonstrated that crews could carry out both planned and unplanned tasks. Having a crew on hand to address problems in real time was an added bonus in maintaining the operational capability of the Great Observatory.

**Tool Time on Hubble**

One of the lighter moments on the mission led to a guest appearance on one of American TV's most popular comedy shows, *Home Improvement*, in which comedian Tim Allen hosts *Tool Time*. It started in the crew quarters prior to the flight, when staying up late to adjust their sleep cycles in preparation to fly a single-shift mission, watching Allen's comedy routine from *Tool Time* and laughing at some of the famous one-liners such as "more power, uh". In 2009 Hoffman recalled that during the flight one of them mentioned that the battery in their power tool was running down and he automatically pointed out, "Oh, I need a battery." Bowersox promptly quipped, "Yes, more power." Someone else chipped in, "Uh-Uh." As a result of hearing this exchange, a person on Allen's show noted that the astronauts were working on Hubble with power tools and invited them to Hollywood to guest on *Home Improvement*. NASA's Public Affairs thought this was a great idea because

Jeff Hoffman displays the selection of EVA tools available on STS-61. Their work earned them a guest appearance on the TV show *Home Improvement*.

the show was one of the most popular on TV at that time and it would be great publicity for the success on Hubble. Due to the re-runs of the show, the now Professor Hoffman at MIT is recognized each year by new students more for appearing on a hit comedy show than for the work as an astronaut at Hubble that got him on the show in the first place!

**Mission success**

With all the attention placed upon the STS-61 mission to succeed, and the importance of acquiring experience in preparation for assembling a space station, the crew were not that aware at the time of just how important their flight would be seen if it were successful. "I don't think anybody was overly surprised that we were successful within the crew, because we really had great trainers, a great plan, and when we went up to do it we didn't have any doubts in our mind, that we weren't going to be successful—unless something completely unforeseen happened," Akers commented. "It's always the unknown that you worry about. We spent hours on 'what if', trying to have a plan for everything, but you know that in the back of your mind you don't have a plan for everything. It's just like the door problem that Story and Jeff had on the first spacewalk; that was something that we hadn't thought of or worried about. Luckily that was the only thing. So no, I don't think we were aware of that 'big picture'. But that's probably good, because we didn't really feel the pressure that I'm sure our bosses and managers felt, going into the mission. We were very confident that we could do that.

"The makeup of the whole Astronaut Office is a 'can do' attitude, once you've done the proper training and planning. Multiple spacewalks were going to be required if we were to build a space station. That was one of the reasons that Kathy Thornton and I were given the spacewalk on STS-49, to get more spacewalking experience in the office. As a result of us working on Hubble, they knew how to do multiple spacewalks back-to-back with four crew members and multiple suits, so there weren't any surprises in getting ready for assembling the space station. The experience from our mission was definitely beneficial there."

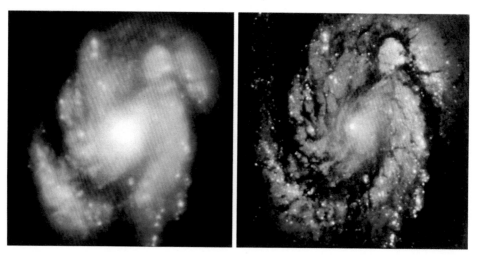

The effects of COSTAR. *Left* an image taken before the corrective optics were fitted, and *right*, the same image afterwards.

A rejuvenated Hubble is placed safely back into orbit to continue its scientific journey around Earth, until its next servicing mission.

## Domestic home improvements

The astronauts were amazed at the public response to the flight. It seems more people had followed it than any other since the Apollo 11 lunar landing. "The warmth of the welcome and the excitement of everybody that this mission had actually gone off so well, was just a wonderful feeling," Hoffman said. "People went overboard in their enthusiasm about how important it was. At one point, people were saying we ought to have a tickertape parade on Wall Street like they did for Apollo. But we never did; that probably would have been a bit over the top."

As with every returning crew, there had to be time after all the debriefing in which they returned to a 'normal life' back home. In his 2009 oral history Hoffman mentioned that his friends complained that he could have rescheduled the EVAs so they didn't have to stay up until 3 am to watch them. His wife mentioned that, having repaired the door on Hubble, he should attend to their own kitchen door, which didn't close properly. Clearly his mission to Hubble was over and it was time to plant his feet firmly on the ground again.

The success of STS-61 indicated the remaining service missions could be prepared with confidence, and made the so-called "Wall of EVA" required to assemble the space station seem less intimidating. It was also a great boost for NASA to have pulled off the mission so successfully. There was talk of restoring the space agency to the "glory days of Apollo". Of course, weeks of testing remained before the telescope could be declared fully operational. While the flag waving certainly raised confidence within NASA, many fingers remained firmly crossed.

As the New Year celebrations wound down in the early hours of January 1, 1994, the Hoffman household received the news from an astronomer friend at the Space Telescope Science Institute that the first new picture had been received and that if Hoffman had any champagne left he should crack open a bottle to celebrate.

Meanwhile it was time to prepare for the next service mission.


## REFERENCES

1. AIS interview with Preston Burch, November 14, 2013
2. NASA News: JSC 92-016
3. NASA News: JSC 92-047
4. Crew Picked for HST Visit, NASA JSC Space News Roundup December 11, 1992, Volume 31 #47 p4
5. NASA News: JSC 93-017
6. Jeffrey A. Hoffman JSC Oral History, November 3, 2009
7. AIS interview with T. Akers November 11, 2013
8. STS-61 Space Shuttle Mission Press Kit, December 1993
9. Hubble Space Telescope First Servicing Mission (SM-01) Training and Operations Media Workshop August 30-September 1, 1993
10. From Engineering Science to Big Science, Pamela E. Mach, Editor, NASA SP-4219, 1998, Chapter 16, the Hubble Space Telescope Servicing Mission. By Joseph N. Tatarewicz
11. *Adventure in Space, The Flight to Fix Hubble*, by Elaine Scott and Margaret Miller, Hyperion Paperbacks, 1995, p20
12. NASA News: JSC 93-038
13. NASA SP-4219
14. From *Engineering Science to Big Science*, p378
15. *We Have Capture*, Tom Stafford and the Space Race, Tom Stafford with Michael Cassutt, Smithsonian Institute Press, 2002 pp233–234 & p239; see also *From Engineering Science to Big Science*, pp378–390

16. KSC Shuttle Status Reports Mission STS-61 (various dates, 1993)
17. Chronology of KSC and KSC Related Events for 1993, by Ken Nail, NASA Technical Memorandum 109196, March 1994, p143 and p145
18. Atlas flies, Shuttle ready, Jim Banke, Florida Today p1A, Nov 29, 1993
19. STS-61 Status Reports, Nos. 1 through 21, NASA Mission Control Center, JSC
20. STS-61 Space Shuttle Mission Report, February 1994, NSTS-08288, Remote Manipulator Systems, p24
21. STS-61 Space Shuttle Mission Report, February 1994, NSTS-08288, Mission Summary, p3
22. Email to D. Shayler from Jonathan McDowell, January 29, 2015
23. AIS interview with Story Musgrave August 22, 2013
24. STS-61 Space Shuttle Mission Report, February 1994, NSTS-08288, p7 and p41, Payload Servicing Tools and Crew Aids

# Closing comments

The time between the launch of Hubble and the landing of STS-61 has been described as a rollercoaster ride for NASA, from the triumphs of success to the pitfalls of setbacks. It had taken years—decades even—to convince the scientists, politicians, and budget planners to commit to a large optical telescope on-orbit, and then to assign it to the space shuttle, which had its own troubled development. As the program evolved, so the costs escalated and the intended launch date slipped, but many people kept the faith, spurred on by the prospect of the great discoveries that the telescope would make once it was in operation high above the Earth's atmosphere.

As the bureaucracy and politics played out, behind the scenes scientists, engineers and designers built the hardware and planned the science. To support the observatory in space a vast infrastructure had to be created to control the telescope, develop the servicing missions and furnish the hardware. All items had to withstand rigorous testing and survive launch on the shuttle. Crews had to be trained—not only for when things went well but also for when they didn't go as planned. There had to be contingencies and backup plans prepared, tested and trained for. These had not only to be feasible, they had also to be safe—as was the case when such actions occasionally proved necessary.

Finally, after recovering from the loss of Challenger in 1986, one of the major setbacks in human space flight, in the spring of 1990, several years later than originally intended, NASA launched the telescope into space. After a few heart-stopping moments during its deployment by STS-31, Hubble was cast off to "do good science". A happy and relieved crew came home knowing that they had placed a telescope into orbit, a concept which had been first proposed some 40 years earlier.

But then the bad news became evident. After all the hard work in ensuring the primary mirror was flawless, it was not. Spherical aberration clouded the expected pristine images, and Hubble went from an example of the latest technology to an object of national ridicule. The costly telescope was flawed by an error made years before during ground processing. From the euphoria of finally placing the telescope into space came the bitter disappointed, indeed embarrassment, that such a naive error could have passed unrecognized. The whole program could well have ended right there.

However, the long term investment in developing techniques for servicing the telescope, and ingenious solutions to the optical problem, resulted in amendments to the planned first servicing mission. It was decided not only to fix the optical system but also to upgrade the solar arrays and several other systems, thereby fully demonstrating the concept of inflight servicing, a capability first suggested over two decades earlier. Even though the remedy to the optical problem was safely installed, there was an anxious wait while the new systems were tested and calibrated, to determine whether the remedy actually worked. After five long weeks, the news came through that the optical performance of the telescope was restored and its science mission was saved. Indeed, NASA too gave a sigh of relief that its reputation had been rescued.

*January 13, 1994*
*Release: 94-7*
*NASA DECLARES HUBBLE SERVICING MISSION SUCCESSFUL*

*NASA Administrator Daniel S. Goldin today declared that last month's Space Shuttle mission to service the Hubble Space Telescope (HST) had been fully successful in correcting the vision of the telescope's optical components. The announcement, accompanied by the first new images from HST, followed the initial 5 weeks of engineering check-out, optical alignment and instrument calibration.*

*Word of the Hubble success came at a press conference at NASA's Goddard Space Flight Center, Greenbelt, Md. Goldin was joined in making the initial announcement by Dr. John H. Gibbons, Assistant to the President for Science and Technology, and Senator Barbara A. Mikulski (Md.), Chair, Appropriations Subcommittee on VA, HUD and Independent Agencies.*

*"This is phase two of a fabulous, two-part success story," Goldin said. "The world watched in wonder last month as the astronauts performed an unprecedented and incredibly smooth series of space walks. Now, we see the real fruits of their work and that of the entire NASA team.*

*"Men and women all across this agency committed themselves to this effort. They never wavered in their belief that the Hubble Space Telescope is a true international treasure," Goldin said.*

*Mikulski, who unveiled two new HST pictures at the press conference, said, "I am absolutely delighted that Hubble is fixed and can see better than ever. This is tremendous news.*

*"Now we are going to look at the origins of our universe," Mikulski said. "What a wonderful victory this is for the Hubble team of astronauts, astronomers, scientists and engineers. Together they are moving American science and technology into the 21st century with exciting new opportunities for scientific and economic progress."*

*Pictures were released from the two cameras that received corrective optics during the servicing mission – the Wide Field/Planetary Camera II and the European Space Agency's Faint Object Camera.*

*In the midst of the news, Program Scientist Ed Weiler was enthusiastic in his excitement, stating, "It's fixed beyond our wildest expectations."*

Two months later, in March 1994, the Robert J. Collier Trophy for 1993 was awarded to the Hubble Space Telescope Recovery Team. The Collier Trophy, established in 1911, is regarded as the highest honor to be awarded in American aviation and is presented annually by the National Aeronautic Association *"for the greatest achievement in aeronautics or astronautics in America, for improving the performance, efficiency, or safety of air or space vehicles, the value of which has been thoroughly demonstrated by its actual use during the preceding year."* The citation on the award that was presented on May 6, 1994 read: *"For outstanding leadership, intrepidity, and the renewal of public faith in America's space program by the successful orbital recovery and repair of the Hubble Space Telescope."*[1]

The HST Recovery Team that received the award was composed of Joseph Rothenberg, previously Associate Director of Flight Projects, Goddard Space Flight Center, Greenbelt, Md.; Randy Brinkley, STS-61 Mission Director, Johnson Space Center (JSC), Houston; James M. "Milt" Heflin, Jr., STS-61 Lead Flight Director, JSC; Brewster H. Shaw, Jr., Director, Space Shuttle Operations, NASA Headquarters, Washington, D.C.; and the members of the STS-61 flight crew, commander Richard O. Covey, pilot Kenneth D. Bowersox, and mission specialists Tom Akers, Jeffrey A. Hoffman, F. Story Musgrave, Claude Nicollier (European Space Agency), and Kathryn C. Thornton.

These eleven individuals were just part of the wider team representing more than 1,200 people who were directly involved in this mission and who were acknowledged in the award. Although NASA did not consider it a "rescue" mission, simply the scheduled first servicing mission, the flight certainly saved the telescope and, as has often been observed, possibly saved the space agency itself. The team behind the mission of STS-61 and its success has often been compared to the effort to return the Apollo 13 crew safely to Earth 23 years earlier. It is certainly celebrated as one of the high points in NASA history.

It was a hard-won battle to prepare the hardware, to ensure that everything would work as intended, and be ready in time for the mission. Frank Cepollina, NASA's manager of space servicing capabilities, recalled that the events had created "great turmoil in checking every socket and bolt". The philosophy of test, re-test, and test again, certainly paid off, and gave enormous confidence not only to the forthcoming follow-on servicing missions in planning but also to the developing International Space Station.

Back in 1946, astronomer Lyman Spitzer had suggested that a large optical telescope orbiting Earth above the restricting layers of the atmosphere would "uncover new phenomena not yet imagined, and perhaps modify profoundly our basic concepts of space and time". Forty-four years later, this became a reality with the deployment of the Hubble Space Telescope, and after a shaky start the first servicing mission was a huge success in restoring the optical system of the telescope to its intended specification, and far beyond what Lyman had dared to dream of.

By 1994, Hubble was finally able to deliver on its promise. The challenge was then to ensure that it remained fully operational for the next decade at least, and hopefully far beyond. Although it could not be known at the time, the once-threatened, even ridiculed space telescope was about to become a glittering prize in the NASA fold, and far beyond the shores of the nation which launched it. Hubble may now be classed as an American national asset, but it is much more than that, and the servicing missions over the next 15 years would ensure it became a global asset. But that is another story, recalled in *Enhancing Hubble's Vision: Service Missions That Expanded Our View Of The Universe*.

**REFERENCE**

NASA News 94-71, May 5, 1994

# Afterword

Hubble was the first major item of hardware designed, from the start, to be serviceable in space. I started with Hubble in 1975. I helped design it; I didn't just go and fix it. I had worked on it for 18 years by the point I got to go fix it on STS-61. There had been some other satellites that had minor design issues, such as Solar Max, but Hubble became the first major satellite designed to be repaired by spacewalking astronauts. It was around 1980 that we started to work out the spacewalks using Hubble mockups in the neutral buoyancy tanks, though these were not rehearsals for the spacewalks, they were research and development activities designed to work out the procedures that I was developing.

In 1975, I was told to find every possible problem that could be imagined on the telescope, and then design tools and procedures to fix them during a spacewalk. At this time there were suggestions to return the telescope to Earth, or utilize automated systems for servicing, but I was tasked not to fix it robotically, I was to fix it by means of a spacewalk. After the Earth-return option was abandoned we did not simply train for the on-orbit servicing mission, we designed the mission.

Back in the 1970s there were no drawings or hardware to work on, only my imagination. I eventually designed about 300 tools for Hubble, some of which were still being used during the final servicing mission in 2009, fully 30 years after we first developed them.

It was a disappointment that, after so many years of working on Hubble, failures happened shortly after it was placed in orbit. Of course there was no question that NASA would put a huge emphasis on fixing the spherical aberration problem of the primary mirror, which ought never to have been there. During preparations for STS-61, the first servicing mission, it was said by people in the media and within NASA, that this mission would define whether there was a future for the space agency. The agency's reputation was tarnished, the capabilities of the shuttle system were in question, and the viability of the space station was in question. An instrument that attempts to bridge the gap between cosmology, theology, and philosophy by seeking to answer the question of our place in the universe, touches people around the world like no other instrument. This added to the responsibility the crew carried to orbit.

We had a promise to do these repairs by spacewalking and this had direct application with the spacewalking required to assemble a space station. During STS-61 we fixed Hubble 100 percent, perhaps even slightly better, because we corrected the whole system, not just the faulty mirror. We corrected all the failures. I believe that with the STS-61 crew, NASA had one of the greatest groups of astronauts that ever flew. We were perhaps one of the most rehearsed missions since the Moon landings, and we demonstrated what NASA was really good at: unbelievably good team work, supported by a tremendous ground support network. We also proved that the on-orbit servicing of a complicated vehicle such as Hubble could be done by spacewalking astronauts, and done well.

Dr. Story Musgrave M.D. (Courtesy Story Musgrave)

The mission was a resounding success in terms of completing the five spacewalks with very few problems, and uninterrupted television coverage of the back-to-back spacewalks was unprecedented since the Apollo era. Hubble clearly touched the general public; I was even stopped during my Christmas shopping shortly after the mission, and congratulated on the success of 'Team NASA'.

I would have loved to return to Hubble on another servicing mission, but it was not to be. The groundwork which we had pioneered was continued to great effect by four more crews between 1997 and 2009, ensuring that today, 25 years after we placed the telescope in orbit and corrected its flaws, it is still flying and generating first class science and spectacular images of our universe, 40 years after we started to figure out how to ensure that longevity.

## Afterword

I have often said that it was very special to work on a spaceship like Hubble. The work was certainly worth it, the passion for Hubble was in that work, and in turn that passion was in ensuring that future generations would be able to enjoy the results of Hubble. It was a long journey from the original concept of servicing a telescope on-orbit to the success in launching the telescope on STS-31 and then restoring its capabilities on our flight. But this was just the start of the Hubble servicing saga.

Dr. Story Musgrave, M.D.
NASA Astronaut 1967–1997
Mission Specialist STS-61/SM-1
Mission Specialist STS-6, -51F, -33, -44 and -80

# Abbreviations

| | |
|---|---|
| A7LB | Apollo spacesuit, 7th model, International Latex Corporation, B variant |
| AAP | Apollo Applications Program |
| AAS | American Astronomical Society |
| AB | Aft Bulkhead |
| ACS | Advanced Camera for Surveys |
| ACS | Attitude Control System |
| AD | Aperture Door |
| AES | Apollo Extension System (AAP) |
| AIAA | American Institute for Aeronautics and Astronautics |
| AMB | Astronomy Mission Board |
| AOA | Abort Once Around |
| APC | Adaptive Payload Carrier |
| APU | Auxiliary Power Unit |
| ART | Anomaly Response Team |
| AS | Aft Shroud |
| Ascan | Astronaut Candidate |
| ASDT | Aft Shroud Door Trainer |
| ASE | Airborne Support Equipment |
| ATDA | Augmented Target Docking Adapter (Gemini) |
| ATL | Advanced Technology Laboratory |
| ATM | Apollo Telescope Mount (Skylab) |
| ATOX | Atomic Oxygen |
| AXAF | Advanced X-ray Astronomical Facility |
| BAPS | Berthing and Positioning System |
| BI | Booster Integration |
| BRT | Body Restraint Tether |
| BSP | BAPS Support Post |
| BUp | Back Up |
| C | Centigrade |

## Abbreviations

| | |
|---|---|
| CAD | Computer Aided Design |
| C&DH | Control & Data Handling |
| CAIB | Columbia Accident Investigation Board |
| CAPCOM | Capsule Communicator |
| CB | NASA Astronaut Office, JSC (Mail Code) |
| CCC | Charge Current Controllers |
| CCD | Charge Coupled Device |
| CCTV | Closed Circuit Tele-Vision |
| CDR | Commander |
| CDR | Critical Design Review |
| CEIT | Crew Equipment Interface Test |
| CGRO | Compton Gamma Ray Observatory |
| CITE | Cargo Integrated Test Equipment |
| cm | Centimeter |
| CMG | Control Moment Gyro |
| COPE | Contingency ORU Protective Enclosure |
| COS | Cosmic Origins Spectrograph |
| COSTAR | Corrective Optics Space Telescope Axial Replacement |
| CSM | Command and Service Module (Apollo) |
| CSS | Course Sun Sensors |
| CXO | Chandra X-ray Observatory |
| DI | Direct Insertion |
| DIU | Data Interface Units |
| DIY | Do It Yourself (Home Improvements) |
| DMS | Data Management System |
| DOD | Department of Defense |
| DSO | Detailed Supplementary Objective |
| DTO | Detailed Test Objective |
| EAFB | Edwards Air Force Base, California |
| ECU | Electronics Control Unit |
| ELV | Expendable Launch Vehicle |
| EMI | Electromagnetic Frequency |
| EMU | Extra Vehicular Mobility Unit |
| EOM | End of Mission |
| EOPTP | EVA Operations Procedures/Training Program |
| EPDSU | Enhanced Powered Distribution and Switching Unit |
| EPS | Electrical Power Subsystem |
| ERO | Early Release Observations |
| ESA | European Space Agency |
| ESF | Exterior Simulator Facility |
| E/STR | Engineering/Science Tape Recorders |
| ESTEC | European Space Technology Center |
| ET | External Tank |
| EUVE | Extreme Ultra-Violet Explorer |
| EURECA | European Retrievable Carrier |

# Abbreviations

| | |
|---|---|
| EV | EVA Astronaut -1, -2, -3 or -4 |
| EVA | Extra Vehicular Activity (spacewalking) |
| F | Fahrenheit |
| FCR | Flight Control Room |
| FD | Flight Day |
| FDF | Flight Data File |
| FGS | Fine Guidance Sensor |
| FHST | Fixed Head Star Tracker |
| FOC | Faint Object Camera |
| FOD | Flight Operations Directorate |
| FOS | Faint Object Spectrograph |
| FOSR | Flexible Optical Solar Reflector |
| FOV | Field of View |
| FPS | Focal Plane Structure |
| FRED | Foot Restraint Equipment Device |
| FRR | Flight Readiness Review |
| FRR | Foot Restraint Receptacle |
| FS | Forward Shell |
| FSS | Flight Support Structure |
| FS&S | Flight Systems & Servicing |
| g | Gravity (force) |
| GAO | Government Accountability Office |
| GHRSR | Goddard High Resolution Spectrograph Kit |
| GPC | General Purpose Computer |
| GRGT | Guam Remote Ground Terminal |
| GSFC | Goddard Space Flight Center (Robert H.) |
| HEAO | High Energy Astronomical Observatory |
| HFMS | High Fidelity Mechanical Simulator |
| HGA | High Gain Antenna |
| HIU | Headset Interface Unit |
| HOST | Hubble Space Telescope Orbiting Systems Test |
| HQ | Headquarters |
| HRS | High Resolution Spectrograph |
| HSP | High Speed Photometer |
| HSPM | Hardware Sun Point Mode |
| HST | Hubble Space Telescope |
| Hz | Hertz (cycles per second) |
| IGY | International Geophysical Year |
| IMAX | Image Maximum |
| INCO | Integrated Communications Officer |
| IR | Infra-Red |
| IS&AG | Image Science & Analysis Group |
| ISS | International Space Station |
| IV | Intra Vehicular |
| IVA | Intra Vehicular Activity |

| | |
|---|---|
| IVT | Interface Verification Test |
| JIS | Joint Integrated Simulation |
| JPL | Jet Propulsion Laboratory (California) |
| JSC | Johnson Space Center (Lyndon B.) |
| JWST | James Webb Space Telescope |
| kg | Kilogram |
| KLSS | Keel Latch Support Structure |
| km | Kilometer |
| KSC | Kennedy Space Center (John F.) |
| LATS | LDEF Assembly Transportation System |
| lbs | Pounds (weight) |
| LCC | Launch Control Center |
| LCMS | Low Cost Modular Spacecraft |
| LDEF | Long Duration Exposure Facility |
| LEO | Low Earth Orbit |
| LGA | Low Gain Antenna |
| $LH_2$ | Liquid Hydrogen |
| LOPE | Large ORU Protective Enclosure |
| LOS | Line of Sight |
| LOT | Large Orbital Telescope |
| LOX | Liquid Oxygen |
| LRR | Launch Readiness Review |
| LSS | Life Support System |
| LST | Large Space Telescope |
| LTA | Lower Torso Assembly |
| M&R | Maintenance & Refurbishment |
| MADWEB | Meteoroids And Debris Website |
| MCC | Mission Control Center |
| MCIU | Manipulator Controller Interface Unit |
| MDD | Mate/Demate Device |
| MDF | Manipulator Development Facility |
| MECO | Main Engine Cut Off |
| MET | Mission Elapsed Time |
| MFR | Multiple Foot Restraint |
| MFR | Manipulator Foot Restraint |
| MHz | Megahertz |
| MILA | Merritt Island Launch Area |
| MISSE | Materials International Space Station Experiments |
| MLI | Multi-Layer Insulation |
| mm | Millimeter |
| MMOD | Micro Meteoroid Orbital Debris |
| MMS | Multi-mission Modular Spacecraft |
| MMT | Mission Management Team |
| MMU | Manned Maneuvering Unit |
| MOCR | Mission Operations Control Room |

# Abbreviations

| | |
|---|---|
| MOL | Manned Orbiting Laboratory |
| MOM | Mission Operations Manager |
| MOR | Mission Operation Room |
| MORL | Manned Orbital Research Laboratory |
| MOT | Manned Orbiting Telescope |
| MOT | Mission Operations Team |
| MoU | Memorandum of Understanding |
| MPS | Main Propulsion System |
| MR | Main Ring |
| MRA | Main Ring Assembly |
| MS | Mission Specialist |
| MSC | Manned Spacecraft Center (Houston) |
| MSE | Mission Safety Evaluation |
| MSFC | Marshall Space Flight Center (Huntsville) |
| MSS | Magnetic Sensing System |
| MTL | Multi-setting Torque Limited |
| MULE | Multi-Use Lightweight Equipment |
| MWS | Mini Work Station |
| NACA | National Advisory Committee for Aeronautics |
| NAS | National Academy of Sciences |
| NASA | National Aeronautics and Space Administration |
| NASC | National Aeronautics and Space Council |
| NASCOM | NASA Communications Network |
| NBL | Neutral Buoyancy Laboratory (Sonny Carter Facility, Houston) |
| NBS | Neutral Buoyance Simulator (Marshall Space Flight Center) |
| NiCd | Nickel Cadmium |
| NICMOS | Near-Infrared Camera and Multi-Object Spectrometer |
| $NiH_2$ | Nickel Hydrogen |
| NOBL | New Outer Blanket Layer |
| NRL | National Research Laboratory |
| NRL | Naval Research Laboratory |
| NSF | National Science Foundation |
| O&C | Operations & Checkout |
| OA | Orbit Adjust |
| OAO | Orbiting Astronomical Observatory |
| OAS | Orbit Adjust Stage |
| OBSS | Orbiter Boom Sensor System |
| OFT | Orbital Flight Test |
| OGO | Orbiting Geophysical Observatory |
| OMDP | Orbiter Maintenance Down Period |
| OMS | Orbiter Maneuvering System |
| OMV | Orbiting Maneuvering Vehicle |
| OPF | Orbiter Processing Facility |
| ORI | Orbital Replacement Instrument |
| ORU | Orbital Replacement Unit |

| | |
|---|---|
| ORUC | Orbital Replacement Unit Carrier |
| OSMQ | Office of Safety and Mission Quality |
| OSO | Orbiting Solar Observatory |
| OSS | Office of Space Science (NASA) |
| OSSA | Office of Space Science and Applications (NASA) |
| OTA | Optical Telescope Assembly |
| OV | Orbital Vehicle |
| PAO | Public Affairs Office |
| PC | Payload Commander |
| PCR | Payload Changeout Room |
| PCS | Pointing Control System |
| PCU | Power Control Unit |
| PCUT | Power Control Unit Trainer |
| PDA | Photon Detector Assembly |
| PDR | Preliminary Design Review |
| PDRS | Payload Deployment Retrieval System |
| PFIP | Post-Flight Investigation Program |
| PFR | Portable Foot Restraint |
| PFRGF | Portable Flight Release Grapple Fixture |
| PGT | Pistol Grip Tool |
| PHA | Payload Hazard Assessment |
| PHSF | Payload Handling Servicing Facility |
| PI | Principal Investigator |
| PIP | Payload Integration Plan |
| PLSS | Portable/Primary Life Support System |
| Plt | Pilot |
| PM | Primary Mirror |
| PMA | Primary Mirror Assembly |
| POHS | Position Orientation Hold Submode |
| POCC | Payload Operations Control Center |
| PRCB | Program Requirements Control Board |
| Prox Ops | Proximity Operations |
| PRT | Power Ratchet Tool |
| PSA | Provisional Stowage Assembly |
| PSEA | Pointing Safing Electronics Assembly |
| PSRP | Payload Safety Review Panel |
| psi | Pounds per square inch |
| RAE | Radio Astronomy Explorer |
| RCC | Reinforced Carbon-Carbon |
| RCS | Reaction Control System |
| RDA | Rotary Drive Actuators |
| RGA | Rate Gyro Assembly |
| RFP | Requests for Proposal |
| RPM | Revolution Per Minute |
| RMGA | Retrievable Mode Gyro Assembly |
| RMS | Remote Manipulator System |

| | |
|---|---|
| RPM | Rendezvous Pitch Maneuver |
| RPSF | Rotation Processing and Surge Facility |
| RSRM | Redesigned Solid Rocket Motor |
| RSS | Rotating Servicing Structure |
| RSU | Rate Sensing Unit |
| RTF | Return To Flight |
| RTG | Radioisotope Thermal Generator |
| RWA | Reaction Wheel Assembly |
| SA | Solar Array |
| SAC | Science Advisory Committee |
| SADE | Solar Array Drive Electronics |
| SAFER | Simplified Aid for EVA Rescue |
| SAIL | Shuttle Avionics and Integration Laboratory |
| SAO | Smithsonian Astrophysical Observatory |
| SAS | Space Adaptation Syndrome |
| SAS | Small Astronomy Satellite |
| SCA | Shuttle Carrier Aircraft (Boeing 747) |
| SCGT | Service Commission Ground Test |
| SCM | Soft Capture Mechanism |
| SDTV | Structural Dynamic Test Vehicle |
| SEP | Scientific Experiment Package |
| SFU | Space Flyer Unit (Japanese) |
| SI | Science Instrument |
| SIC&DH | Science Instrument Control & Data Handler |
| SIHM | Software Internal Hold Mode |
| SIP | Science Instrument Package |
| SIRTF | Shuttle/Space Infra-Red Telescope Facility |
| SLF | Shuttle Landing Facility (KSC) |
| SLIC | Super Lightweight Interchangeable Carrier |
| SM | Secondary Mirror |
| SM | Service Mission |
| SMA | Secondary Mirror Assembly |
| SMGT | Systems Management Ground Test |
| SMIT | Service Mission Integrated Timeline |
| SMS | Shuttle Mission Simulator |
| SOPE | Small ORU Protective Enclosure |
| SPSM | Sun Point Safe Mode |
| SRB | Solid Rocket Booster |
| SRL | Space Radar Laboratory |
| SSAT | S-Band Single Access Transmitter |
| SSB | Space Science Board |
| SSE | Space Support Equipment |
| SSHR | Space Shuttle Hazard Reports |
| SSM | Support System Module |
| SSME | Space Shuttle Main Engines |
| SSPM | Software Sun Point Mode |

| | |
|---|---|
| SSRF | Shell/Shield Replacement Fabric |
| SSS | Star Selector Servos |
| SST | Spitzer Space Telescope |
| ST | Space Telescope |
| STA | Shuttle Training Aircraft |
| STAR | Space Telescope Axial Replacement |
| STDN | Space (flight) Tracking and Data Network |
| STIS | Space Telescope Imaging Spectrograph |
| STOCC | Space Telescope Operations Control Center |
| STS | Space Transportation System |
| STScI | Space Telescope Science Institute |
| Systems SIG | Systems Special Interest Group |
| TAL | Trans-Atlantic Abort |
| TCDT | Terminal Countdown Demonstration Test |
| TCS | Thermal Control Subsystem |
| TDRSS | Tracking and Data Relay Satellite System |
| TPAD | Trunnion Pin Attachment Device |
| TRR | Test Readiness Review |
| TSB | Temporary Stowage Brackets (for solar arrays) |
| TV | Television |
| UARS | Upper Atmosphere Research Satellite |
| UHF | Ultra High Frequency |
| US | United States |
| USA | United States of America |
| USA | United States Army |
| USAF | United States Air Force |
| USN | United States Navy |
| UV | Ultra Violet |
| V | Volt |
| V1, V2, V3 | HST Axis |
| VAB | Vehicle Assembly Building (KSC) |
| VDAS | Video Digital Analysis System (Laboratory) |
| VEST | Vehicle Electrical Systems Test |
| VESS | Vehicle Electrical Support Structure |
| VHF | Very High Frequency |
| VPF | Vertical Processing Facility |
| VR | Virtual Reality |
| VRL | Virtual Reality Laboratory |
| VTIK | Voltage/Temperature Improvement Kit |
| W | Watt |
| WBKL | Wide Body Keel Latch |
| WETF | Weightless Environment Training Facility |
| WFPC | Wide Field/Planetary Camera |
| WSC | White Sands Complex |
| WX | Weather |
| ZGSP | Zero Gyro Sun Point |

# Bibliography

In a project such as this, countless reference was made to a huge variety of publications over many years and from many sources, too many to list here in their entirety. In addition to those references cited in the main text, the following were frequently used in the compilation of the resulting two books.

**Interviews**

Several personal interviews were conducted during the compilation of these books, and the details of these are below:

| Name | Date |
| --- | --- |
| Akers, Thomas | November 11, 2013 |
| Burch, Preston | November 14, 2013 |
| Carey, Duane, | January 28, 2013 |
| Clervoy, Jean-François | December 9, 2011 |
| Covey, Richard | September 9, 1994 |
| Crippen, Robert | February 5, 2013 |
| Foale, Michael | June 28, 2000 |
| Gerlach, Lothar | February 12, 2013 |
| Hawley, Steven | March 1, 2012 |
| Hoffman, Jeff | September 8, 1994; August 1996 |
| King, Joyce | December 2013 |
| McCandless, Bruce | August 17, 2006 |
| Musgrave, Story | August 22, 2013 |
| Nelson, George | July 23, 2013 |
| Newman, James | December 6, 2013 |
| Reed, Ben | October 25, 2013 |
| Richards, Paul | November 24, 2013 |

(continued)

# Bibliography

(continued)

| | |
|---|---|
| Smith, Steven | February 15, 2013 |
| Tanner, Joseph | February 28, 2012 |
| Vernacchio, Al | October 24, 2013 |
| Werneth, Ross | November 11, 2013 |

**Official Oral Histories**

| | | |
|---|---|---|
| Charles F. Bolden | January 6, 2004 | JSC Oral History Project |
| Loren J. Shriver | December 18, 2002 | JSC Oral History Project |
| Randy H. Brinkley | January 25, 1998 | JSC Oral History Project |
| Richard O. Covey | March 28, 2007 | JSC Oral History Project |
| Steven A. Hawley | December 4, 17 2002 January 14, 2003 | JSC Oral History Project |
| Kathryn D. Sullivan | May 28, 2009 | JSC Oral History Project |
| Jeffrey A. Hoffman | November 3, 12, 17, 2009 | JSC Oral History Project |
| Ceppolina Frank | June 11, 2013 | NASA HQ Oral History Project |
| Lennard A. Fisk | September 8, 2010 | NASA HQ Oral History Project |
| Nancy Roman | September 15, 2000 | NASA HQ Oral History Project |

**Periodicals**

Aviation Week and Space Technology
Capcom
Countdown
ESA Bulletin
Flight International
Journal of the British Interplanetary Society
NASA Activities
Orbiter
Space World
Spaceflight
World Spaceflight News

**Newspapers**

Florida Today
Houston Chronicle
Houston Post
The Daily Telegraph
The Times, London
Washington Post

**NASA Publications**

| | |
|---|---|
| 1961–2009 | *Astronautics and Aeronautics, A Chronology*, NASA SP various editions |
| 1990–2009 | *Chronology of KSC and KSC Related Events*, NASA TM various editions |
| 1969 | *Long-Range Program in Space Astronomy*, July 1969, NASA SP-213 |
| 1976 | *The Space Telescope*, NASA SP-392 |
| 1982 | *Space Telescope*, Joseph J. McRoberts, NASA EP-166 |
| 1988 | *NASA Historical Data Book, Volume III Programs and People 1969–1978*, Linda Neuman Ezell, NASA SP-4012 |

(continued)

| | |
|---|---|
| (continued) | |
| 1993 | *Suddenly Tomorrow Came... A History of the Johnson Space Center*, Henry C. Dethloff, NASA SP-4307 |
| 1993 | *EVA Tools and Equipment Reference Book* NASA TM-109350/ JSC-24066 Rev-B, November 1993 |
| 1995 | *Exploring the Unknown, Selected Documents in the History of the U.S. Civil Space Program, Volume I, Organising for Exploration*; edited by John M. Logsdon, NASA SP-4407 |
| 1996 | *Exploring the Unknown, Selected Documents in the History of the U.S. Civil Space Program, Volume II, External Relations*; edited by John M. Logsdon, NASA SP-4407 |
| 1997 | *Walking to Olympus: An EVA Chronology*, David S.F. Portree and Robert C. Treviño, NASA Monographs in Aerospace History #7 |
| 1998 | From Engineering Science to Big Science, Edited by Pamel E. Mack, NASA SP-4219 |
| 1999 | *The Space Shuttle Decision, NASA's Search for a Reusable Space Vehicle*, T.A. Heppenheimer, NASA SP-4221 |
| – | *Power to Explore, a History of Marshall Space Flight Center*, Andrew J. Dunar and Stephen P. Waring, NASA SP-4313 |
| 2001 | *Exploring the Unknown, Selected Documents in the History of the U.S. Civil Space Program, Volume V, Exploring the Cosmos*; edited by John M. Logsdon, NASA SP-4407 |

**Media Publications**

| | |
|---|---|
| 1982 | (March) *Space Shuttle Trasnspotation Stesme Press Information*, Rockwell International |
| 1984 | (January) *Space Shuttle Trasnspotation Stesme Press Information*, Rockwell International |
| 1990 | *Hubble Space Telescope, Media Reference Guide*, Lockheed Missiles & Space Company, Inc. |

**Other Books**

| | |
|---|---|
| 1968 | *Telescopes in Space*, Zedenek Kopal, Hart Publishing Company, Inc. |
| 1995 | *Adventure in Space, the Flight to Fix the Hubble*, Elaine Scott, Margaret Miller, Hyperion Paperbacks (Juvenile literature) |
| 1997 | *Astronomy from Space, the Design and Operation of Orbiting Observatories*, John K. Davis, Wiley-Praxis |
| 1998 | *The Hubble Wars, Astrophysics Meets Astropolitics in the Two-Billion-Dollar Struggle over the Hubble Space Telescope*, Eric J. Chaisson, Harvard University Press, 2nd edition |
| 1999 | *Who's Who In Space, the International Space Station Edition*, Michael Cassutt, Macmillan |
| 2001 | *Space Shuttle, The History of the National Space Transportation System, The First 100 Missions*, Dennis R. Jenkins, Midland Publishing |
| – | *Skylab, America's Space Station*, David J. Shayler, Springer-Praxis |

(continued)

(continued)

| | |
|---|---|
| 2002 | *Apollo, The Lost and Forgotten Missions*, David Shayler, Springer-Praxis |
| – | *History of the Space Shuttle, Volume 1, The Space Shuttle Decision 1965–1972*, T.A. Heppenheimer, Smithsonian Institute Press, |
| – | *History of the Space Shuttle, Volume 2, Development of the Space Shuttle 1972–1981, Volume 1*, T.A. Heppenheimer, Smithsonian Institute Press |
| 2004 | *The Story of the Space Shuttle*, David M. Harland, Springer-Praxis |
| – | *Walking in Space*, David J. Shayler, Springer-Praxis |
| 2005 | *Space Shuttle Columbia, Her Missions and Crews*, Ben Evans, Springer-Praxis |
| – | *Women in Space, Following Valentina*, David J. Shayler and Ian Moule, Springer-Praxis |
| 2006 | *Hubble Space Telescope, Pocket Space Guide*, Steve Whitfield, Apogee Books |
| 2007 | *NASA's Scientist Astronauts*, David J. Shayler and Colin Burgess, Springer-Praxis |
| – | *Praxis Manned Spaceflight Log, 1961–2006*, Tim Furniss, David J. Shayler with Michael D. Shayler, Springer-Praxis |
| 2008 | *How Spacecraft Fly, Spaceflight without Formulae*, Graham Swinerd, Copernicus Books-Praxis |
| 2009 | *Serving the Hubble Space Telescope, Space Shuttle Atlantis—2009*, Dennis R. Jenkins and Jorge R. Frank, Speciality Press |
| 2010 | *The Universe in a Mirror, The saga of the Hubble Space Telescope and the visionaries who built it*, Robert Zimmerman, Princeton University Press, 4th printing |
| 2012 | *U.S. Spacesuits*, Kenneth S. Thomas, Harold J. McMann, Springer-Praxis, 2nd Edition |
| – | *Tragedy and Triumph in Orbit—the Eighties and Early Nineties*, Ben Evans, Springer-Praxis |
| 2013 | *Manned Spaceflight Log II*, David J. Shayler and Michael D. Shayler, Springer-Praxis |
| 2014 | *Partnership in Space, The Mid to Late Nineties*, Ben Evans, Springer-Praxis |
| – | *To Orbit and Back Again, How the Space Shuttle Flew in Space*, Davide Sivolella, Springer-Praxis |
| – | *Hubble's Legacy, Reflections by Those Who Dreamed It, Built It, and Observed With It*, Edited by Roger D. Launius and David H. DeVorkin, Smithsonian Institute Scholarly Press |

# About the author

Spaceflight historian David J. Shayler, F.B.I.S. (Fellow of the British Interplanetary Society or, as he likes to call it, Future Briton In Space!), was born in England in 1955. His lifelong interest in space exploration began by drawing rockets aged 5, but it was not until the launch of Apollo 8 to orbit the Moon in December 1968 that an interest in human space exploration became a passion. He fondly recalls staying up late with his grandfather one wonderful night in July 1969 to watch the Apollo 11 moonwalk.

Dave joined the British Interplanetary Society as a Member in January 1976, becoming an Associate Fellow in 1983, and Fellow in 1984. He was elected to the Council of the society in 2013. His first articles were published by the society in the late 1970s and in 1982 he created Astro Info Service (www.astroinfoservice.co.uk) to focus his research efforts.

Dave's first book was published in 1987 and has been followed by over 20 other titles featuring the American and Russian space programs, spacewalking, women in space, and the human exploration of Mars. His authorized biography of astronaut Jerry Carr was published in 2008. In 1989 Dave applied as a prospective cosmonaut candidate for the UK Project Juno program in cooperation with the Soviet Union (now Russia). The mission was to spend 7 days on the Mir space station. He didn't reach the final selection, but progressed farther than he had expected. The mission was flown by Helen Sharman in May 1991. In undertaking his research, Dave has visited NASA field centers in Houston and Florida in the United States, and the Yuri Gagarin Cosmonaut Training Center in Russia. It was during these trips that he was able to interview many space explorers and workers, tour training facilities, and handle real space hardware. He also gained a valuable insight into the activities of a space explorer and the realities of not only flying and living in space but also what goes into preparing for a mission and planning future programs.

Dave is on friendly terms with many former and current astronauts and cosmonauts, some of whom have accompanied him on visits to schools across the UK. For over 30 years he has delivered space-themed presentations and workshops to children and social groups to inform ordinary members of the public and interested individuals about the history and development of human space exploration and, wherever possible, to help youngsters develop an interest in science and technology and the world around them.

Dave lives in the West Midlands region of the UK and enjoys spending time with his wife Bel and a rather large white German Shepherd that answers to the name Jenna, and indulging in his fondness for cooking, fine wines, and classical music. His other interests are in reading about military history, visiting historical sites and landmarks, and following Formula 1 motor racing.

# Other works by the author

**Other space exploration books by David J. Shayler**
Challenger Fact File (1987), ISBN 0-86101-272-0
Apollo 11 Moon Landing (1989), ISBN 0-7110-1844-8
Exploring Space (1994), ISBN 0-600-58199-3
All About Space (1999), ISBN 0-7497-4005-X
Around the World in 84 Days: The Authorized Biography of Skylab Astronaut Jerry Carr (2008), ISBN 9781-894959-40-7

**With Harry Siepmann**
NASA Space Shuttle (1987), ISBN 0-7110-1681-X

**Other books by David J. Shayler in this series**
Disasters and Accidents in Manned Spaceflight (2000), ISBN 1-85233-225-5
Skylab: America's Space Station (2001), ISBN 1-85233-407-X
Gemini: Steps to the Moon (2001), ISBN 1-85233-405-3
Apollo: The Lost and Forgotten Missions (2002), ISBN 1-85233-575-0
Walking in Space (2004), ISBN 1-85233-710-9
Space Rescue (2007), ISBN 978-0-387-69905-9

**With Rex Hall**
The Rocket Men (2001), ISBN 1-85233-391-X
Soyuz: A Universal Spacecraft (2003), ISBN 1-85233-657-9

**With Rex Hall and Bert Vis**
Russia's Cosmonauts (2005), ISBN 0-38721-894-7

**With Ian Moule**
Women in Space: Following Valentina (2005), ISBN 1-85233-744-3

# Other works by the author

**With Colin Burgess**
NASA Scientist Astronauts (2006), ISBN 0-387-21897-1

**Other books by David J. Shayler and Michael D. Shayler in this series**
Manned Spaceflight Log II—2006-2012 (2013), ISBN 978-1-4614-4576-0

**With Andy Salmon**
Marswalk One: First Steps on a New Planet (2005), ISBN 1-85233-792-3

**With Tim Furniss**
Praxis Manned Spaceflight Log: 1961–2006 (2007), ISBN 0-387-34175-7

# Index

**A**

Abbey, George, 13, 14, 17, 19, 21, 47
Abt, Helmut A., 77
Administrator (NASA), 95, 103, 321, 331, 346, 347, 383
Advanced Camera for Surveys (ACS), 150, 177, 179, 286, 389
Advanced X-Ray Astronomical Facility (AXAF), 155
Akers, Thomas D., ("Tom"), xiii, 240, 241, 333–336, 338, 340, 342, 355, 356, 358, 360, 361, 363–364, 367–372, 374–376, 378, 384, 397, 400
Aldrin, Edwin E. ("Buzz"), 55
Allen, Joseph P., 346
Allen, Lew, 45
Allen, Tim (actor), 377
Altman, Scott D., 261, 315
American Astronomical Society (AAS), 56, 389
American Institute of Aeronautics and Astronautics (AIAA), 83, 92
American Optical Company, 66
Ames Research Center (ARC), California (NASA), 47
Anomaly Response Team (ART), 291, 292, 307, 389
Apollo (program), ii, ix, xi, xv, xxiii, 14, 50, 59, 60, 64–66, 69, 71–73, 78–81, 84, 86, 111, 117, 124, 141, 192–193, 197–198, 212, 219–220, 242, 271, 284, 346, 380, 387, 389, 390, 400–401, 403
   Apollo 11, 55, 74, 379
   Apollo 12, 55
   Apollo 13, 84, 384
   Apollo 17, 372
   Apollo Soyuz, 111

Apollo (spacecraft), 64, 65, 390
Apollo Applications Program (AAP), 65, 66, 79, 80, 103, 389
Apollo Telescope Mount (ATM-Skylab), 66, 90, 92, 389
Apt, Jay, 239, 240, 261
Armstrong, Neil A., 55
Associate Administrator (NASA), 40, 45, 65, 102, 311
*Astounding Science Fiction* (magazine), 51
*Astra* (Advanced OAO), 77
*Astro* (shuttle astronomy payload)
   Astro 1, 79, 334
   Astro 2, 79
Astronaut Office (Code CB, JSC), 198, 281, 312
Astronomy Experiment Module, 123, 124
Astronomy Mission Board, 56, 77, 389
*Atlantis* (space shuttle), 15, 26, 120, 155
Atlas (launch vehicle), 61, 63, 64, 120, 155
Atlas-Agena D (launch vehicle), 61
Atlas-Centaur (launch vehicle), 61, 63
Atomic Oxygen (ATOX), 167, 168, 170, 389
Austin Bryan P., 315
Aviation Week (magazine), 398

**B**

Bantle, Jeffrey W., 315
Batteries, 21, 28–29, 33, 34, 69, 92, 112, 124, 138, 146, 148, 156, 166, 170–171, 202, 207, 224, 237, 242, 265, 266, 270, 305–306, 364, 367, 374, 375, 395
   nickel-cadmium (NiCd), 138, 393
   nickel-hydrogen (NiH$_2$), 28, 138, 170, 393
Bellcomm Inc., 74, 76, 77

Beltsville Space Center, Maryland (currently GSFC), 57
Berthing and Positioning System (BAPS), 201, 202, 256, 257, 389
Big Bird (USAF reconnaissance satellite), 102
Boe, Eric, 315
Boeing Company, 44, 66–68, 137, 395
Bolden, Charles F., ("Charlie"), 1, 7, 9–11, 14, 15, 18, 19, 28, 29, 31–32, 34, 36, 47, 398
Bowersox, Kenneth D. ("Sox" or "Ken"), 314, 333–335, 345, 352, 361, 373, 377, 384
Brandenstein, Daniel, 15, 16
Brinkley, Randy, 339, 346, 348, 384, 398
Briscoe, A. Lee, 314, 315
British Aerospace Space Systems Ltd., England, 17–19, 121, 122, 168, 179
Brown, Curtis L., 315
Bullock, Sandra (actress), 261
Burch, Preston, xiv, 284–288, 290, 332, 397
Burner II (upper stage), 135
Bush, George W. (U.S. President)

C

Canadarm (shuttle RMS), 106
Cape Canaveral Air Force Station, Florida, 117, 310, 348
Carey, Duane G. ("Digger"), xiii, 397
Cassegrainian (telescope), 69
Cassutt, Michael (author), 346
Castle, Jr. Robert E., 315
Ceccacci, Anthony J., 315, 327
Cepollina, Frank ("Cepi"), 112, 191–194, 197, 224, 248, 259, 281, 290–292, 384
Cernan, Eugene A., 372
*Challenger* (space shuttle), 12–15, 17, 20, 22–23, 47, 78, 83, 85, 93, 95, 102, 107, 112, 119, 120, 155, 169, 237, 240, 245, 272, 310–312, 334, 348, 382, 403
*Chandra* X-Ray Observatory, 155, 390
Charged Current Controllers (CCC), 166, 170, 390
Chief of Astronomy (NASA), 86, 87
Clervoy, Jean-François, xiii, 107, 121, 397
Clifford, Richard, 239, 261
Closed Circuit TV, 106, 390
Closeout Mission, 313
Coarse Sun Sensors (CSS), 165, 236, 390
Code, Arthur D., 77
Cold War, 54, 83
Collins, Michael, 55
Columbia (Apollo 11 Command Module), 55
*Columbia* (space shuttle), 12, 23, 24, 26, 107, 115, 126, 155, 167, 220, 233, 348, 400
*Compton* Gamma Ray Observatory (CGRO), 21, 155, 194, 240, 390

Control Moment Gyros (CMG), 145, 146, 390
Copeland, Aaron, 355
*Copernicus* (satellite), 61
Copernicus, Nicolaus., 50
*Corona* (USAF reconnaissance satellite), 102
Corrective Optics Space Telescope Axial Replacement (COSTAR), xxiii, 161, 175, 177, 179–181, 256, 257, 286, 290, 336–338, 340, 342, 344, 348, 349, 368–372, 376, 378, 383, 390
Cosmic Background Explorer (project), 86
Cosmic Origins Spectrograph (COS), 177, 179, 181, 286, 390
Covey, Richard O., ("Dick"), xiii, 333–335, 339, 352, 353, 355, 360, 361, 376, 384, 397, 398
Crippen, Robert L., ("Bob"), xiii, 111, 112, 310, 311, 347, 397
Critical Design Review (CDR), 95, 132, 228, 390
Crocker, James, 338
Currie (née Sherlock), Nancy J., 107, 241

D

Data Interface Units (DIU), 163, 175, 176, 214, 374, 390
Data Management System (DMS), 137, 163, 175, 294, 305, 306, 390
Davis, John, xiv, 124, 399
Department of Defense (DOD), 16, 23, 40, 99, 102, 194, 311, 312, 322, 390
Deployment Controls Electronics Unit (DCEU), 166
Deployment Mission (STS-31), xi, xv, xxii, 1–23, 26, 30, 32, 34, 37, 39, 40, 42, 44, 45, 47, 107, 154, 236, 237, 248, 250, 260, 265, 282, 311, 314, 316, 320, 322, 323, 344, 349, 355, 375, 382, 388, 396
Deputy Administrator (NASA), 93, 192
Detailed Test Objective (DTO), 240, 248, 390
Die Rakete zu den Planetenräumen, 51
*Discovery* (space shuttle), 1, 4, 6–8, 12, 14, 23, 25–29, 31–34, 37, 39–44, 120, 154, 155, 242, 267
Dittemore, Ronald D., 314
Docking, 66, 69, 81, 108–110, 135, 137, 138, 144, 151, 195, 270, 389
Dornier, Germany, 121, 179
Douglas (McDonnell), 64, 152, 346
Downey, James A., 88
Drift Orbit, 110
Dryden Flight Research Center, California (Hugh L.), 117, 313
Dye, Paul F., 315

**E**

*Eagle* (Apollo 11 Lunar Module), 55
Eisenhower, Dwight (U.S. President), 56
Electrical Power Subsystem (EPS), 166, 176
Electronics Control Unit (ECU), 166, 336, 356, 390
*Endeavour* (space shuttle), 120, 240, 330, 347–349, 351–353, 355, 360, 361, 374–376
Engineering/Science Tape Recorder, 146, 148, 164, 209, 214, 390
Engle, Joe H., 346
Essex Corporation, Alabama, 213–215, 218, 235, 236, 238
Equipment Section, 137, 155, 158, 161, 163, 164, 166, 170, 172–175, 181, 196, 197, 209, 213–215, 223, 227, 236
EURECA (free flying satellite, ESA), 112, 347, 390
European Space Agency (ESA), xiv, 94–96, 98, 121, 122, 124, 169, 170, 179, 228, 278, 308–309, 334, 353, 360, 383, 384, 390
European Space Technology Center (ESTEC), 121, 170, 390
Expendable Launch Vehicle (ELV), 86, 100–102, 127, 135, 190, 194, 390
Explorer (program)
    *Cosmic Background Explorer*, 86
    *Explorer 1*, 62
    *Explorer 38*, 62
    *Explorer 42*, 62
    *Explorer 48*, 62
    *Explorer 49*, 62
    *Explorer 53*, 62
Extra Vehicular Activity (EVA)/spacewalks, ix, x, xiv, xv, xvii, xxiii, 21, 66, 90, 152, 161, 198, 203, 207, 220, 264, 278, 333, 340, 346, 358, 361, 364, 378, 386, 387, 391
    Apollo, 197, 212, 220, 242, 271, 372, 379, 380
    Gemini, 212, 213, 229, 372
    ISS, 205, 213, 220, 240–244, 247, 264, 269
    Skylab, x, 14, 66, 90, 103, 115, 197, 200, 212, 213, 248, 399, 403
    Space Shuttle, 21, 152, 153, 204, 205, 207, 213, 240, 243, 248, 251, 252, 259, 340
    EDFT (EVA Demonstration Flight Test), 242, 243
    Extra Vehicular Activity (Service Missions)
        STS-31 (deployment mission), 6–7, 9, 11, 13, 17–21, 34, 37
        STS-61 EVA's, 331, 333–336, 338–340, 344, 346, 353
        (EVA-1), 354–360
        (EVA-2), 360–365
        (EVA-3) 365–368
        (EVA-4) 368–372
        (EVA-5) 372–374
*Extreme Ultra-Violet Explorer* (EUVE), 194

**F**

Faget, Max, 247
Faint Object Camera (FOC), 45, 46, 94, 95, 121, 177, 179, 308, 346, 383, 391
Faint Object Spectrograph (FOS), 136, 146, 177, 180, 391
Farrow, John, 124
Feustel, Andrew J., 367
Fine Guidance Sensor (FGS), 161, 165, 166, 176, 177, 180, 223, 227, 228, 236, 237, 253, 288, 305, 307, 345, 367, 391
Fixed Head Star Tracker (FHST), 151, 165, 221, 223, 227, 229, 232, 253, 257, 391
Fisher, Anna L, 198, 223, 261
Fisk, Lenard A., 45, 398
Fitts, Richard, 346
Flexible Optical Solar Reflector (FOSR), 172, 173, 391
Flight Readiness Review (FRR), 27, 28, 40, 132, 347, 351, 391
Flight Support Structure (FSS), 200–202, 237, 246, 256, 257, 293, 303, 348, 349, 391
*Florida Today* (newspaper), 352
Foale, C. Michael. ("Mike"), xiii, 121, 190, 209, 242, 397
Focal Plane Structure (FPS), 161, 165, 175, 176, 391
Foot Restraint Equipment Device (FRED), 391

**G**

Gagarin, Yuri A., 55, 401
Galilei, Galileo, xv, xxi, 50
*Galileo* (Jupiter probe), 23
Gemini (program), 14, 60, 79, 84, 111, 212, 213, 229, 389, 403
General Electric, 66, 193
General Purpose Computer (STS), 3, 107, 321, 391
Gerlach, Lother, xiv, 121, 131, 168, 308, 309, 397
Gilbert, Allen C., 64
Goddard High Resolution Spectrograph Redundancy Kit, 372, 373
Goddard Space Flight Center (GSFC), Maryland (Robert H.), xiii, xxiii, 7, 23, 26, 33, 36, 37, 57, 59, 62, 86, 88–90, 94–96, 109, 112, 116–119, 121, 174, 177, 180, 191–194, 197, 200, 202, 203, 207–209, 248, 250, 254, 255, 257–266, 276, 280–292, 294–296, 301–304, 306, 307, 311, 313, 319, 320, 323, 325, 326, 332, 340, 342, 345, 346, 348, 349, 353–356, 360, 363, 365, 367, 369, 372–374, 383, 384, 391
Goldin, Daniel S. (NASA administrator), 331, 346, 347, 383

408  Index

Good, Michael T., 255
Gore, Al, 375
Grapple, 32, 34, 106–114, 143, 161, 196, 233, 253, 256, 257, 270, 348, 353–355, 360, 375, 394
*Gravity* (movie), 261
Great Observatory, NASA's, 1, 154, 275, 372, 377
Greenford, Michael, 346
Griffin, Gerald, 47
Grumman Aerospace, 59, 193, 284
Grunsfeld, John M., 255, 261
Gyros, 39, 43, 90, 92, 112, 145, 165, 174, 223, 270, 305, 307, 308, 335, 336, 356, 361

**H**

Hale, Jr. N. Wayne, 314, 315
Hamilton Sundstrand, 205, 281
Ham, Linda J., 315
Hanley, Jeffrey M., 315
Hansen, Grant, 102
Hansen, James R., 16
Harbaugh, Gregory J., 240, 315, 334, 342, 364
Hardware Sun Point (HSP), 113, 174
Hartford, James J., 83, 92
Hartsfield, Henry W., 32
Hart, Terry, 112
Hawker Siddley Dynamics, Stevanage, England, xiv, 123, 124
Hawley, Steven A., ("Steve"), xii, xiv, xv, xvii, 1–4, 7, 9–17, 19, 21, 22, 29, 31–34, 36, 37, 42, 47, 107, 155, 277, 397, 398
Heaton, Jim, 124
Heflin, Jr. J. Milton, 314–315, 384
Herbig, George H., 77
High Energy Astronomical Observatory (HEAO) (program), 135, 141, 142, 149
  *HEAO-1*, 63
  *HEAO-2*, 63
  *HEAO-3*, 63
High Gain Antenna (HGA), 4–7, 17, 21, 42, 44, 113, 118, 156, 158, 159, 161, 163, 202, 229, 232, 236, 253, 305, 354, 374, 391
High Resolution Spectrograph (HRS), 136, 177, 180, 336, 345, 348, 372, 373, 391
High Speed Photometer (HSP), 177, 180, 256, 257, 286, 336, 338, 340, 342, 353, 369, 370, 376, 391
Hoffman, Jeffrey A., ("Jeff"), xiii, 208, 241, 265, 331, 333–336, 339–344, 353, 355–358, 360, 364–368, 372, 373, 376–380, 384, 397, 398
*Home Improvements, Tool Time* (T.V. program), 377

Honeycutt, Jay, 27
Houson, Carl M., 64
Hubble, Edwin P., 96, 97
  Hubble Constant, 93
*Hubble Legacy* (book), 89
Hubble Space Telescope (HST)
  budget, 12, 46, 63, 70, 73, 77, 78, 84, 86, 87, 92, 93, 100, 116, 121, 128, 129, 193, 217, 220, 231, 287, 291, 300, 302, 303, 382
  cost, 3, 12, 23, 46, 60, 66, 67, 69–72, 79, 86, 88, 89, 91–93, 95, 96, 116, 121, 131, 132, 135, 138–142, 144, 149, 151–153, 155, 194–197, 215, 227, 239, 251, 266, 286, 345, 346, 382
Hurricanes, 124, 296

**I**

IMAX Camera, 9–11, 16, 43, 349, 355, 376, 391
International Geophysical Year (IGY), 55, 391
International Space Station (ISS), xv, 2, 32, 50, 103, 107, 125, 127, 190, 191, 205, 213, 220, 240–244, 246, 247, 254, 261, 264, 269, 314, 316, 324, 328, 330, 344, 346, 384, 391–392
Itek Corporation, 66

**J**

Jackson, Jr. Richard D., 315
Johnson, Gregory C., 315
Johnson (Lyndon B.) Space Flight Center (JSC), Houston, Texas, xiv, 13, 16, 26, 29, 47, 88, 90, 116, 117, 152, 198, 200, 207, 208, 210–213, 220, 223, 228, 235, 238–240, 247–250, 254, 259–262, 264–266, 270, 276, 278, 281–283, 286–288, 290, 291, 294–296, 298, 302, 305–306, 312–314, 320, 323, 325, 327, 328, 334, 340, 341, 346, 349, 355, 356, 367, 384, 390, 392, 398–399

**K**

Kelly Air Force Base, San Antonio, Texas, 44
*Kennan* (USAF reconnaissance satellite), 102
Kennedy, John F. (U.S. President), 55, 64
Kennedy Space Flight Center (KSC), Florida (John F.), xxiii, 23, 29, 96, 117, 276, 280, 292, 309, 310, 314, 323, 392
King, Joyce A., xiv, 292–296, 304
Kitt Peak National Laboratory, 66
Knight, Norman D., 315

## L

Lang, Watson, 124
Laika, 56
Landsat (program)
  *Landsat* 4, 194, 197
  *Landsat* 5, 194
Langley Memorial Aeronautical Laboratory (NACA), 58
Langley Research Center (LRC), Virginia (NASA), 66
Large Orbital Telescope (LOT), 54, 63, 64, 97, 181, 392
Large Space Telescope (LST), x, xi, xxiii, 52, 60, 64, 72, 73, 76–81, 83, 85, 86, 88–93, 97, 101, 121, 124, 131–153, 181, 190, 194, 195, 212, 270–271, 292
Launch Complex 39, KSC, 26, 117, 312, 314
*Leasat* (satellite), 111, 204
Leckrone, David, 372
Ledbetter, Kenneth, 372
Lee, Mark C., 16, 239, 242
Leinbach, Mike, 352
Leonov, Alexei A., 219
Lockheed Missile & Space Company, xiv, 18, 23, 94–96, 102, 117, 118, 182, 186, 195, 196, 198, 200, 213, 215, 248, 250, 262, 281, 282, 299, 301
*Long Duration Exposure Facility* (LDEF), (satellite), 32, 111, 112, 126, 311, 348, 392
Low-Z (rendezvous approach mode), 111, 114–115
Low, George M., 93, 192
Lucid, Shannon W., 198, 199, 223, 261
Lunar (program), ix, xxiii, 14, 56, 59, 60, 64, 81, 84, 99, 242
Lunar Module (Apollo), 55, 66, 284
Lunney, Bryan C., 315

## M

Magnetic Sensing System (MSS), 165, 336, 393
Manned Orbital Research Laboratory (MORL), 64–66, 71, 75, 393
Manned Orbiting Laboratory (MOL), 102, 393
Manned Orbiting Telescope (MOT), 66–71, 75, 79, 124, 181, 393
Manned Space Flight Center (MSC, later JSC), 65, 66, 69, 393
Mariner (planetary probe), 60, 84
Mars (program), ix, 60, 81, 84, 99
Marshall Space Flight Center (MSFC), Huntsville, Alabama (George C.), 12, 19, 33, 63, 88–90, 93, 96, 103, 116, 117, 126, 132, 196, 198–200, 209, 213–215, 218–220, 223–226, 228, 234–239, 250, 261, 276, 278, 280–282, 288, 297, 301, 334, 335, 340, 346, 350, 393, 399

Martin Marietta, 47, 103, 180, 268
Massimino, Michael J., 255, 261
Matra, France, 121, 179
McArthur, K. Megan, 107
McCandless II, Bruce, xiii, 1, 2, 4, 6, 7, 9, 10, 13–21, 29, 33, 34, 37–39, 42, 43, 47, 198, 199, 223, 224, 226, 227, 233, 237, 248, 250, 268, 270, 272, 277, 337, 397
McCullough, John A., 315
McDonnel Douglas Corporation, 64, 152, 346
Mercury (NASA spacecraft), 247, 312, 314, 316
Mikulski, Barbara A. (U.S. Senator), 383
Mir (Soviet/Russian space station)
Mission Control Centre, JSC, xiv, xiii, 4, 10, 18, 29, 31, 34, 117, 205, 260, 276, 295, 212, 213, 282, 314-322, 325-328, 350, 392
Mission Operations Manager (MOM), 295, 303, 304, 306, 307, 393
Mission Operation Room (MOR), 301, 304, 306, 307
Moore, Jesse, 47
Motorola, 215
Mountain, Matt, 284
Mueller, George, 65
Multi-Layer Insulation (MLI), 172, 173, 257, 368, 372, 391
Multi-Mission Modular Spacecraft (MMS), 193–195, 392
Muratore, John F., 315,
Musgrave, F. Story, xiii, 18, 34, 37, 42, 198, 200, 239, 241, 314, 331, 333–336, 340–342, 355, 356, 359, 360, 365, 367, 368, 372–374, 384, 387, 388, 397
Myers, Dale, 102

## N

National Academy of Sciences (NAS), 59, 72, 92, 393
National Advisory Committee for Aeronautics (NACA), 56, 58, 393
National Aeronautics and Space Administration (NASA)
  Budget, 12, 46, 63, 73, 77, 78, 84, 86, 87, 100, 116, 220, 330, 382
  Commercial Satellites, 12, 46, 63, 73, 77, 78, 84, 87, 100, 116, 140, 220, 330, 382
  Chief of Astronomy, 86, 87
  Deputy Administrator, 93, 192
  Associate Administrator, 40, 45, 65, 102, 311
  Office of Space Science (OSS), 58, 86, 346, 394
National Aeronautics and Space Council (NASC), 56, 393

National Air and Space Museum, Washington DC, 344
National Research Laboratory (NRL), 57, 393
National Science Foundation (NSF), 56, 393
Naugebauer, Gerry, 77
Naval Research Laboratory (NRL), 86, 393
Near Infra-Red Camera and Multi-Object Spectrometer (NICMOS), 177, 180, 203, 286, 393
Nelson, George D ("Pinky"), xiii, 16, 111, 223, 224, 226, 261, 268, 271, 397
Nickel cadmium (batteries), 138, 393
Nickel hydrogen (batteries), 28, 138, 170, 393
Nicollier, Claude, xiv, 107, 121, 333–335, 339, 345, 352, 353, 355, 361, 363, 365, 369, 373–375, 384
Nimbus (program), 73–77
Nimbus (servicing), 73–77
Nixon, Richard M. (U.S. President), 63, 84, 102, 103

## O

Oberth, Herman, 51
O'Dell, Charles Robert., 77, 91
Office of, (NASA)
    Space Science (OSS), 86, 131, 346, 394
    Space Science and Applications (OSSA), 115, 394
    Space Tracking and Data Systems, 117
O'Keefe, John, 57
Oliver, Jean, 88
Optical Astronomy Panel, 77
Optical Telescope Assembly (OTA), 33, 89, 95, 118, 135, 152, 156, 160, 175, 184, 213, 214, 228, 236, 301, 307, 394
Orbit Adjust Stage (OAS), 133, 135, 393
Orbital Flight Test (OFT), 111, 248, 393
Orbital Maneuvering System (OMS), 31, 39, 40, 42, 134, 135, 319, 320, 352, 355
Orbital Replacement Unit (ORU), 90, 122, 129, 153, 157, 181, 194, 196, 200, 202, 209, 213–217, 223, 225, 227–229, 236, 237, 242, 246, 248–251, 253, 256, 263, 266, 278, 286, 287, 293, 303, 307, 344, 349, 393
Orbital Replacement Unit Carrier (ORUC), 122, 200, 202, 203, 236, 237, 246, 256, 293, 303, 348, 349, 356, 394
Orbital Replacement Unit Protective Enclosure, 202, 286, 348, 392, 395
Orbiter Boom Sensor System (OBSS), 107, 393
Orbiter (shuttle)
    *Atlantis* (OV-104), 15, 26, 120, 155
    *Challenger* (OV-099), 12–15, 17, 20, 22–23, 47, 78, 83, 85, 93, 95, 102, 107, 112, 119, 120, 155, 169, 237, 240, 249, 272, 310–312, 334, 348, 382, 403
    *Columbia* (OV-102), xxiii, 12, 23, 26, 55, 107, 115, 126, 155, 167, 220, 233, 348, 390
    *Discovery* (OV-103), 1, 4, 6–8, 12, 14, 23, 25–29, 31–34, 37, 39–44, 120, 154, 155, 242, 267
    *Endeavour* (OV-105), 120, 240, 247, 249, 330, 348, 351–353, 355, 360, 361, 374–376
Orbiting Astronomical Observatory (OAO program), xiv, 57, 58, 393
    *OAO-1*, 59–61, 63, 149, 191, 284
    *OAO-2*, 61
    *OAO-B*, 61
    *OAO-Copernicus (OAO (C) )*, 61, 284
    *OAO-D* through *J*, 77
Orbiting Geophysical Observatory (OGO program), 59, 62, 191, 393
Orbiting Solar Observatory (OSO program), 59, 62, 87, 394
Oval Office (The White House), 375

## P

*Palapa* (satellite), 111, 248
Palomar, Mount. (telescope), 52
Payload Bay (space shuttle), xxii, 2, 4, 6, 9, 13, 17, 19, 21, 23, 25–29, 31, 34, 36, 37, 44, 80, 85, 92, 98, 99, 101–103, 106–109, 114, 122, 128–129, 131, 134, 135, 143, 147, 151, 155, 159, 161, 166, 195–196, 198, 200, 201, 206, 207, 216, 220–221, 224, 229, 233, 237, 240–243, 248, 250, 256, 259, 272, 282, 283, 286, 288, 320, 328, 336, 339, 340, 347–349, 354, 355, 357, 360, 361, 365, 369, 370, 373–374
Payload Deployment Retrieval System (PDRS), 10, 106, 321, 394
Perkin Elmer Corporation, 54, 94, 95, 118, 195, 198, 215, 238
Perry, Richard U., 347
Pioneer-Venus (program), 60, 93
Phase (program), 66, 131, 132
    Phase A, 88, 131–145, 151
    Phase B, 110, 111, 132, 141, 150, 151
Pointing Control System (PCS), 305, 307, 397
Pointing Safing Electronics Assembly (PSEA), 173, 394
Power Tools, 33, 216, 242, 253, 254, 261, 265–267, 271, 377
Pistol Grip Tool (PGT), 254, 261, 262, 265, 266, 268, 269, 394
Power Ratchet Tool (PRT), 254, 257, 263–266, 369, 394

President (U.S.), 55, 56, 63, 64, 84, 87, 102, 103, 375
Principle Investigator (PI), 118, 394
Princeton University, 52
Project Scientist (LST/HST), 91, 301–303, 372
Proximity Operations (Prox Ops), 32, 107, 109–112, 114, 128, 129, 278, 319, 394
Purcell, Joseph, 191

**Q**
Quale, Dan (U.S. Vice President), 46

**R**
Race (Arms), 54, 83
Race (Moon), 55, 83
Race (Space), 55, 83
Radio Astronomy Explorer (RAE), 62
Radioisotope Thermoelectric Generator (RTG), 75, 76, 395
RAND Corporation, 52
Ranger (lunar probe program), 60
Rate Gyro Assembly (RGA), 43, 146, 166, 174, 348, 394
Rate Sensing Unit (RSU), 165, 166, 221, 223, 227, 263, 335, 356, 395
R-Bar (rendezvous approach mode), 110
Reaction Control System (RCS), 31, 111, 112, 115, 150, 319, 320, 342, 355, 374, 394
Reaction Wheel Assembly (RWA), 112, 158, 166, 395
Reboost, 12, 202, 319
Reed, Benjamin, ("Ben"), xiv, 397
Reeves, William, ("Bill"), 37, 314
Remote Manipulator System (RMS) (Shuttle Canadarm), xxiii, 1–5, 8–10, 13, 16–17, 19, 21, 22, 26, 31–36, 80, 103–107, 109, 110, 112–114, 121, 128, 129, 143–146, 190, 196, 200, 202, 205–207, 211–213, 217, 220, 227, 229, 233–237, 241, 242, 246, 248, 270, 271, 277, 278, 281, 282, 319–321, 332–334, 339, 344, 345, 348, 353, 355, 356, 358, 361, 362, 367, 371, 373–376, 394
  Orbiter Boom Sensor System (shuttle), 107, 393
  Sepentutor (AAP), 103
Rendezvous, xiii, xvii, xxii, 4, 12, 20, 40, 42, 69, 85, 98, 107–113, 128, 129, 132, 134, 143, 196, 200, 277, 278, 281, 304, 315, 319, 320, 324, 330, 344, 352–355, 395
Resnik, Judith A., 32

Restraints
  Body, 206, 389
  Foot, 8, 158, 159, 161, 190, 200, 206, 208, 216, 217, 221, 223, 227, 229, 231–233, 236, 237, 239, 241, 242, 248, 253, 254, 257, 271, 361, 364, 365, 367, 369, 391–329, 394
Retrievable Mode Gyro Assembly (RMGA), 112, 174, 394
Richardson, R.S., 51
Richards, Paul W., xiii, 263–269, 274, 397
Ritchey-Chretien (optics), 124
Robotic servicing, 190, 212
Roman, Nancy Grace, 57, 58, 77, 86, 87, 89, 96, 398
Ross, Jerry, 240, 241, 350, 364
Rotary Drive Actuators (RDA), 161, 394
Rothenberg, Joe, 346, 364, 384

**S**
Salt Lake City International Airport, Utah
Sanders, Fred, 228
Saturn IB (launch vehicle), 65, 218
Saturn V (launch vehicle), 66, 69
S-Band Single Access Transmitters (SSAT), 158
Schilling, Gerhardt, 57
Science Advisory Committee (SAC), 56, 395
Science instrument, xv, xxiii, 69, 95, 96, 127, 129, 138, 143, 154, 156, 163, 172, 174, 175, 177, 179, 181, 194, 223, 225, 232, 236, 246, 257, 286–288, 295, 302–305, 395
Science Instrument Control & Data Handling (SIC&DH), 150, 156, 158, 163, 172, 174, 176, 181, 296, 305, 395
Science Instrument Package (SIP), 135, 136, 138, 395
Science and Technology Advisory Committee, 100
Scientific Experiment Package (SEP), 156
Scout (launch vehicle), 62
Service Mission
  SM-1, 107, 126, 175, 177, 180, 241, 254, 256, 294, 308, 316, 323, 334, 340, 344–348, 388
  SM-2, 107, 177, 242, 290, 293, 308, 316, 322, 323
  SM-3A, 107, 126, 177, 242, 280, 285, 314, 316
  SM-3B, 107, 126, 177, 180, 211, 241, 280, 294, 308, 314, 316
  SM-4, 107, 126, 171, 177, 180, 243–244, 249, 257, 275, 280, 294–296, 300, 314, 316, 323, 328, 367

Service Mission Integrated Timeline (SMIT), 305–306, 328, 386, 395
Shannon, John P., 315
Shaw, Charles W. ("Chuck"), xiv, 275, 315, 322–328
Sheffield, Ron, 248–250, 262, 266, 274
Shriver, Loren J., 1, 2, 4, 7, 10, 15, 16, 18, 19, 22, 29, 31, 33, 34, 43, 47, 349, 353, 398
Shuttle/Space Infra-Red Telescope Facility (SIRTF), 152, 155, 395
Skylab (program), x, xxiii, 14, 32, 65, 66, 79, 80, 84, 88, 90, 92, 103, 141, 149, 150, 192, 197, 200, 212, 213, 219, 220, 248, 333, 399
Small Astronomy Satellite (SAS), 62, 161
Smithsonian Astrophysical Observatory (SAO), 57, 395
Smith, Helen, 77
Software Sun Point Mode (SSPM), 174, 395
Solar Array (SA), xv, 75, 90, 113, 114, 121, 124, 129, 142–144, 148–150, 153, 155, 156, 170, 173, 174, 192, 203, 214, 223, 229, 233, 236, 237, 251, 256, 257, 305–308, 335–336, 348, 349, 372, 383, 395, 396
  first generation, 4–7, 10, 11, 17–19, 21, 34, 36, 37, 43, 44, 94–96, 158, 159, 161, 166–169, 228, 240, 244, 245, 353, 355, 358, 360–364, 376
  second generation, 228, 308, 331, 340, 344, 345, 358, 360, 374, 375
  third generation, 170, 292
Solar Array Drive Electronics (SADE), 166, 214, 336, 349, 372, 373, 395
Solar cycle, 12, 42, 134
*Solar Max* (satellite), xiii, 13, 23, 111, 112, 126, 194, 197, 201, 204, 220, 224, 248, 268, 270–272, 348, 386
Solid State Recorder, 164
Soyuz (Russian spacecraft), xi, 111, 334, 403
*Spaceflight* Magazine (British Interplanetary Society), 34
Space Act, 56
Space Age, ix, x, 51, 54, 55, 319
Spacehab, 240, 347
Spacelab, 14, 79, 80, 122, 123, 128, 152, 198, 202, 236, 334, 348
Space Science Board (SSB of the NAS), 59, 60, 72, 395
Space Shuttle (orbiter)
  See Atlantis and Orbiter OV-104
  See Challenger and Orbiter OV-099
  See Columbia and Orbiter OV-102
  See Discovery and Orbiter OV-103
  See Endeavour and Orbiter OV-105

Space Shuttle Missions
  STS-1, 205
  STS-2, 106, 233
  STS-4, 32, 205
  STS-7, 111
  STS-26, 23, 120, 311, 334
  STS-27, 311
  STS-28, 311
  STS-29, 120, 311
  STS-30, 311
  STS-31, xi, xxi, 1, 3, 4, 12, 15–17, 21–23, 26, 30, 32, 34, 37, 39, 40, 42, 44, 47, 107, 115, 154, 248, 265, 311, 314, 316, 320, 322, 323, 344, 349, 355, 382, 388
  STS-32, 32, 112, 311
  STS-33, 23, 37, 40, 311, 333–335
  STS-34, 311
  STS-35, 25, 26
  STS-36, 311
  STS-37, 21, 155, 240
  STS-41, 26, 44, 334
  STS-41B, 13, 14, 204, 239, 268, 272, 282
  STS-41C, 13, 111, 112, 201, 204, 248, 268, 270, 272, 282
  STS-41D, 14, 32, 124
  STS-41G, 14
  STS-49, 112, 204, 206, 240, 271, 334, 335, 338, 339, 346, 360, 369, 378
  STS-51, 112, 241, 347, 348
  STS-51A, 111, 248, 268, 270
  STS-51C, 16
  STS-51D, 111, 204, 333, 340, 368
  STS-51I, 111, 204, 334
  STS-51L, 15, 120
  STS-54, 120, 240
  STS-57, 112, 240, 347
  STS-61, xiii, xxiii, 37, 107, 115, 121, 180, 200, 208, 241, 243, 264, 265, 294, 311, 314–316, 331–333, 335, 336, 338, 339, 343–352, 356, 369, 372, 374–378, 380, 382, 384, 386–388
  STS-61B, 10, 204
  STS-61C, 14, 15, 17, 32
  STS-61J, 12, 13, 15–17, 21
  STS-64, 204
  STS-72, 243
  STS-82, xiii, 107, 115, 180, 210, 242, 243, 268, 280, 315, 316, 376
  STS-102, 267–269
  STS-103, 107, 115, 121, 180, 209, 242, 280, 285, 315, 316
  STS-109, xiii, 107, 115, 179, 180, 206, 241, 280, 294, 315, 316
  STS-114, 204, 325

STS-125, 107, 115, 179, 180, 230, 243–244, 249, 255, 257, 277, 280, 284, 294, 315–318, 327, 367
Space Stations, ix, xv, 46, 50, 64, 66, 69–75, 78, 80, 81, 84, 85, 88, 90, 99–101, 103, 107, 108, 111, 123–125, 127, 190, 192, 197, 204, 220, 240, 241, 259, 263, 264, 266, 271, 314, 330, 331, 333–335, 344, 346, 376, 378, 380, 384, 386, 387, 391, 392, 401, 403
Space Support Equipment (SSE), 200, 306, 348, 395
Space Telescope (ST), ix, x, xi, xiii, xv, xvii, xix, xxi-xxiii, 1, 4, 13, 29, 33, 44, 45, 47, 50–52, 54, 56–61, 64, 69, 72–73, 76, 77, 79, 83, 85, 86, 88, 90, 91, 93, 94, 96–99, 101, 102, 106, 107, 115–119, 121, 124, 125, 128, 129, 131–133, 135–137, 139, 140, 151–156, 159, 163, 174, 177, 180, 181, 190, 191, 194, 195, 197, 198, 201, 202, 207, 213, 214, 216, 218–220, 222, 223, 226, 235, 236, 239, 246, 247, 250, 259, 270, 271, 275, 276, 282, 286, 290, 292, 300, 301, 304, 308, 330, 336, 338, 339, 344, 352, 353, 364, 368, 380, 382–384, 386, 390–392, 396
Space Telescope Imaging Spectrograph (STIS), 177, 180, 286, 396
Space Telescope Operations Control Center (STOCC), 4, 6, 33, 42, 44, 95, 96, 112, 113, 117, 118, 174, 276, 282, 287, 301–307, 319, 328, 353, 360, 372, 374, 375, 396
Space Telescope Science Institute (STSI), 47, 117, 118, 121, 207, 286, 290, 304, 308, 380, 396
Space Transportation System (program), 98, 100, 103, 127, 143, 193
 Payloads, 100–102, 122, 128, 152
Spar Aerospace, Toronto, Canada, 106
Spitzer, Jr. Lyman, 1, 30, 52–54, 59, 60, 77, 155, 181, 384
Spitzer Space Telescope (SST), 155, 396
Sputnik (satellite program), 55, 56
 *Sputnik 1*, 55, 56
 *Sputnik 2*, 55, 56
Stafford, Thomas P., 111, 346
State University of Iowa, 59
Stone, B. Randy, 314, 315
Stratoscope, Project (balloon program), 54, 59
 Stratoscope I, 54
 Stratoscope II, 54
Structural Dynamic Test Vehicle (SDTV Space Telescope), 344, 395

Sullivan, Kathryn D., 1, 4, 6, 7, 9, 10, 13–21, 29, 33, 34, 37, 39, 40, 42, 43, 47, 198, 237, 250, 273, 398
Support Systems Module (SSM), 94, 95, 118, 135, 137, 138, 141–145, 147–148, 150, 152, 156, 158, 161, 163, 166, 171, 172, 176, 181, 213, 221, 223, 227, 233, 236, 395

**T**
Tanner, Joseph R., ("Joe"), xiii, 398
Test Program
 Vehicle Electrical Systems Test (VEST), 209, 286
 Vehicle Electrical Support Structure (VESS), 294, 396
 Service Commission Ground Test (SCGT), 294, 395
Thermal Blankets (TB)
 Multi-Layer Insulation (MLI), 172–176, 392
 Flexible Optical Solar Reflector (FOSR), 172–173, 391
Thermal Control Subsystem (TCS), 172, 174, 202, 396
Thornton, Kathryn C., 40, 240, 241, 314, 333–336, 338–340, 342, 352, 358, 360–365, 367–371, 378, 384
Tools (EVA), x, xi, xiii, xxiii, 33, 88, 90, 122, 124, 147–148, 151, 190–192, 196, 200, 206–207, 212, 215, 216, 220, 223, 224, 227–229, 233, 234, 236–239, 241–243, 246–248, 250–254, 256–266, 268, 270, 271, 273, 274, 280, 283, 286, 288, 289, 303, 320, 322, 328, 339, 341, 342, 344, 358, 367, 373, 375, 377, 386, 399
Trevino, Robert, xiv, 200, 239, 247–248, 273, 274, 281, 282
Titan (launch vehicle)
 Titan 4, 47
 Titan II, 64
 Titan III, 88, 102, 132, 135
 Titan III-C, 100
 Titan III-D, 135
 Titan III-E, 133, 135, 141
Tracking and Data Relay Satellite System (TDRSS), 117–120, 128, 163, 375, 396
Training, xvii, 3, 4, 9, 10, 13–20, 22, 27, 32, 47, 66, 69, 80, 89, 90, 94, 116, 117, 152, 207–213, 216–218, 220, 221, 228, 230, 235–238, 240, 241, 243, 244, 248–250, 259, 260, 262, 264, 268, 270, 274, 276–282, 284–286, 288, 291, 294, 296, 297, 300, 303, 312–314, 320, 322, 324, 327, 331–335, 338–340, 342–347, 352, 356, 361, 364, 368, 374–376, 378
TRW, 46, 193

**Index**

**U**

*Uhuru* (satellite), 62
*Ulysses* (solar probe), 26, 44
Underwater (EVA training), xxiii, 19, 52, 89, 213–239, 243, 262, 263, 265, 275, 303, 313
United States Air Force (USAF)
  Assistant Secretary for Research and Development, 102
  Reconnaissance satellites, 101, 102
United States Congress, 63, 72, 73, 85–87, 89, 92, 93, 96, 101, 116, 121, 193, 199, 220, 331, 337
United States President, 55, 56, 63–64, 84, 87, 102, 103, 375
United States Vice President, 46, 375
*Upper Atmosphere Research Satellite* (UARS), 194, 197, 396

**V**

V-2 (German rocket), 54
Vandenberg Air Force Base (California), 101, 102
V-Axis, 113, 114, 158, 227
V-Bar (rendezvous approach mode), 110
Vernacchio, Al, xiv, 289–291, 398
Vietnam War, 84, 248
Virtual Reality (VR), 208, 209, 260, 282, 343, 344, 396
Virutal Reality Laboratory (VRL), 208, 344, 396
Voss, James S.("Jim"), 239, 242, 261, 314, 315
Vostok (Soviet spacecraft), 55

**W**

Warner Robbins Air Force Base, Macon, Georgia, 44
Water tanks (for EVA simulations), 129, 213, 220, 238, 243, 248, 261, 281, 282, 303, 344
  Neutral Buoyancy Laboratory (NBL)–JSC (Sonny Carter Facility, Houston), 208, 213, 230, 243, 244, 250, 257, 260–262, 288, 327, 393
  Neutral Buoyancy Simulator (NBS)-MSFC, 19, 214–223, 225, 226, 231, 233–239, 248, 250, 253, 264, 266, 278, 393
  Weightless Environment Training Facility (WETF-JSC), 208–210, 220, 235, 240, 241, 278, 396
*We Have Capture* (book), 346
Weiler, Edward, 383
Werneth, Ross, xiv, 257–263, 268, 274, 398
*Westar* (satellite), 111, 248
*When We Left The Earth* (documentary), 42
Whipple, Fred, 57, 77
White Sands Test Facility, New Mexico, 29, 117, 118, 396
Wide Field/Planetary Camera (WFPC), 95, 161, 177, 180, 223, 227, 233, 237, 257, 283, 295, 335–337, 342, 348, 349, 353, 365, 367, 396
  WFPC-1, 177, 180, 342
  WFPC-2, 177, 286, 337, 342, 348, 349, 367
  WFPC-3, 177, 180, 286, 295, 367
Wilkins, Chris, 293
Woods Hole (Study Group), 50, 59, 63–64, 72, 92

**X**

X-Axis, 113

**Y**

Yale University, 52
Young, John W., 14–16, 346

**Z**

Z-Axis, 36, 114
Z-Bar, 110
Zimmerman, Robert (author), 89, 194, 195,